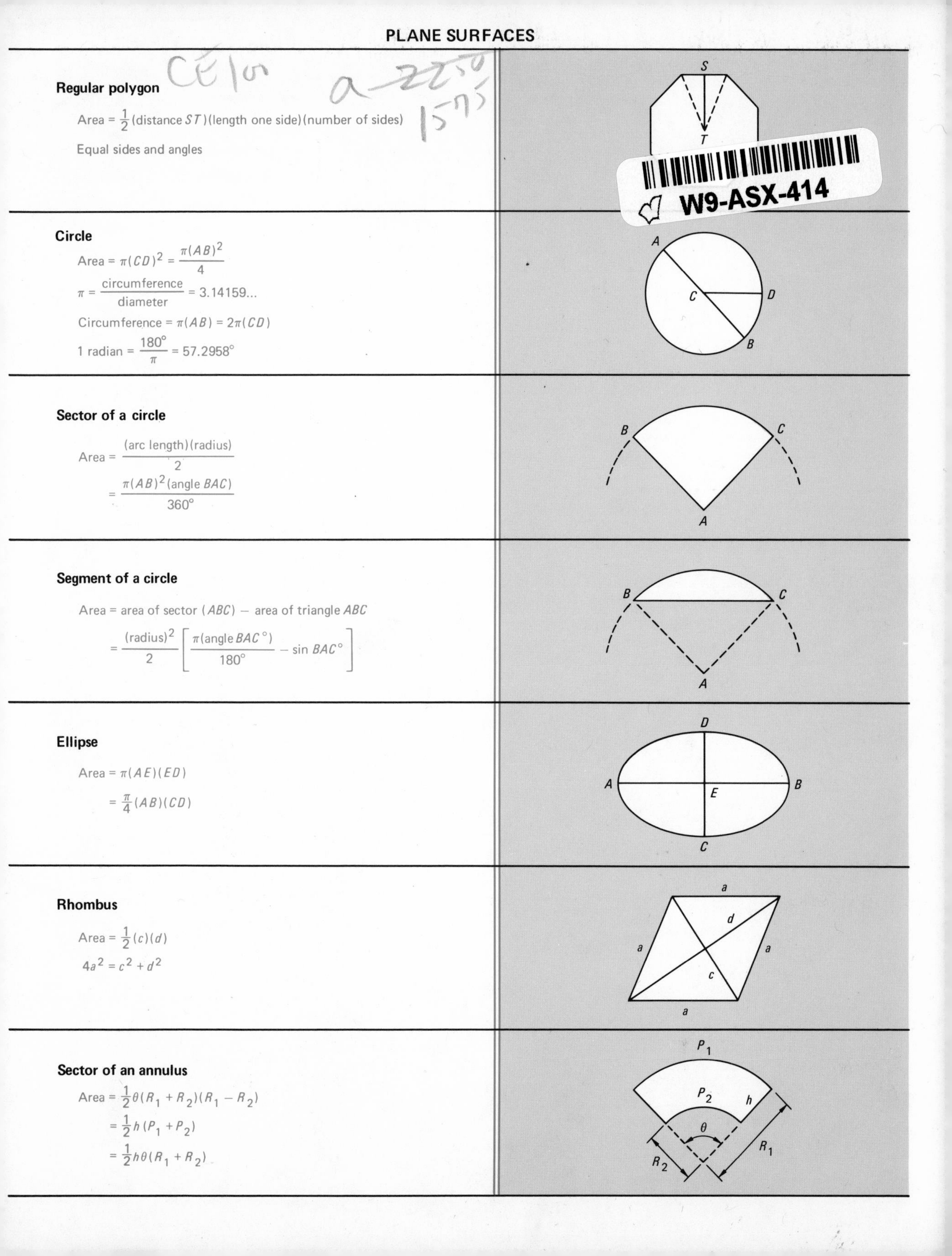

Regular polygon

Area = $\frac{1}{2}$(distance ST)(length one side)(number of sides)

Equal sides and angles

Circle

$$\text{Area} = \pi(CD)^2 = \frac{\pi(AB)^2}{4}$$

$$\pi = \frac{\text{circumference}}{\text{diameter}} = 3.14159...$$

$$\text{Circumference} = \pi(AB) = 2\pi(CD)$$

$$1 \text{ radian} = \frac{180^\circ}{\pi} = 57.2958^\circ$$

Sector of a circle

$$\text{Area} = \frac{(\text{arc length})(\text{radius})}{2} = \frac{\pi(AB)^2(\text{angle } BAC)}{360^\circ}$$

Segment of a circle

Area = area of sector (ABC) − area of triangle ABC

$$= \frac{(\text{radius})^2}{2}\left[\frac{\pi(\text{angle } BAC^\circ)}{180^\circ} - \sin BAC^\circ\right]$$

Ellipse

$$\text{Area} = \pi(AE)(ED) = \frac{\pi}{4}(AB)(CD)$$

Rhombus

$$\text{Area} = \frac{1}{2}(c)(d)$$

$$4a^2 = c^2 + d^2$$

Sector of an annulus

$$\text{Area} = \frac{1}{2}\theta(R_1 + R_2)(R_1 - R_2) = \frac{1}{2}h(P_1 + P_2) = \frac{1}{2}h\theta(R_1 + R_2)$$

Engineering fundamentals and problem solving

ENGINEERING FUNDAMENTALS AND PROBLEM SOLVING

Arvid R. Eide
Roland D. Jenison
Lane H. Mashaw
Larry L. Northup

Professors of Engineering
Iowa State University

McGraw-Hill Book Company
New York St. Louis San Francisco
Auckland Bogotá Düsseldorf Johannesburg
London Madrid Mexico Montreal New Delhi
Panama Paris São Paulo Singapore
Sydney Tokyo Toronto

Engineering fundamentals and problem solving

4 5 6 7 8 9 0 V H V H 8 3 2 1 0

Library of Congress Cataloging in Publication Data

Main entry under title:

Engineering fundamentals and problem solving.

Bibliography: p.
Includes index.
1. Engineering. I. Eide, Arvid R.
TA147.E52 620 78-25613
ISBN 0-07-019123-9

This book was set in Century Expanded by Typothetae Book Composition.
The editors were Julienne V. Brown and J. W. Maisel;
the designer was Ben Kann;
the production supervisor was Dennis J. Conroy.
The drawings were done by Felix Cooper.
Von Hoffmann Press, Inc., was printer and binder.

Contents

Preface *vii*

Part 1 An introduction to engineering

1. The Engineering Profession *2*
2. Engineering Solutions *34*
3. Representation of Technical Information *50*
4. Engineering Estimations and Approximations *78*

Part 2 Metric (SI) units in engineering

5. Dimensions, SI Units, and Conversions *98*

Part 3 Computers and calculators in engineering

6. Introduction to Computing Systems *122*
7. Preparation for Computer Solutions *148*

Part 4 Applied engineering concepts

8. Mechanics *164*
9. Chemistry—Concepts and Calculations *192*
10. Material Balance *214*
11. Electrical Theory *230*
12. Energy *254*
13. Engineering Economy *280*
14. Statistics *304*

Part 5 Foundations of design

15. Engineering Design—A Process 324

Appendix

A. Selected Topics from Algebra 378
B. Trigonometry 388
C. Graphics 402
D. General 425

Unit conversions 437

Answers to selected problems 443

Selected bibliography 448

Index 451

Preface

To the student

As you begin the study of engineering you will no doubt be filled with enthusiasm, curiosity, and a desire to succeed. Your first year will be spent primarily establishing a solid foundation in mathematics, physical sciences, and communications. You may at times question what the benefits of this background material are and when the real engineering work and experience will begin. We believe that they begin now. We hope that the material in this book will motivate you in your educational pursuits as well as provide you with a basis for understanding how the engineer functions in today's technological world.

To the teacher

During the past decade, engineering courses for freshman students have been in a state of transition. The traditional engineering drawing and descriptive geometry courses as well as formal training in the use of the slide rule have been pared considerably and in turn supplemented or replaced by courses in computations, computer programming, design, and career orientation. Of course, the emphasis placed on each of these courses varies considerably among engineering schools.

In 1974, an engineering faculty committee at Iowa State University was assigned the task of developing a proposal for a computations course that would include the use of programmable calculators. The course, which eventually replaced the course on the slide rule, furnished a great deal of material for improving the problem-solving skills of the student. This book evolved from the authors' experiences with that engineering computations course.

The book has four broad objectives: (1) to motivate engineering students in their first year, when exposure to engineering subject matter is limited; (2) to provide them with experience in solving problems (using the new international SI units) and presenting solutions in a logical manner; (3) to introduce students to subject areas common to most engineering disciplines, which require the application of fundamental engineering concepts; and (4) to develop their basic skills for solving open-ended problems through a design process.

The material in this book is presented in a manner that allows the instructor to emphasize certain areas more than others without loss of continuity. The problems that follow most chapters vary in difficulty, so that students can experience success rather quickly and still be challenged as problems become more complex.

There is sufficient material in the first fourteen chapters for a three-credit semester course. By omitting selected chapters and/or varying the coverage from term to term, you can present a sound computations course for a two- to four-credit quarter course or a two- to three-credit semester course. Expanded efforts in computer programming and coverage of Chapter 15 on design would provide sufficient material for a one-year sequence.

The book is conveniently divided into five parts. Part One, An Introduction to Engineering, begins with a description and breakdown of the engineering profession that can be expanded to the instructor's liking. If a formal orientation course is given elsewhere, Chapter 1 can be simply a reading assignment. Chapters 2, 3, and 4 provide procedures for approaching an engineering problem, determining the necessary data and method of solution, and presenting the results. The authors have found from experience that emphasis in this area will reap benefits when the material and problems become more difficult later on.

Part Two, although only one chapter in length, is important because it emphasizes the SI metric units. Throughout the book, discussions and example problems are primarily SI metric, so that coverage of Chapter 5 is advisable. Other unit systems do appear in some of the discussion and many problems contain nonmetric units, so the students are exposed to conversions and to units that are still commonly used.

Part Three is an introduction to computers and calculators. Unless the instructor plans to provide some programming experience on available programmable calculators or computers, this material can be omitted. Because of the variety of computational equipment available to students on most engineering campuses, the authors felt that treatment of the many computer languages available would not be practical. This in no way implies that programming is not an important part of the engineering students' course of study. They should be able to obtain calculated results in the most economical, efficient manner. Chapters 6 and 7 bring the student to the point of computation by computer. The instructor need only introduce the language and available equipment for the student to complete the solution. At Iowa State, we introduce the programmable pocket calculators in the computations course and all engineering students are introduced to FORTRAN, PL 1, or other languages in later courses.

Part Four allows a great deal of flexibility for the instructor. The time available and the instructor's personal interests dictate to what depth any or all of these chapters can be covered. For example, the authors have found that discussion of engineering economy (Chapter 13) is popular with the students and provides a good introduction to an area in which engineers need to be more knowledgeable. Material in Part Four may be covered in any order, since no chapter depends on another for background material.

The design process, Part Five, is a logical extension of the fundamental problem-solving approach in engineering. A nine-step design process is explained and supplemented with an actual preliminary design performed by a freshman student team. The process as described allows the instructor to supplement the text material with specific examples and bring design experience into the classroom.

Mathematical expertise beyond algebra, trigonometry, and analytical geometry is not required for any material in the book. The authors have found, however, that providing additional experience in precalculus mathematics is important at this stage of the student's education.

Acknowledgments

The authors are indebted to many who assisted in the development of this textbook. First we would like to thank the faculty in the College of Engineering at Iowa State University who have taught the Engineering Computations course since its inception in 1974. They have contributed many ideas and problems to this textbook; but most importantly, they have made the course a success with their efforts. We want to thank the hundreds of students who by their evaluations of the course prompted improvements and refinements in our initial manuscripts. Thanks is owing to the teaching assistants who demanded excellence on homework assignments. A special thanks goes to Professor Gordon Sanders for his review of the graphics portion of the Appendix. We also express grateful appreciation to Jodie Brown and Vicky Bice who sacrificed many lunch hours and evenings to produce clear copy from sometimes cloudy input. Finally, we thank our families for their unfailing support of our efforts.

Arvid R. Eide
Roland D. Jenison
Lane H. Mashaw
Larry L. Northup

PART ONE

An introduction to engineering

The engineering profession

CHAPTER 1

1.1 Introduction

The rapidly expanding sphere of science and technology may seem overwhelming to the individual seeking a career in a technological field. A technical specialist today may be called either engineer, scientist, technologist, or technician, depending upon education, industrial affiliation, or specific work. For example, more than 200 colleges and universities offer engineering programs accredited by the Engineer's Council for Professional Development (ECPD). Included are such traditional specialties as aerospace, agricultural, ceramic, chemical, civil, electrical, industrial, and mechanical engineering; as well as the expanding areas of computer, energy,

Fig. 1.1 The space shuttle, a modern engineering feat, will help mankind learn more about space by ferrying cargo into and out of earth orbit. (*Honeywell, Inc.*)

environmental, materials, and nuclear engineering. Programs in construction engineering, engineering science, mining engineering, and petroleum engineering add to a lengthy list of career options in engineering alone. Coupled with thousands of programs in science and technical training offered at hundreds of other schools, the task of choosing the right field no doubt seems formidable.

Since you are reading this book, we assume that you are interested in studying engineering or at least are trying to decide whether or not to do so. Up to this point in your academic life, you have probably had little experience with engineering and have gathered your impressions of engineering from advertising materials, counselors, educators, and perhaps a practicing engineer or two. Now you must investigate as many careers as you can as soon as possible to be sure of making the right choice.

The study of engineering requires a strong background in mathematics and the physical sciences. Section 1.4 discusses typical areas of study within an engineering program that leads to the bachelor's degree. You should also consult with your counselor about specific course requirements. If you are enrolled in an engineering college but have not chosen a specific discipline, consult with an adviser or someone on the engineering faculty about particular course requirements in your areas of interest.

When considering a career in engineering or any closely related fields, you should explore the answers to several questions. What is engineering? What is an engineer? What are the functions of engineering? Where does the engineer fit into the technical spectrum? How are engineers educated? What is meant by professionalism and engineering ethics? What have engineers done in the past? What are engineers doing now? What will engineers do in the future? Finding answers to such questions will assist you in assessing your educational goals and obtaining a clearer picture of the technological sphere.

Brief answers to some of these questions are given in the remainder of this chapter. By no means are they intended to be a complete discussion of engineering and related fields. You can find additional and more detailed technical career information in the reference materials listed in the bibliography at the end of the book.

1.2 The technology team

In 1876, 15 men led by Thomas Alva Edison gathered in Menlo Park to work on "inventions." By 1887, the group had secured over 400 patents, including ones for the electric light bulb and the phonograph. Edison's approach typified that used for early engineering developments. Usually one person possessed nearly all the knowledge in one field and directed the research, development, design, and manufacture of new products in this field.

Today, however, technology has become so advanced and sophisticated that one person cannot possibly be aware of all the intricacies of a single device or process. The concept of systems engineering has thus evolved: that is, technological problems are studied and solved by a technology team.

Scientists, engineers, technologists, technicians, and craftsmen form the technology team. The abilities of the team range across what is often called the technical spectrum. At one end of the spectrum are individuals with an understanding of scientific and engineering principles. They possess the ability to apply these principles for the benefit of mankind. At the other end of this technical spectrum are persons skilled in the use of their hands. They are the individuals who bring the actual ideas into reality.

Each of the technology team members has a specific function in the technical spectrum, and of utmost importance is that each specialist understands the role of all team members. It is not difficult to find instances where the education and tasks of team members overlap. For any engineering accomplishment, successful team performance requires cooperation that can be realized only through an understanding of the functions of the technology team. We will now investigate each of the team specialists in more detail.

1.2.1 Scientist

Scientists have as their prime objective increased knowledge of nature (see Fig. 1.2). In the quest for new knowledge, the scientist conducts research in a systematic manner. The research steps referred to as the scientific method are often summarized as follows.

1. Formulate a hypothesis to explain a natural phenomenon.
2. Conceive and execute experiments to test the hypothesis.
3. Analyze test results and state conclusions.
4. Generalize the hypothesis into the form of a law or theory if experimental results are in harmony with the hypothesis.
5. Publish the new knowledge.

An open and inquisitive mind is an obvious characteristic of a scientist. Although the scientist's primary objective is that of obtaining an increased knowledge of nature, many scientists are also engaged in the development of their ideas into new and useful creations. But to differentiate quite simply between the scientist

Fig. 1.2 Many scientists perform their work in a laboratory. Here a chemist investigates the effect of a ruthenium catalyst on the reaction of hydrogen and carbon dioxide. (*Ames Laboratory, U.S. Department of Energy.*)

and engineer, we might say that the true scientist seeks to understand more about natural phenomena, whereas the engineer primarily engages in applying new knowledge.

1.2.2 Engineer

In the 1963 Annual Report of ECPD, the following definition of engineering appears.

Engineering *is the profession in which a knowledge of the mathematical and natural sciences gained by study, experience, and practice, is applied with judgment to develop ways to utilize, economically, the materials and forces of nature for the benefit of mankind.*

In the National Council of Engineering Examiners' Model Law, the following statement is found.

Engineer *shall mean a person who, by reason of his special knowledge and use of mathematical, physical, and engineering sciences and the principles and methods of engineering analysis and design, acquired by education and experience, is qualified to practice engineering.*

Both the engineer and scientist are thoroughly educated in the mathematical and physical sciences, but the scientist primarily uses this knowledge to acquire new knowledge, whereas the engineer applies the knowledge to design and develop usable devices, structures, and processes. In other words, the scientist seeks to know, the engineer aims to do.

You might conclude that the engineer is totally dependent on the scientist for the knowledge to develop ideas for human benefit. Such is not always the case. Scientists learn a great deal from the work of engineers. For example, the science of thermodynamics was developed by a physicist from studies of practical steam engines built by engineers who had no science to guide them. On the other hand, engineers have applied the principles of nuclear fission discovered by scientists to develop nuclear power plants and numerous other devices and systems requiring nuclear reactions for their operation. The scientist's and engineer's functions frequently overlap, leading at times to a somewhat blurred image of the engineer. What distinguishes the engineer from the scientist in broad terms, however, is that the engineer often conducts research, but with a definite purpose in mind.

The end result of an engineering effort—generally referred to as design—is a device, structure, or process which satisfies a need. A successful design is achieved when a logical procedure is followed to meet a specific need. The procedure, called the design process, is similar to the scientific method with respect to a step-by-step routine, but it differs in objectives and end results. The design process encompasses the following activities, all of which must be completed.

1. Identification
2. Definition
3. Search
4. Establishment of criteria
5. Consideration of alternatives
6. Analysis
7. Decision
8. Specification
9. Communication

In the majority of cases, designs are not accomplished by an engineer's simply completing the nine steps shown in the order given. As the designer proceeds through each step, new information may be discovered or new objectives may be specified for the design. If so, the designer must backtrack and repeat steps. For example, if none of the alternatives appear to be economically feasible when the final solution is to be selected, the designer must redefine the problem or possibly relax some of the criteria to admit less expensive alternatives. Thus, because decisions must frequently be made at each step as a result of new developments or unexpected outcomes, the design process becomes iterative.

The engineering design process is discussed in detail in Chap. 15. Before you can meaningfully study it, you must become acquainted with problem analysis: that is, with finding a solution to a problem when the necessary data are available and the governing laws of nature are known. You will note that analysis is but one of the steps of the design process; but in itself it is indeed a powerful tool for decision making in an engineering design. You will solve many problems by application of the engineering method of problem analysis introduced in Chap. 2. But for now, an example will further illustrate the relation of analysis to design.

Example problem 1.1 For the beam shown in Fig. 1.3 we must know the reactions at points *A* and *B* in order to have knowledge of the total external loading on the beam. The beam has a mass of 1.0×10^3 kg shown acting at the center of mass.

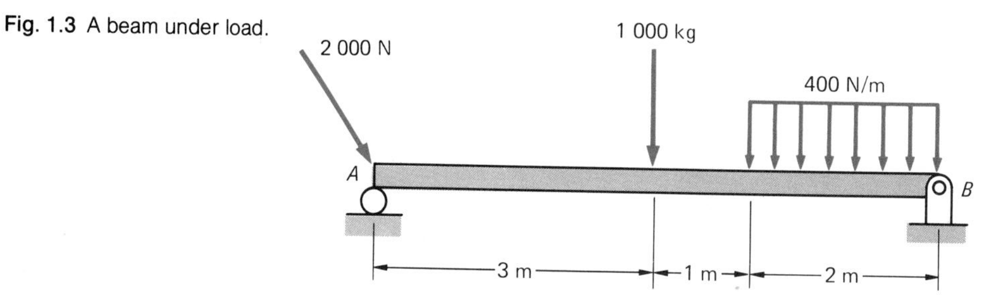

Fig. 1.3 A beam under load.

Solution Finding the reactions is a matter of analysis. As an engineer you begin the analysis by replacing the supports at *A* and *B* with appropriate force vectors of unknown magnitude. Then with knowledge from statics (Newton's second law with zero acceleration), you apply the equations of equilibrium to find the unknown forces.

Suppose that you are next given the task of finding a beam that will support the loading determined in the previous analysis, and will furthermore have a factor of safety of 3. If the factor of safety is the maximum allowable stress divided by the maximum existing stress in the beam, then you must continue the analysis of the problem to find the value of the maximum stress in the beam. The maximum allowable stress will depend on the material of which the beam is made. You can see that the analysis cannot be completed until a beam shape and material are specified. This is design. Once a shape is chosen, say an I beam or T beam, the overall dimensions must be assumed, the stresses computed, and a factor of safety calculated. If the factor of safety is not close to the desired value of 3, then the process must be repeated until a satisfactory solution is found. Other important factors enter into the design considerations, such as time boundaries, economic limitations, space requirements, and availability of materials. Decisions are made as the solution is developed. The analysis may thus be performed several times before the best solution is obtained for the specified conditions. Analysis is fundamental to the design process, so you should carefully develop your ability to solve problems by analysis before attempting to carry the design process to a successful conclusion.

1.2.3 Technologist and technician

Much of the actual work of converting the ideas of scientists and engineers into tangible results is performed by technologists and technicians (see Fig. 1.4). A technologist generally possesses a baccalaureate degree and a technician, an associate degree. Technologists are involved in the direct application of their education and

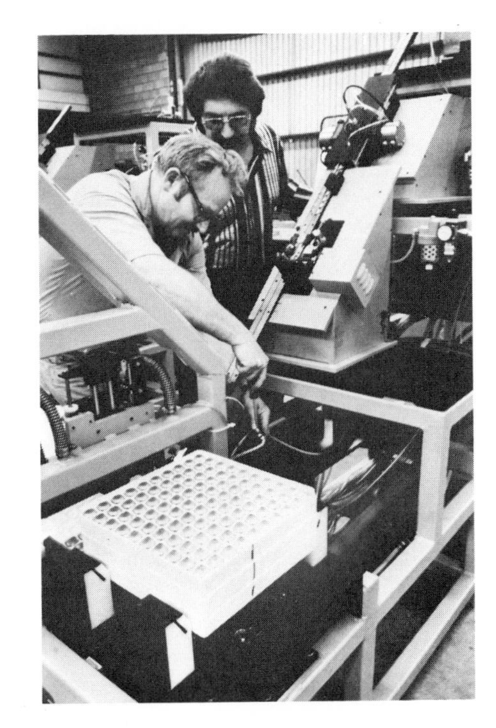

Fig. 1.4 Two members of a technology team, an engineer and a technician, are performing modifications on an automatic packaging machine for electronic components. (*FMC Corporation.*)

experience to make appropriate modifications in designs as the need arises. Technicians primarily perform routine computations and experiments and prepare design drawings as requested by engineers and scientists. Thus, technicians (typically) are educated in mathematics and science but not to the depth required of scientists and engineers. Technologists and technicians obtain a basic knowledge of engineering and scientific principles in a specific field and develop certain manual skills that enable them to communicate technically with all members of the technology team. Some tasks commonly performed by technologists and technicians include drafting, estimating, model building, data recording and reduction, troubleshooting, servicing, and specification. Often they are the vital link between an idea on paper and the idea in practice.

1.2.4 Craftsman

The craftsman possesses manual skills necessary to produce parts specified by scientists, engineers, technologists, and technicians. Craftsmen need not be particularly concerned with the principles of science and engineering incorporated in a design (see Fig. 1.5). They are usually trained on the job, serving an apprenticeship during which the skills and abilities to build and operate specialized equipment are developed. Some of the specialized jobs of craftsman are welder, machinist, electrician, carpenter, plumber, and mason.

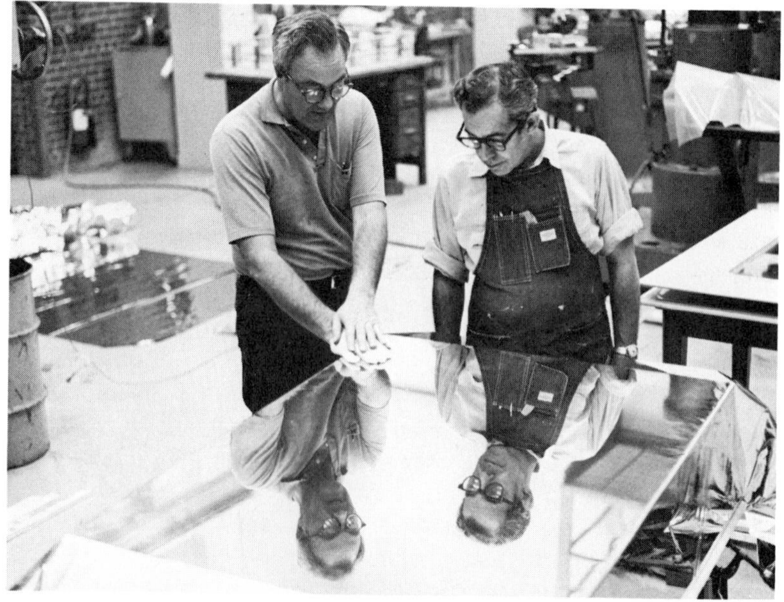

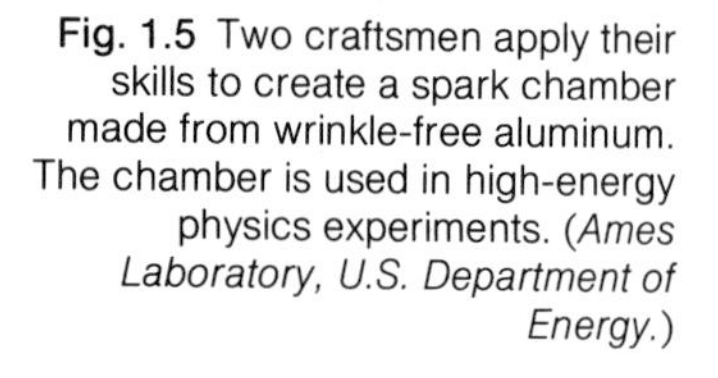

Fig. 1.5 Two craftsmen apply their skills to create a spark chamber made from wrinkle-free aluminum. The chamber is used in high-energy physics experiments. (*Ames Laboratory, U.S. Department of Energy.*)

1.3 The functions of the engineer

As we alluded to in the previous section, engineering feats accomplished from earliest recorded history up to the Industrial Revolution could best be described as one-man shows. The pyramids of Egypt were usually designed by one individual, who directed tens of thousands of laborers during construction. The person in charge called every move, made every decision, and took the credit if the project was successful or the consequences if the project failed.

With the Industrial Revolution, there was a rapid increase in scientific findings and technological advances. The one-man engineering team was no longer practical or desirable. We know that today no single aerospace engineer is responsible for the jumbo jets; and no one civil engineer completely designs a bridge. Automobile manufacturers assign several thousand engineers to the design of a new model. So we not only have the technology team as described earlier, but we have engineers from many disciplines working together on single projects.

To describe specific aspects of each engineering discipline would require repeating much information. A more enlightening approach is to discuss the different types of work that engineers do. For example, in civil, electrical, industrial, etc., engineering, all engineers become involved in design, which is an engineering function. The engineering functions, which we will discuss briefly here, are research, development, design, production, testing, construction, operations, sales, management, consulting, and teaching.

To avoid confusion between the meaning of engineering disciplines and engineering functions, let us consider the following. Normally a student selects a curriculum (aerospace, chemical, mechanical, etc.) either before or soon after admission to an engineering college. When and how the choice is made varies with each school. The point is, the student does not choose a function, but a discipline. To illustrate further, consider a student who has chosen mechanical engineering. This student will, during an undergraduate education, learn how mechanical engineers are involved in the engineering functions of research, development, design, etc. Some program options allow a student to pursue an interest in a specific subdivision within the curriculum, such as energy conversion in a mechanical engineering program. Most other curricula have similar options.

Upon graduation, when you accept a job with a company, you will be assigned to a functional team performing in a specific area such as research, or design, or sales. Within some companies, particularly smaller ones, you may become involved in more than one function, design *and* testing, for example. It is important to realize that regardless of your choice of discipline, you may become involved in one or more of the functions to be discussed in the following paragraphs.

1.3.1 Research

Successful research is one catalyst for starting the activities of a technology team or, in many cases, the activities of an entire industry. The research engineer seeks new findings, as does the scientist; but it must be kept in mind that the research engineer also seeks a way to use the discovery.

Some qualities of a successful research engineer are intelligence, perceptiveness, cleverness, patience and self-confidence. Most students interested in research will pursue the master's and doctor's degrees in order to develop their intellectual abilities and the necessary research skills. An alert and perceptive mind is needed to recognize nature's truths when they are encountered. When attempting to reproduce natural phenomena in the laboratory, cleverness and patience are prime attributes. Research often involves tests, failures, retests, etc., for long periods of time. Research engineers are therefore often discouraged and frustrated and must strain their abilities and rely on their self-confidence in order to sustain their efforts to a successful conclusion.

Billions of dollars are spent each year on research at colleges and universities, industrial research laboratories, government installations, and independent research institutes. The team approach to research is predominant today primarily because of the need to incorporate a vast amount of technical information into the research effort. Individual research is also carried out but not to the extent it was several years ago. In the late 1970s, a large share of the research money has been concentrated in the areas of energy, environment, and health; whereas a few years earlier, a large portion was spent on defense and space exploration. Research funding from federal funds is very sensitive to national and international priorities. During a career as a research engineer, you might expect to work in many diverse, seemingly unrelated, areas, but your qualifications will allow you to adapt to many different research efforts. See Fig. 1.6.

1.3.2 Development

Using existing knowledge and new discoveries from research, the development engineer attempts to produce a device, structure, or process that is functional. Building and testing scale or pilot models is the primary means by which the development engineer evaluates ideas. A major portion of the development work requires use of well-known devices and processes in conjunction with established theories. Thus reading of available literature and a solid back-

Fig. 1.6 A research engineer in a laboratory uses a sophisticated electron-probe microanalyzer to study surface qualities of metals. (*Pratt and Whitney Aircraft.*)

ground in the sciences and engineering principles are necessary for the development engineer's success.

Many people who suffer from heart irregularities are able to function normally today because of the pacemaker, an electronic device that maintains a regular heartbeat. The pacemaker is an excellent example of the work of development engineers.

The first model, conceived by medical personnel and developed by engineers at the Electrodyne Company, was an external device that sent pulses of energy through electrodes to the heart. However, the large power requirement for stimulus was so great that patients got severe burns on their chests. As improvements were being studied, research in surgery and electronics enabled development engineers to devise an external pacemaker with electrodes through the chest attached directly to the heart. Although more efficient from the standpoint of power requirements, the devices were uncomfortable and patients suffered infection where the wires entered the chest. Finally two independent teams developed the first internal pacemaker, 8 years after the first pacemaker had been tested. Their experience and research with tiny pulse generators for spacecraft that could supply adequate power led to this achievement. But the very fine wire used in these early models proved to be inadequate and quite often failed, forcing patients to have the entire pacemaker replaced. In 1965 a team of engineers at General Electric developed a pacemaker that incorporated a new wire, called a helicable. The helicable consisted of 49 strands of wire coiled together and then wound into a spring. The spring diameter was about 46 μm, half the diameter of a human hair. Thus, with doctors and development engineers working together, an effective, comfortable device was perfected in 13 years that has enabled many heart patients to enjoy a more active life. Today pacemakers have been developed that operate at more than one speed, enabling the patient to speed up or slow down heart rate depending on physical activity.

We have discussed the pacemaker in detail to point out that changes in technology can be in part owing to development engineers. Only 13 years to develop an efficient, dependable pacemaker; 5 years to develop the transistor; 25 years to develop the digital computer indicates that modern engineering methods generate and improve products nearly as fast as research generates new knowledge.

Successful development engineers are ingenious and creative. Astute judgment is often required in devising models that can be used to determine whether a project will be successful in performance and economical in production. Obtaining an advanced degree is helpful but not as important as it is for an engineer who will be working in research. Practical experience more than anything else produces the qualities necessary for a career as a development engineer. See Fig. 1.7.

Development engineers are often asked to demonstrate that an idea will work. Within certain limits, they do not work out the exact specifications that a final product should possess. Such matters are usually left to the design engineer if the idea is deemed feasible.

1.3.3 Design

The development engineer produces a concept or model that is passed on to the design engineer for converting into a device, process, or structure (see Fig. 1.8). The designer relies on education and experience to evaluate many possible solutions, keeping in mind cost of manufacture, ease of production, availability of mate-

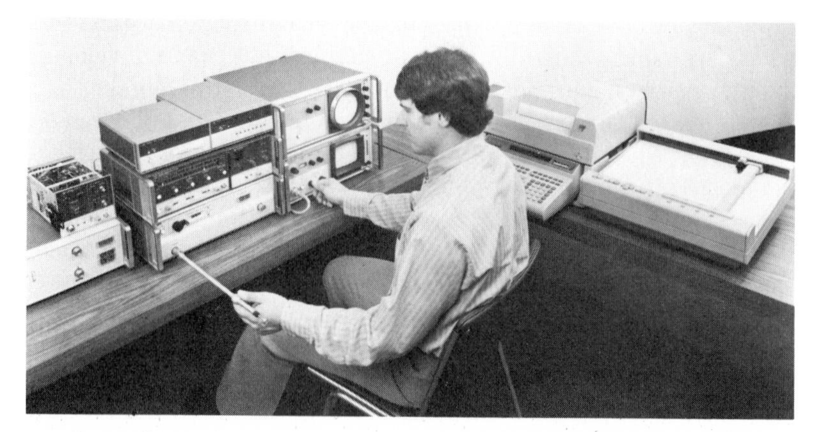

Fig. 1.7 A development engineer investigates new electronic devices with equipment such as this semiautomatic network analyzer. *(Hewlett-Packard.)*

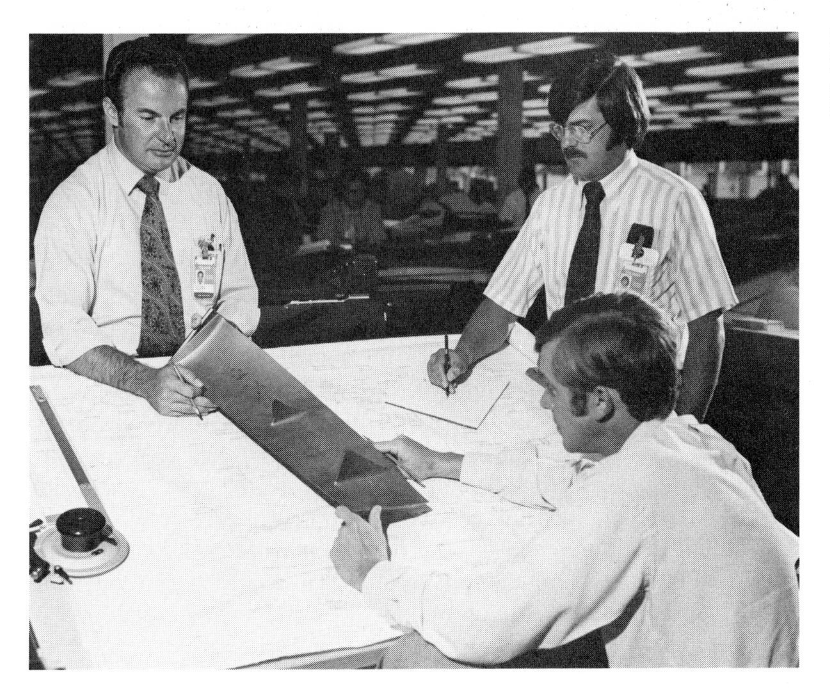

Fig. 1.8 Three design engineers view the part that is the result of their efforts to take an idea and make it into a functional product. (*Pratt and Whitney Aircraft.*)

rials, and performance requirements. Usually several designs and redesigns will be undertaken before the product is brought before the general public.

To illustrate the role the design engineer plays, we will discuss the development of the over-the-shoulder seat belts for added safety in automobiles, which created something of a design problem. Designers had to decide where and how the anchors for the belts would be fastened to the car body. They had to determine what standard parts could be used and what parts had to be designed from scratch. Consideration was given to passenger comfort inasmuch as awkward positioning could deter usage. Materials to be used for the anchors and the belt had to be selected. A retraction device had to be designed that would give flawless performance.

From one such part of a car, one can extrapolate the numerous considerations that must be given to the approximately 12 000 other parts that form the modern automobile: optimum placement of engine accessories, comfortable design of seats, maximization of trunk space, and aesthetically pleasing body design all require thousands of engineering man-hours to be successful in a highly competitive industry.

Like the development engineer, the designer is creative. However, unlike the development engineer, who is usually concerned only with a prototype or model, the designer is restricted by the state of the art in engineering materials, production facilities, and perhaps most important, economic considerations. An excellent design from the standpoint of performance may be completely impractical when viewed from a monetary point of view. To make the necessary decisions, the designer must have a fundamental knowledge of many engineering specialty subjects as well as an understanding of economics and people.

1.3.4 Production and testing

When research, development, and design have created a device for use by the public, the production and testing facilities are geared for mass production (see Figs. 1.9 and 1.10). The first step in production is to devise a schedule that will efficiently coordinate materials and manpower. The production engineer is responsible for such tasks as ordering raw materials at the optimum times, setting up the assembly line, and handling and shipping the finished product. The individual who chooses this field must possess the ability to visualize the overall operation of a particular project as well as know each step of the production effort. Knowledge of design, economics, and psychology is of particular importance for production engineers.

Test engineers work with a product from the time it is conceived by the development engineer until such time as it may no longer be manufactured. In the automobile industry, for example, test engi-

Fig. 1.9 Test engineers use a laboratory model to establish a design basis for a new production unit. (*Procter and Gamble.*)

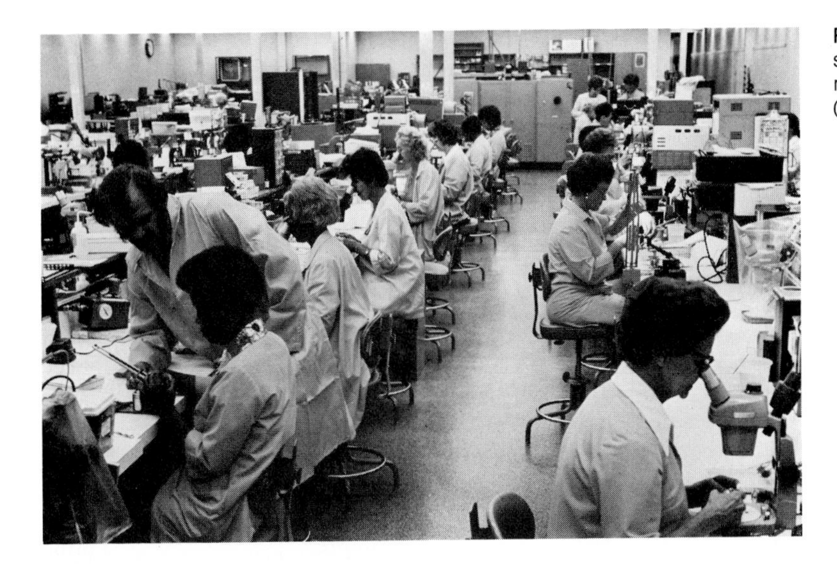

Fig. 1.10 A production engineer supervises employees assembling miniature electronic components. (*Bourns, Inc.*)

neers evaluate new devices and materials that may not appear in automobiles until several years from now. At the same time, they test component parts and completed cars currently coming off the assembly line. They are usually responsible for quality control of the manufacturing process. In addition to the education requirements of the design and production engineers, a fundamental knowledge of statistics is beneficial.

1.3.5 Construction

The counterpart of the production engineer in manufacturing is the construction engineer in the building industry (see Fig. 1.11). When an organization bids on a competitive construction project,

Fig. 1.11 Engineers in the construction industry become involved in estimating, site planning, and designing new structures. This engineer is inspecting new construction for a nuclear-energy installation. (*Union Carbide.*)

the construction engineer begins the process by estimating material, labor, and overhead costs. If the bid is successful, a construction engineer assumes the responsibility of coordinating the project. On large projects, a team of construction engineers may supervise the individual segments of construction such as mechanical (plumbing), electrical (lighting), and civil (building). In addition to a strong background in engineering fundamentals, the construction engineer needs on-the-job experience and an understanding of labor relations.

1.3.6 Operations

Up to this point, discussion has centered around the results of engineering efforts to discover, develop, design, and produce products that are of benefit to human beings. For such work, engineers obviously must have offices, laboratories, and production facilities in which to accomplish it. The major responsibility for supplying such facilities falls on the operations engineer (see Fig. 1.12). Sometimes called a plant engineer, this individual selects sites for facilities, specifies the layout for all facets of the operation, and selects the fixed equipment for climate control, lighting, and communication. Once the facility is in operation, the plant engineer is responsible for maintenance and modifications as requirements demand. Because this phase of engineering comes under the economic category of overhead, the operations engineer must be very conscious of cost and keep up with new developments in equipment so that overhead is maintained at the lowest possible level. A knowledge of

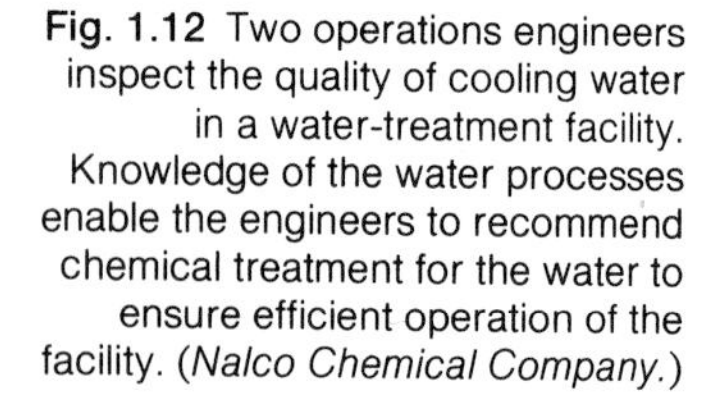

Fig. 1.12 Two operations engineers inspect the quality of cooling water in a water-treatment facility. Knowledge of the water processes enable the engineers to recommend chemical treatment for the water to ensure efficient operation of the facility. (*Nalco Chemical Company.*)

basic engineering, industrial engineering principles, economics, and law are prime educational requirements of the operations engineer.

1.3.7 Sales

In many respects, all engineers are involved in selling. To the research, development, design, production, construction, and operations engineers, selling means convincing management that money should be allocated for development of particular concepts or expansion of facilities. This is, in essence, selling one's own ideas. Sales engineering, however, means finding or creating a market for a product. The complexity of today's products requires an individual thoroughly familiar with materials in and operational procedures for consumer products to demonstrate to the consumer in layman's terms how the products can be of benefit. The sales engineer is thus the liaison between the company and the consumer, a very important means of influencing a company's reputation. An engineering background plus a sincere interest in people and a desire to be helpful are the primary attributes of a sales engineer. The sales engineer usually spends a great deal of time in the plant learning about the product to be sold. After a customer purchases a product, the sales engineer is responsible for coordinating service and maintaining customer satisfaction. As important as sales engineering is to a company, it still has not received the interest from new engineering graduates that other engineering functions have. See Fig. 1.13.

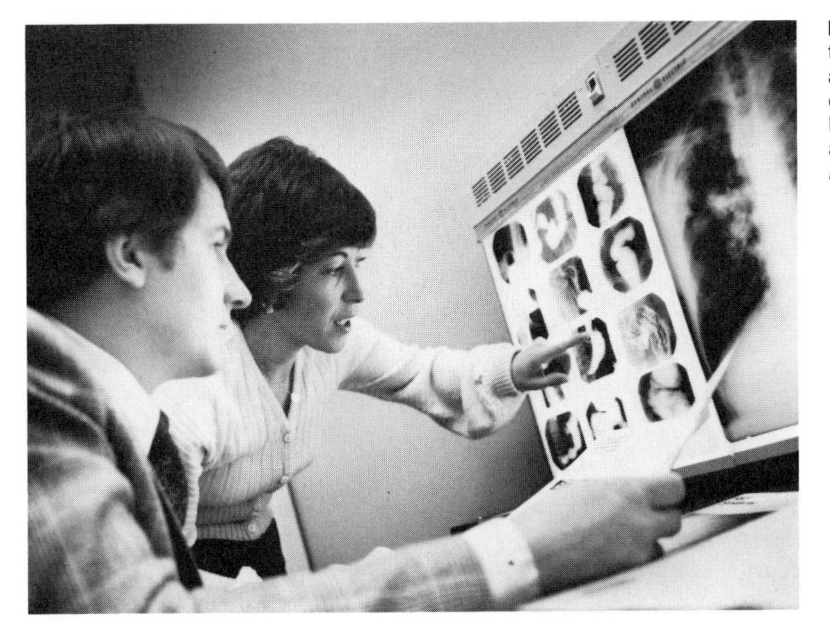

Fig. 1.13 A sales engineer points out features of a new product to a prospective customer. Sales engineering requires thorough knowledge of company operations and product performance. (*General Electric Company.*)

1.3.8 Management

Traditionally, management has consisted of individuals trained in business and groomed to assume positions leading to the top of the corporate ladder. However, with the influx of scientific and technological data being used in business plans and decisions, there has been a need for people in management with knowledge and experience in engineering and science. Recent trends indicate that a growing percentage of management positions are being assumed by engineers and scientists. Inasmuch as one of the principal functions of management is to use company facilities to produce an economically feasible product, and decisions must often be made that may affect thousands of people and involve millions of dollars over periods of several years, a balanced education of engineering or science and business seems to produce the best managerial potential. See Fig. 1.14.

At some time during your career as an engineer, a decision must be made about whether to remain with the technical functions of research, development, design, etc., or to obtain business education or experience and pursue the managerial route. This will require an honest self-evaluation and a commitment to abide totally by the decision. Your future success will depend on this.

1.3.9 Consulting

For someone interested in self-employment, a consulting position may be an attractive one (see Fig. 1.15). Consulting engineers

Fig. 1.14 Many engineers are involved in management, a great deal of which is communication with people. Here a supervisor discusses the progress of an engineering project with a young engineer. (*Dravo Corporation.*)

Fig. 1.15 The complexity of problems faced by consulting engineers requires a coordinated effort by persons from many disciplines. Meetings to discuss project status are commonplace. (*Stanley Consultants, Inc.*)

operate alone or in partnership, furnishing specialized help to clients who request it. Of course, as in any business, risks must be taken. Moreover, a sense of integrity and a knack for correct engineering judgment are primary necessities in such work.

A consulting engineer must possess a professional engineer's license before beginning practice. Consultants usually spend many years in a specific area before going on their own. A successful consulting engineer maintains a business primarily by being able to solve unique problems for which other companies have neither the time nor capability. In many cases, large consulting firms maintain a staff of engineers of diverse background so that a wide range of engineering problems can be contracted.

1.3.10 Teaching

Individuals interested in a career that involves helping others to become engineers will find teaching very rewarding (see Fig. 1.16). The engineering teacher must possess an ability to communicate abstract principles and engineering experiences in a manner that young people can understand and appreciate. By merely following general guidelines, the teacher is usually free to develop his or her own method of teaching and means of evaluating its effectiveness. In addition to teaching, the engineering educator can also become involved in student advising and research.

Engineering teachers today must have a mastery of fundamental engineering and science principles and a knowledge of applications. Customarily, they must obtain an advanced degree in

Fig. 1.16 The ability to communicate is one of the assets of a good teacher and not all teaching takes place at schools. Here an engineer uses a blackboard to explain a nuclear process to a coop student and a mathematician. (*Union Carbide.*)

order to improve their understanding of basic principles, to perform research in a specialized area, and perhaps to gain teaching experience on a part-time basis.

1.4 Education of the engineer

The amount of information coming from the academic and business world is increasing exponentially; and at the current rate, it will double in 20 years. More than any other group, engineers are using this knowledge to shape civilization. To keep pace with a changing world, engineers must be educated in their lifetimes to solve problems that are unheard of at the present time. A large share of the responsibility for this mammoth education task falls on the engineering colleges and universities. But the completion of an engineering program is only the first step toward a lifetime of education. The engineer must continue to study, with the assistance of employers and universities. See Fig. 1.17.

Logically, then, an engineering education should provide a broad base in scientific and engineering principles, some study in humanities and social sciences, and specialized studies in a chosen engineering curriculum. Some specific questions concerning engineering education might also require answers. We will deal here with the following questions that are frequently asked by students. What are the desirable characteristics for success in an engineering program? What knowledge and skills should be acquired in college? What is meant by continuing education with respect to an engineering career?

Fig. 1.17 Classroom situations such as this one are a major portion of the engineer's education in college.

1.4.1 Desirable characteristics

Years of experience have enabled engineering educators to analyze the performance of students in relation to abilities and desires possessed by students entering college. The most important characteristics in this respect can be summarized as

1. A strong interest in and ability to work with mathematics and science taken in high school
2. An ability to think through a problem in a logical manner
3. A knack for organizing and carrying through to conclusion the solution to a problem
4. An unusual curiosity about how and why things work

Although such attributes are desirable, having them is no guarantee of success in an engineering program. Simply a strong desire for the job has made successful engineers of some individuals who did not possess any of these characteristics; and, conversely, many who possessed them did not complete an engineering degree. Moreover, an engineering education is not easy, but it can offer a rewarding career to anyone who accepts the challenge.

1.4.2 Knowledge and skills that should be acquired

As indicated previously, over 200 colleges and universities offer programs in engineering that are accredited by the Engineer's Council for Professional Development (ECPD). It is safe to say that for any given discipline, no two schools will have identical offerings. However, close scrutiny will show a framework within which most courses can be placed, with differences occurring only in textbooks used, topics emphasized, and sequences followed. Figure 1.18 depicts this framework and some of the courses that fall within each of the areas. The approximate percent of time spent on each course grouping is indicated.

The sociohumanistic block is a small portion of most engineering curricula, but it is important because it helps the engineering student understand and develop an appreciation for the potential impact of engineering undertakings on the environment and general society. When the location of a nuclear power plant is being considered, the engineers involved in this decision must respect the concerns and feelings of all individuals who might be affected by the location. Discussions of the interaction between engineers and the general public is done in few engineering courses; sociohumanistic courses are thus needed to furnish engineering students with an insight into the needs and aspirations of society.

Chemistry and physics are almost universally required in engineering. They are fundamental to the study of engineering science.

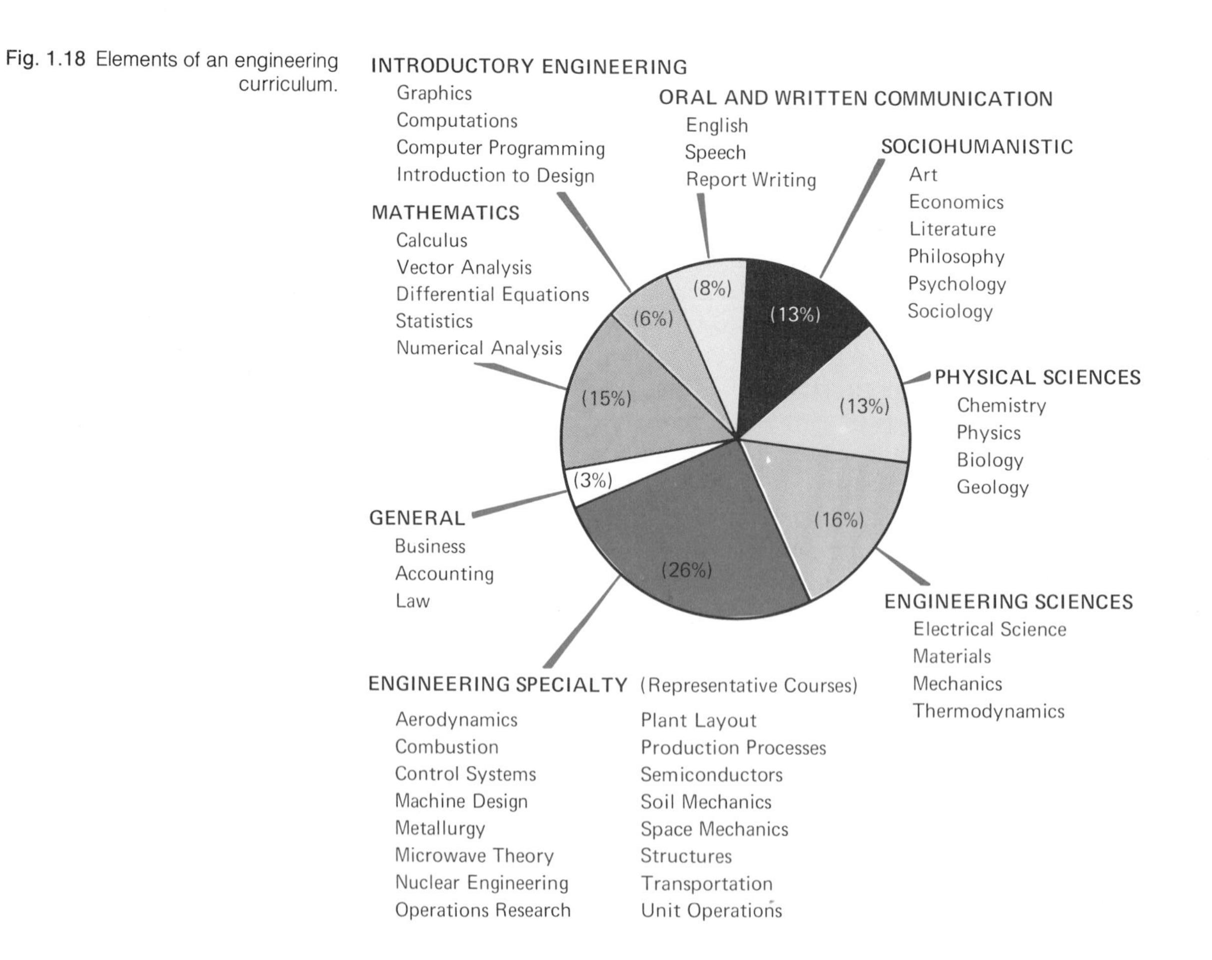

Fig. 1.18 Elements of an engineering curriculum.

The mathematics normally required for college chemistry and physics is more advanced than that for the corresponding high school courses. Higher-level chemistry and physics may also be required, depending on departmental structure. Finally, other physical science courses may be required in some programs or may be taken as electives.

An engineer cannot be successful without the ability to communicate ideas and the results of work efforts. The research engineer writes reports and orally presents ideas to management. The production engineer must be able to converse with craftsmen in understandable terms. And all engineers have dealings with the public and must be able to communicate on a nontechnical level.

Engineers have been accused of not becoming involved in public affairs. The reason often given for their not becoming involved is that they are not trained sufficiently in oral and written communications. However, the equivalent of one-third of a year is spent on formal courses in these subjects and additional time is spent in design presentations, written laboratory reports, and the like. But a conscious effort by the student engineer must be made to improve his or her abilities in oral and written communication to overcome this nonactivist label.

Mathematics is the most powerful tool the engineer uses to solve problems. The amount of time spent in this area is indicative of the importance. Calculus, vector analysis, and differential equations are common to all programs. Statistics, numerical analysis, and other mathematics courses support some engineering specialty areas. Students desiring an advanced degree may want to take mathematics courses beyond the baccalaureate level requirements.

In the early stages of an engineering education, introductory courses in graphical communication, computational techniques, design, and computer programming are taken. Engineering schools vary somewhat in their emphasis on these areas, but the general intent is to develop skills in the application of theory to practical problem solving and familiarity with engineering terminology. Design is presented from a conceptual point of view to aid the student in creative thinking. Graphics develops the visualization capability and assists the student in transferring mental thoughts into well-defined concepts on paper. The tremendous potential of the computer to assist the engineer has led to computer programming as a requirement of almost all curricula. Use of the computer to perform many tedious calculations has increased the efficiency of the engineer and allows more time for creative thinking.

With a sound background in mathematics and physical sciences, you can begin study of engineering sciences, courses that are fundamental to all engineering specialties. Electrical science includes study of charges, fields, circuits, and electronics. Materials

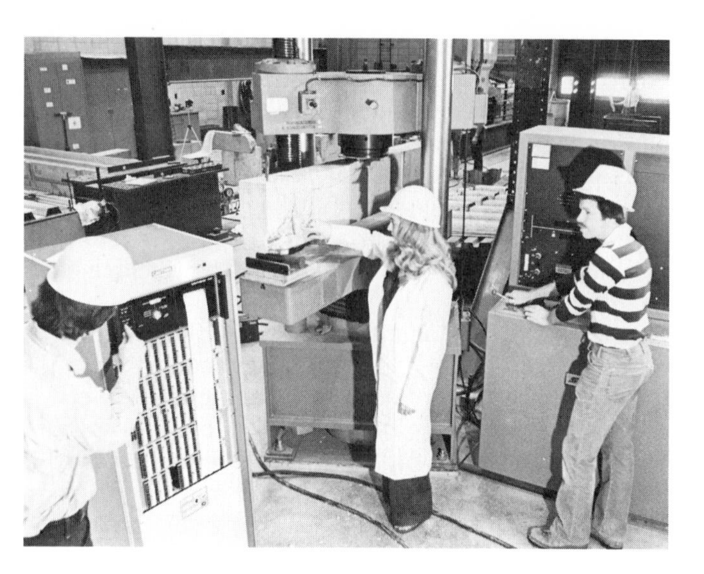

Fig. 1.19 A student team conducts a test in a laboratory on a concrete beam to demonstrate failure characteristics.

science courses involve training in the properties and chemical compositions of metallic and nonmetallic substances (see Fig. 1.19). Mechanics includes study of statics, dynamics, fluids, and mechanics of materials. Thermodynamics is the science of heat and is the basis for the study of all types of energy and energy transfer. A sound understanding of the engineering sciences is most important for anyone interested in pursuing post-graduate work and research.

Figure 1.18 shows only a few examples of the many specialized engineering courses given. Scanning course descriptions in a college general bulletin or catalog will provide a more detailed insight into the specialized courses required in the various engineering disciplines.

Most curricula allow a student flexibility in selecting a few courses in areas not previously mentioned. For example, a student interested in management may take some courses in business and accounting. Another may desire some background in law or medicine, with the intent of entering a professional school in one of these areas upon graduation from engineering.

1.5 Professionalism and ethics

Engineering is a learned vocation, demanding an individual with high standards of ethics and sound moral character. When making judgments that may create controversy and affect many people, the engineer must keep foremost in mind a dedication to the betterment of humanity.

The engineering profession has attempted for many years to become a unified profession. It is presently an open profession, as compared with such closed professions as law, medicine, and theology. In addition, technical societies representing individual engineering disciplines have grown strong and have tended to keep the various engineering factions separated.

1.5.1 Professionalism

Professionalism is a way of life. A professional person is one who engages in an activity that requires a specialized and comprehensive education and is motivated by a strong desire to serve humanity. He or she thinks and acts in a manner that brings favor upon the individual and the entire profession. Developing a professional frame of mind begins with your engineering education.

The professional engineer can be said to have the following:

1. Specialized knowledge and skills used for the benefit of humanity
2. Honesty and impartiality in engineering service
3. Constant interest in improving the profession
4. Support of professional and technical societies that represent the professional engineer

It is clear that these characteristics include not only technical competence but also a positive attitude toward life that is continually reinforced by educational accomplishments and professional service.

A primary reason for the rapid development in science and engineering is the work of the technical societies. The fundamental service provided by a society is the sharing of ideas, which means that technical specialists can publicize their efforts and assist others in promoting excellence in the profession. When information is distributed to other society members, new ideas evolve and duplicated efforts are minimized. The societies conduct meetings on international, national, and local bases. Students of engineering will find a technical society in their specialty that may operate as a branch of the regular society or as a student chapter on the campus. The student organization is an important link with the professional workers, providing motivation and the opportunity to make acquaintances that will help students to formulate career objectives.

The technical societies presently functioning are too numerous to mention here. Some are quite strong and influential in the engineering profession. To unify the entire engineering community will require cooperation of all the societies. But the individual societies have more than satisfied the professional needs for many engineers, so that no pressing desire to unify is apparent. Nonetheless to preserve the advantages of technical societies while unifying the entire profession is a major concern of engineers today.

1.5.2 Professional registration

The power to license engineers rests with the states. Since the first registration law in Wyoming in 1907, all states have developed legislation specifying requirements for engineering practice. The purpose of registration laws is to protect the public. Just as one would expect a doctor to provide competent medical service, an engineer can be expected to provide competent technical service. However, the laws of registration for engineers are quite different from those for lawyers or doctors. An engineer does not have to be registered to practice engineering. Legally, only the chief engineer of a firm need be registered for that firm to perform engineering services. In some instances, the practice of engineering is allowed as long as the individual does not advertise as an engineer.

Registration does have many advantages. Most public employment positions, all expert witness roles in court cases, and some high-level company positions require the professional engineer's license. However, less than one-half the eligible candidates are currently registered. You should give serious consideration to becoming registered as soon as possible upon graduation. Satisfying the requirements for registration can be started even before graduation from an ECPD accredited curriculum.

1.5.3 Professional ethics

Ethics is the guide to personal conduct of a professional. Most of the technical societies have a written code of ethics for their members. Because of this, some variations exist; but a general view of ethics for engineers is provided here from two of the technical societies. Figure 1.20 is a code endorsed by the Engineer's Council for Professional Development. Appendix D gives the most widely endorsed code of ethics, that of the National Society of Professional Engineers. As you read both codes, note the many similarities. Figure 1.21 is the Engineers' Creed as published by the National Society of Professional Engineers.

1.6 Challenges of the future

The world continues to undergo rapid and sometimes tumultuous changes. As an engineer, you will occupy center stage in many of these changes and will become even more involved in the future. The huge tasks of providing energy; maintaining a clean, safe world; exploring the universe; providing for the subsistence and

Fig. 1.20 Code of Ethics for Engineers. (*Engineers' Council for Professional Development.*)

CODE OF ETHICS OF ENGINEERS

THE FUNDAMENTAL PRINCIPLES

Engineers uphold and advance the integrity, honor and dignity of the engineering profession by:

I. using their knowledge and skill for the enhancement of human welfare;

II. being honest and impartial, and serving with fidelity the public, their employers and clients;

III. striving to increase the competence and prestige of the engineering profession; and

IV. supporting the professional and technical societies of their disciplines.

THE FUNDAMENTAL CANONS

1. Engineers shall hold paramount the safety, health and welfare of the public in the performance of their professional duties.

2. Engineers shall perform services only in the areas of their competence.

3. Engineers shall issue public statements only in an objective and truthful manner.

4. Engineers shall act in professional matters for each employer or client as faithful agents or trustees, and shall avoid conflicts of interest.

5. Engineers shall build their professional reputation on the merit of their services and shall not compete unfairly with others.

6. Engineers shall associate only with reputable persons or organizations.

7. Engineers shall continue their professional development throughout their careers and shall provide opportunities for the professional development of those engineers under their supervision.

Approved by the Board of Directors, October 1, 1974

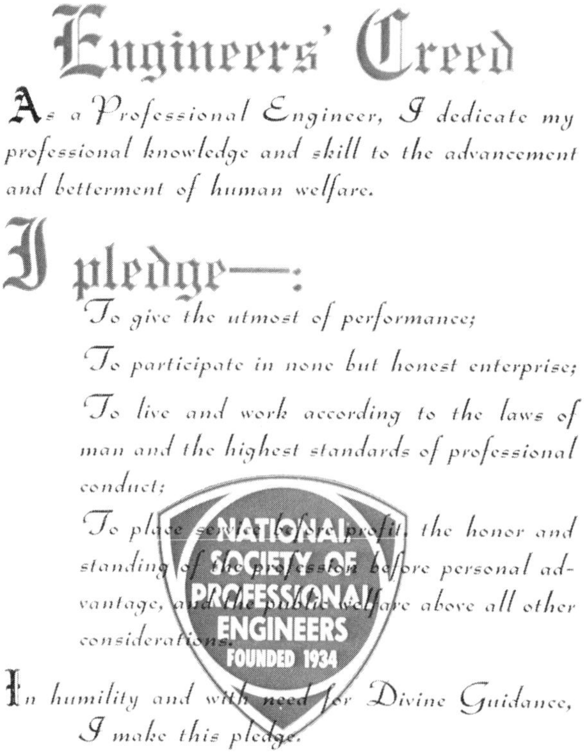

Fig. 1.21 Engineers' Creed. (*National Society of Professional Engineers.*)

Engineers' Creed

As a Professional Engineer, I dedicate my professional knowledge and skill to the advancement and betterment of human welfare.

I pledge—:

To give the utmost of performance;

To participate in none but honest enterprise;

To live and work according to the laws of man and the highest standards of professional conduct;

To place service before profit, the honor and standing of the profession before personal advantage, and the public welfare above all other considerations.

In humility and with need for Divine Guidance, I make this pledge.

NATIONAL SOCIETY OF PROFESSIONAL ENGINEERS
FOUNDED 1934

health of billions of people; and solving or coping with many other problems will challenge the technical community beyond anyone's imagination. With knowledge doubling every 20 years and with even the rate increasing, you must plan to be in a lifetime educational program.

Fig. 1.22 What lies in store for mankind as fossil energy supplies dwindle? This model of a liquid metal fast breeder reactor depicts one potential source for additional energy. (*Ames Laboratory, U.S. Department of Energy.*)

Fig. 1.23 As the population increases, solid waste disposal becomes increasingly difficult. Unique methods of separating combustible and noncombustible waste for recycling are currently in operation. This illustration shows a conveyor bringing solid waste to a shredder before the separation process begins. (*Iowa State University Engineering Research Institute.*)

You must work hard during your college years in obtaining a knowledge base that is necessary for your first technical position. Do not rest on your laurels after graduation when beginning your first job. If you do, you will find your supply of knowledge obsolete in a very short time. Challenges await you. Prepare to meet them well.

Problems

1.1 Compare definitions of an engineer and scientist from at least three different books that discuss engineering career opportunities.

1.2 Compare definitions of an engineer and technologist from at least two sources.

1.3 Find three textbooks that introduce the "design process." Write down from each the steps in the process. Note similarities and differences and write a paragraph describing your conclusions.

1.4 Find the name of a pioneer engineer in the field of your choice and write a brief paper on the accomplishments of this individual.

1.5 Find the name of an engineer from your state, province, or country and write a brief paper on the accomplishments of this individual.

1.6 Select a specific field of engineering and list at least 20 organizations that utilize engineers from this field.

1.7 Select one organization from your list in Prob. 1.6 and write a summary of the functions performed in this organization by engineers in your specified field.

1.8 For a particular traditional branch of engineering, for example, mechanical engineering, find the program of engineering courses for the first 2 years of study and compare it with the program at your school 25 years ago. Comment on major differences.

1.9 Do Prob. 1.8 for the last 2 years of study of a particular branch.

1.10 List five characteristics of yourself and compare with the list in Sec. 1.4.1.

1.11 Interview a professor in your department and ask how each of the following courses is used in his or her particular branch of engineering.

(*a*) Physics
(*b*) Chemistry
(*c*) Graphics
(*d*) Computer programming

1.12 Research and write a brief paper on the registration laws in your state, province, or country.

1.13 Prepare a 5-min talk to give to your class describing one of the technical societies and how it can benefit you while you are a student.

1.14 Choose one of the following topics (or one suggested by your instructor) and write a paper that discusses technological changes that have occurred in this area during the period 1970 to the present. Justify the changes from the viewpoint of engineering.

(*a*) Passenger automobile
(*b*) Electric power generating plants
(*c*) Television
(*d*) Air-pollution control systems
(*e*) Surgery
(*f*) National defense systems

Engineering
solutions

CHAPTER 2

2.1 Introduction

This chapter provides a basic guide to problem analysis, solution, and presentation. Early in your education, you must develop an ability to solve and present a range of complex problems in an orderly, logical, and systematic way. The material presented here will be clarified by examples throughout the text. Problems at the end of many of the subsequent chapters will reinforce what is presented in the text.

2.2 Problem analysis

A distinguishing characteristic of a qualified engineer is the ability to solve problems. Mastery involves a combination of an art and a science. By science we mean a knowledge of the principles of mathematics, chemistry, physics, mechanics, and other technical subjects that must be learned so that they can be applied correctly when appropriate. By art is meant the proper judgment, common sense, and know-how that must be used to reduce a real-life problem to such a form that science can be applied to its solution. To know when and how rigorously science should be applied and whether the resulting answer reasonably satisfies the original problem is an art.

Fig. 2.1 Engineers often spend much of their time solving problems, many times in different surroundings, such as in the office of a consulting firm as depicted here. (*Stanley Consultants.*)

Much of the science portion of successful problem solving comes from formal training in school or from continuing education after graduation. But most of the art of problem solving cannot be learned in a formal course but is rather a result of experience and common sense. Its application can be more effective, however, if problem solving is approached in a logical and organized method.

To clarify the distinction let us suppose that a manufacturing engineer working for an electronics company is given the task of recommending whether or not a new calculator can be profitably produced. At the time the engineering task is assigned, the competitive selling price has already been established by the marketing division. Also, the design group has developed working models of the calculator with specifications of all components, which means that the cost of these components is known. The question of profit thus rests on the cost of assembly. The theory of engineering economy (the science portion of problem solving) is well-known by the engineer and is applicable to the cost factors and time frame involved. Once the production methods have been established, the costs of assembly can be computed using standard techniques, such as methods time measurement analysis. Selection of production methods (the art portion of problem solving) depends strongly on the experience of the engineer. Knowing what will or will not work in each part of the manufacturing process is a must in the cost estimate, but that data cannot be found in a handbook. It is in the head of the engineer. It is an art originating from experience, common sense, and good judgment.

All problem solving requires some type of reasoning. Reasoning can be categorized as *deductive* (reasoning from the general to the specific) and *inductive* (reasoning from the particular to the general).

The most typical type of reasoning used in solving academic problems is deductive reasoning, because the majority of problems presented to students are of the closed-form or one-answer variety. Briefly stated, deductive reasoning has a major premise, a minor premise, and a conclusion. The *major premise* is a statement that is regarded as a general law; the *minor premise* is a statement that places a specific problem within the scope of the general law; and the *conclusion* is the statement that applies the general law to the specific case. A simple example of deductive reasoning is:

Major premise: The area of a triangle is $\frac{1}{2}$(base)(height).

Minor premise: A plot of land is triangular in shape.

Conclusion: The area of the plot of land can be calculated from the formula, $\frac{1}{2}$(base)(height).

A somewhat more difficult example of deductive reasoning from the field of fluid mechanics is the following:

Major premise: Pressure and velocity are related by Bernoulli's equation [pressure + $\frac{1}{2}$(density)(velocity)^2 = constant] for incompressible inviscid fluids.

Minor premise: Water flowing in a venturi acts approximately as an incompressible, inviscid fluid.

Conclusion: The velocity of the flow can be calculated from Bernoulli's equation if the pressure can be measured.

It could be said that deductive reasoning allows application of a general law to a specific case if you can recognize that the specific problem meets the criteria for application of the law. On the other hand, *inductive reasoning* involves observing specific cases and generalizing these observations to develop new laws. This is done in some of the following ways: if characteristics of two or more members of a class are the same, it is likely that other members of the class will share the same characteristics; observations at a particular time will likely be true in similar situations at other times; if some characteristics of members of a group are the same, other characteristics are also likely to be the same.

When reasoning, whether deductive or inductive, is used to solve a problem, the result is often some type of an idealized model of the real object or process. Most often, it is impossible to duplicate exactly every characteristic of a system in a mathematical or physical model, so you must judge which characteristics are to be included and which can be ignored because they do not materially affect the desired answer. The most common models are mathematical or physical ones. However, other types of models such as free-body diagrams in mechanics, control volumes in flow analysis, schematics in electronics, etc., are also frequently encountered in engineering work. These idealized models are amenable to analysis using mathematical, scientific, and engineering principles. It should be remembered that the solution found from the model is *not* the solution to the real problem, so its accuracy and usefulness as a solution to the real problem must be judged by the engineer who tries to apply the result.

The design process was introduced in Chap. 1 and is amplified in Chap. 15. One of the design steps involves analysis of alternative solutions. The goal of the next several chapters of this text is to introduce basic topics from engineering in order to provide experience in problem solving, which is necessary in the analysis of alternative solutions. This analysis procedure will hereinafter be called the *engineering method.* It can be stated in five basic steps:

1. Recognize and understand the analysis problem.

   Many times, the most difficult part of solving a problem is being able to recognize and define it precisely. The academic problems that you must solve generally have this step completed by the instructor. For example, if your instructor asks you to solve a quadratic algebraic equation but provides you with all the coefficients, the problem has been completely defined before it is given to you, so little doubt remains about what the problem is.

   If the problem is not totally defined, considerable effort must be expended in studying the problem, eliminating the things that are unimportant, and zeroing in on the root problem. Effort at this step pays great dividends by eliminating or reducing false trials and thereby shortening the time taken to complete later steps.

2. Accumulate facts.

   All pertinent physical facts such as sizes, temperatures, voltages, currents, costs, concentrations, weights, times, etc., must be ascertained. Some problems require that steps 1 and 2 be done simultaneously. In others, step 1 might automatically produce some of the physical facts. Do not mix or confuse these details with data that is suspect or only assumed to be accurate. Deal only with items that can be verified. Sometimes it will pay to actually verify data which you believe to be factual but that may actually be in error.

3. Make necessary assumptions and select appropriate theory or principle.

   Real problems cannot always be solved in their complete forms because they are sometimes too complicated to lend themselves to workable solutions. Assumptions may be made that do not greatly affect the solution or reduce its accuracy but do allow the application of known scientific or engineering principles. These assumptions must always be clearly stated and their effects understood so that a valid interpretation of the resultant solution can be made. At this step, experience and ingenuity are extremely important to have. Here existing methods or solutions can possibly be adapted to a new application. However, you must be capable of recognizing the possibility of new applications while guarding against overextending the use of a theory.

4. Solve the problem.

   If step 3 has resulted in a mathematical equation (model) to be solved, it can be attacked by trial and error, graphically, with mathematical theories, graphs, etc. The result will be an algebraic or numerical answer that must be interpreted in terms of the physical problem. It is also possible that the solution may be in terms of a laboratory model from which answers can be found experimentally.

5. Verify and check.

The work isn't finished merely because a solution has been obtained. It must still be checked to see if it is mathematically correct and the units must be specified. Correctness can be checked by reworking the problem, by using a different technique or by performing the calculations in a different order to be certain that the numbers agree in both trials. The units can be examined to see that all equations are dimensionally correct.

The answer must then be examined to see if it makes sense. You will generally have a good idea of the order of magnitude to expect. If the answer doesn't seem reasonable, there is probably an error in the mathematics or in the assumptions and theory used. For example, suppose that you are asked to compute the monthly payment required to repay a car loan of $5 000 over a 3-year period at an annual interest rate of 12 percent. After solving this problem, your calculated answer is $11 000 per month. Even if you are inexperienced in engineering economy, this answer is not reasonable, so you should reexamine your theory and computations. Examination and evaluation of the reasonableness of an answer is a habit that you should strive to acquire. Your employer and instructor alike will find it unacceptable to be given results which you have indicated to be correct but which are obviously incorrect by a significant percentage. Otherwise the employer or instructor might conclude that you have not developed good judgment or, worse yet, have not taken the time to do the necessary checking.

2.3 Problem presentation

Once a problem has been solved and checked, it is necessary to present the solution to others who either must know about the results or are responsible for implementing them. Most often, the solution must be presented to another individual who is technically trained but who may not have an intimate knowledge of the problem itself or of the solution technique used. Presenting technical information to laymen may require different methods from those used when communicating with another engineer to be certain that key bits of information are understood.

But the objective is usually to furnish a technical presentation designed for an instructor or supervisor who does understand such data. A mark of an engineer is the ability to present information with great clarity in a neat, careful manner. In short, the information must be communicated wholly to the reader. (Discussion of drawings or simple sketches will not be included in this chapter although they are important in many presentations. See App. C.)

Employers insist on carefully done presentations that completely document all work involved in solving the problems. Thorough documentation may be important in the event of a lawsuit, for which the details of the work might be introduced into the court proceedings as evidence. Lack of such documentation may result in the loss of a case that might otherwise have been won. Moreover internal company use of the work is easier and more efficient if all aspects of it have been carefully supported and substantiated by data and theory.

Each industrial company, consulting firm, governmental agency, or university has established standards for presenting technical information. These standards vary slightly, but all fall into a basic pattern, which will be discussed below. Each organization expects its employees to follow its standards. Details can be easily learned in a particular situation once you are familiar with the general pattern that exists in all of these standards.

2.4 Problem layout

It is not possible to specify a single problem layout or format that will accommodate all types of engineering problem solutions. Such a wide variety of solutions exists that the technique used must be adapted to fit the information to be communicated. In all cases, however, one must lay out a given problem in such a fashion that it can be easily grasped by the reader. No matter what technique is used, it must be logical and understandable.

Guidelines for problem presentation are suggested below. Acceptable layouts for problems in engineering are also illustrated. The guidelines are not intended as a precise format that must be followed but rather as a suggestion that should be considered and incorporated whenever applicable.

1. The most common type of paper used is that which is ruled horizontally and vertically on the reverse side, with only heading and margin rulings on the front. It is often called engineering problems paper. The rulings on the reverse side which are faintly visible through the paper, help one maintain horizontal lines of lettering and provide guides for sketching and simple graph construction. Moreover, the lines on the back of the paper will not be lost as a result of erasures.

2. The completed top heading of the problems paper should include such information as name, date, course number, and sheet number. The upper right-hand block should normally contain a notation such as *a*/*b*, where *a* is the page number of the sheet and *b* is the total number of sheets in the set.

3. Work should ordinarily be done in pencil with a lead that is hard enough (approximately H or 2H) that the linework is crisp and unsmudged. Erasures should always be complete, with all eraser particles removed.

4. Either vertical or slant letters may be selected as long as they are not mixed. Care should be taken to produce good, legible lettering but without such care that little work is accomplished. (See App. C for more information about lettering.)

5. Spelling should be checked for correctness. There is no reasonable excuse for incorrect spelling in a properly done problem solution.

6. Work must be easy to follow and uncrowded. Making an effort to keep it so contributes greatly to readability and ease of interpretation.

7. If several problems are included in a set, they must be distinctly separated, usually by a horizontal line drawn completely across the page between problems. Never begin a second problem on the same page if it cannot be completed there. It is usually better to begin each problem on a fresh sheet except in cases where two or more problems can be completed on one sheet. It is not necessary to use a horizontal separation line if the next problem in a series begins at the top of a new page.

8. Diagrams that are an essential part of a problem presentation should be carefully rendered and of excellent quality. (Refer to App. C for details of graphical techniques.) Often a good rough sketch is adequate but using a straightedge can greatly improve the appearance and accuracy of the sketch. A little effort in preparing a sketch to approximate scale can pay great dividends when it is necessary to judge the reasonableness of an answer, particularly if the answer is a physical dimension that can be seen on the sketch.

9. The proper use of symbols is always important, particularly when the International System (SI) of units is used. It involves a strict set of rules that must be followed so that absolutely no confusion of meaning can result. (Details concerning the use of units can be found in Chap. 5.) There are also symbols in common and accepted use for engineering quantities that can be found in most engineering handbooks. These symbols should be used whenever possible. It is important that symbols be consistent throughout a solution and that all are defined for the benefit of the reader and also for your own reference.

The physical layout of a problem solution logically follows steps similar to those of the engineering method. You should attempt to present the process by which the problem was solved in addition to the solution, so that any reader can readily understand all aspects of the solution. The following steps should be observed when planning and executing a problem layout. (Figure 2.2 illustrates the placement of the information.)

1. State the problem to be solved. The statement can often be simply a summary of the problem, but it should contain all the essential information, including what is to be determined.

2. Prepare a diagram (sketch) with all pertinent dimensions, flow rates, currents, voltages, weights, etc. A diagram is a very efficient method of showing given and required information. It also is a simple way of showing the

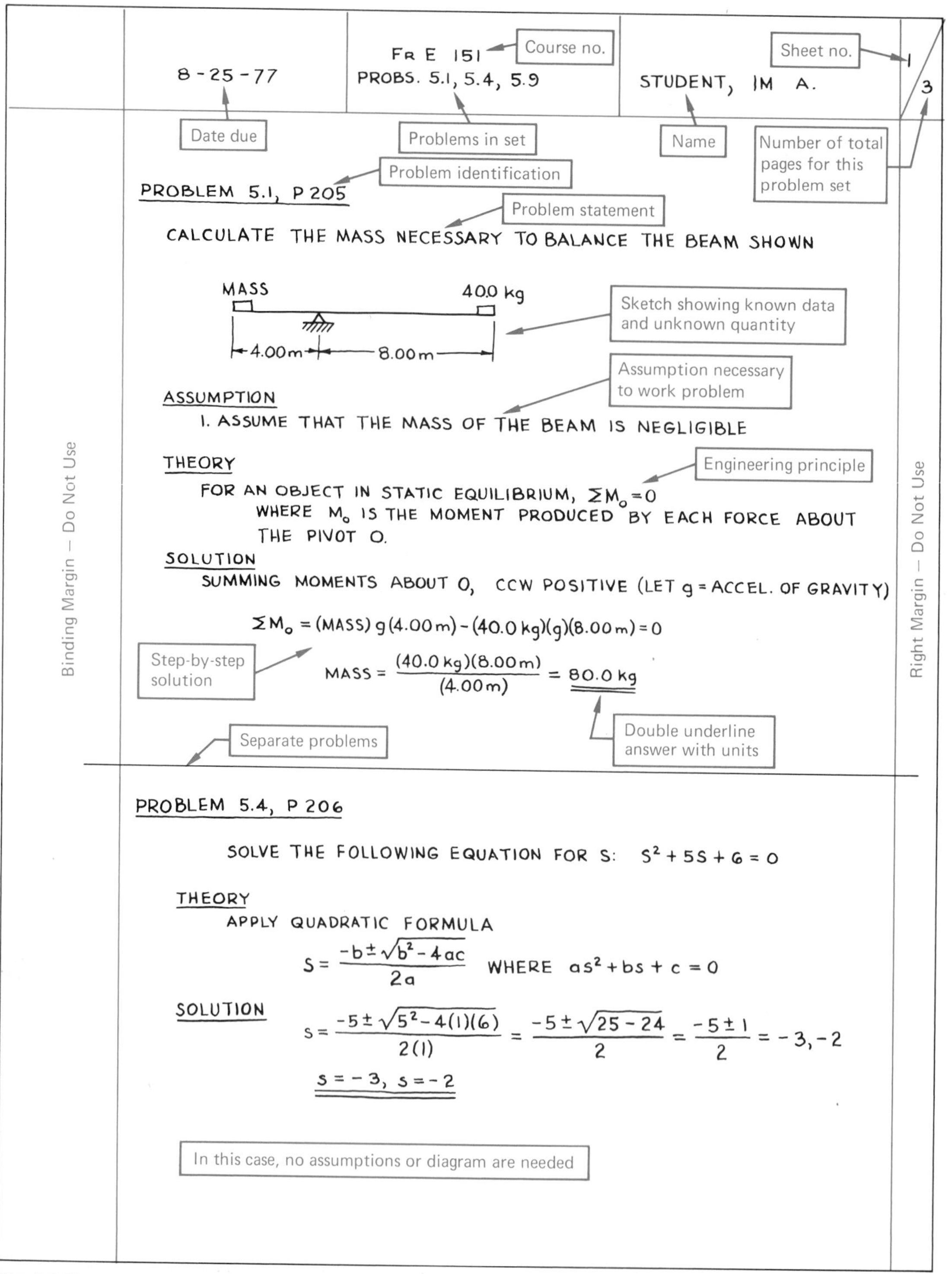

Fig. 2.2 Elements of a problem layout.

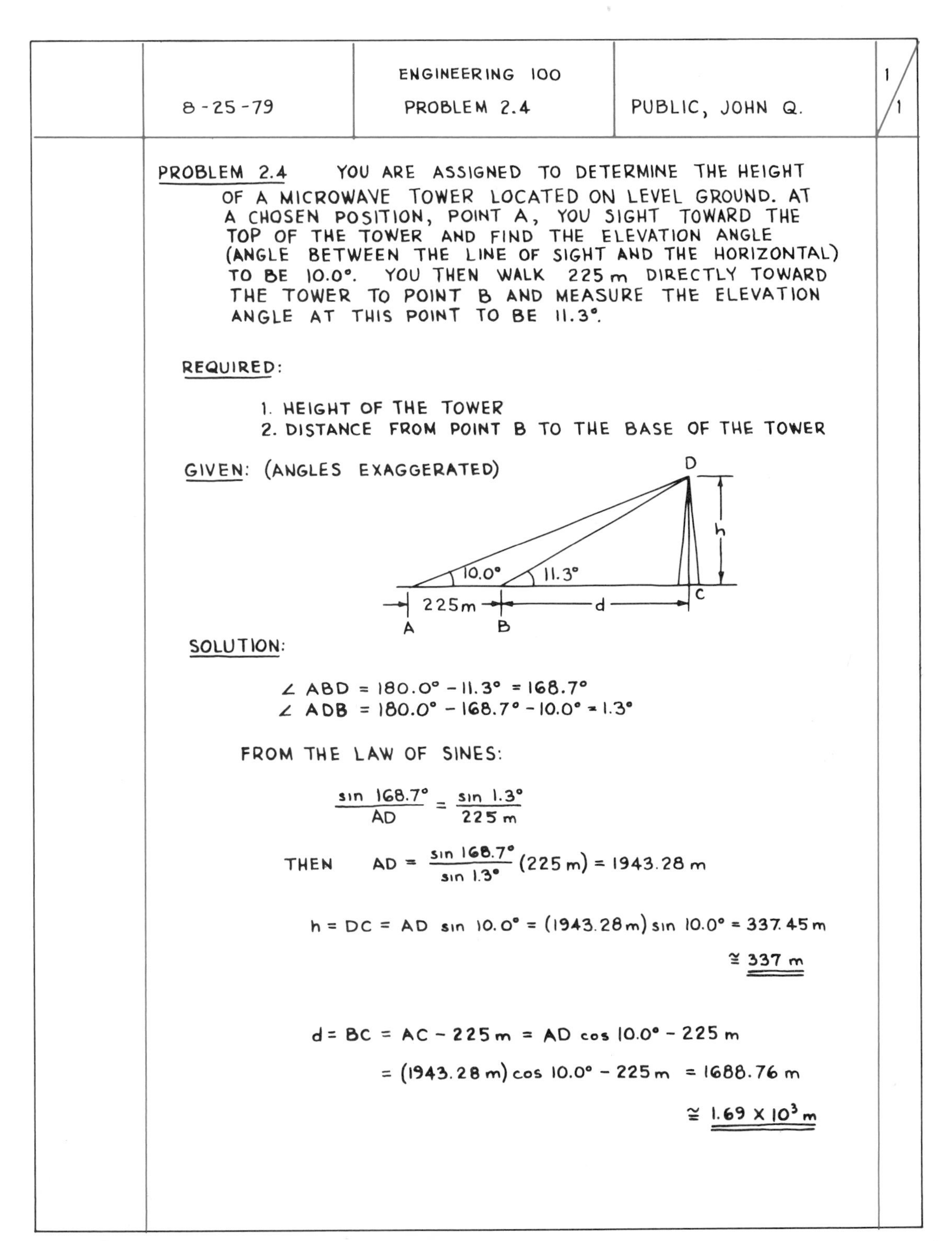

	8-25-79	ENGINEERING 100 PROBLEM 2.4	PUBLIC, JOHN Q.	1/1

PROBLEM 2.4 YOU ARE ASSIGNED TO DETERMINE THE HEIGHT OF A MICROWAVE TOWER LOCATED ON LEVEL GROUND. AT A CHOSEN POSITION, POINT A, YOU SIGHT TOWARD THE TOP OF THE TOWER AND FIND THE ELEVATION ANGLE (ANGLE BETWEEN THE LINE OF SIGHT AND THE HORIZONTAL) TO BE 10.0°. YOU THEN WALK 225 m DIRECTLY TOWARD THE TOWER TO POINT B AND MEASURE THE ELEVATION ANGLE AT THIS POINT TO BE 11.3°.

REQUIRED:

1. HEIGHT OF THE TOWER
2. DISTANCE FROM POINT B TO THE BASE OF THE TOWER

GIVEN: (ANGLES EXAGGERATED)

SOLUTION:

$\angle ABD = 180.0° - 11.3° = 168.7°$
$\angle ADB = 180.0° - 168.7° - 10.0° = 1.3°$

FROM THE LAW OF SINES:

$$\frac{\sin 168.7°}{AD} = \frac{\sin 1.3°}{225\ m}$$

THEN $$AD = \frac{\sin 168.7°}{\sin 1.3°}(225\ m) = 1943.28\ m$$

$$h = DC = AD \sin 10.0° = (1943.28\ m)\sin 10.0° = 337.45\ m$$
$$\cong \underline{\underline{337\ m}}$$

$$d = BC = AC - 225\ m = AD \cos 10.0° - 225\ m$$
$$= (1943.28\ m)\cos 10.0° - 225\ m = 1688.76\ m$$
$$\cong \underline{\underline{1.69 \times 10^3\ m}}$$

Fig. 2.3 Sample problem presentation.

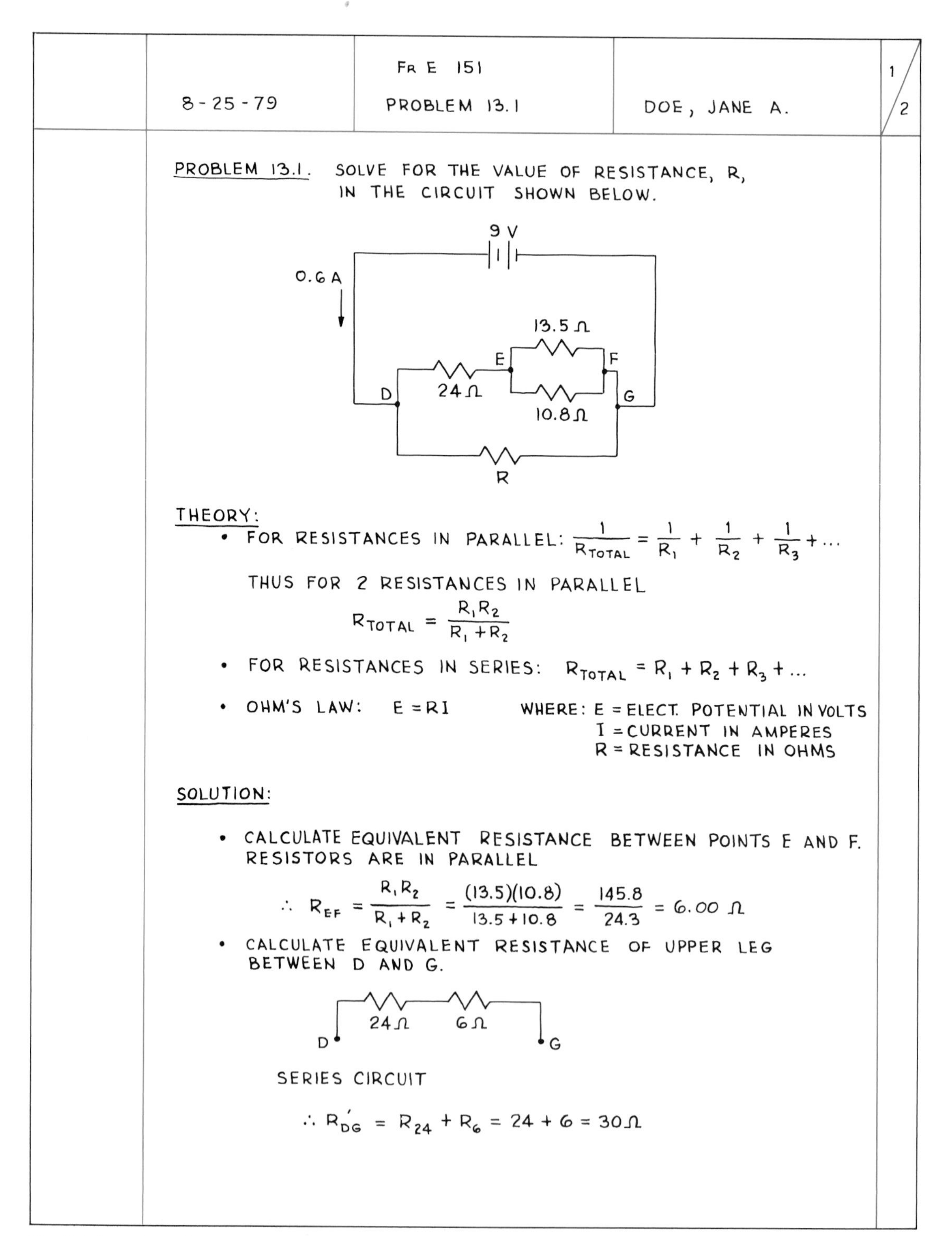

8-25-79 | FR E 151 PROBLEM 13.1 | DOE, JANE A. | 1/2

PROBLEM 13.1. SOLVE FOR THE VALUE OF RESISTANCE, R, IN THE CIRCUIT SHOWN BELOW.

THEORY:

- FOR RESISTANCES IN PARALLEL: $\frac{1}{R_{TOTAL}} = \frac{1}{R_1} + \frac{1}{R_2} + \frac{1}{R_3} + \ldots$

  THUS FOR 2 RESISTANCES IN PARALLEL

$$R_{TOTAL} = \frac{R_1 R_2}{R_1 + R_2}$$

- FOR RESISTANCES IN SERIES: $R_{TOTAL} = R_1 + R_2 + R_3 + \ldots$
- OHM'S LAW: $E = RI$ WHERE: E = ELECT. POTENTIAL IN VOLTS
  I = CURRENT IN AMPERES
  R = RESISTANCE IN OHMS

SOLUTION:

- CALCULATE EQUIVALENT RESISTANCE BETWEEN POINTS E AND F. RESISTORS ARE IN PARALLEL

$$\therefore R_{EF} = \frac{R_1 R_2}{R_1 + R_2} = \frac{(13.5)(10.8)}{13.5 + 10.8} = \frac{145.8}{24.3} = 6.00\ \Omega$$

- CALCULATE EQUIVALENT RESISTANCE OF UPPER LEG BETWEEN D AND G.

SERIES CIRCUIT

$$\therefore R'_{DG} = R_{24} + R_6 = 24 + 6 = 30\ \Omega$$

Fig. 2.4 Sample problem presentation.

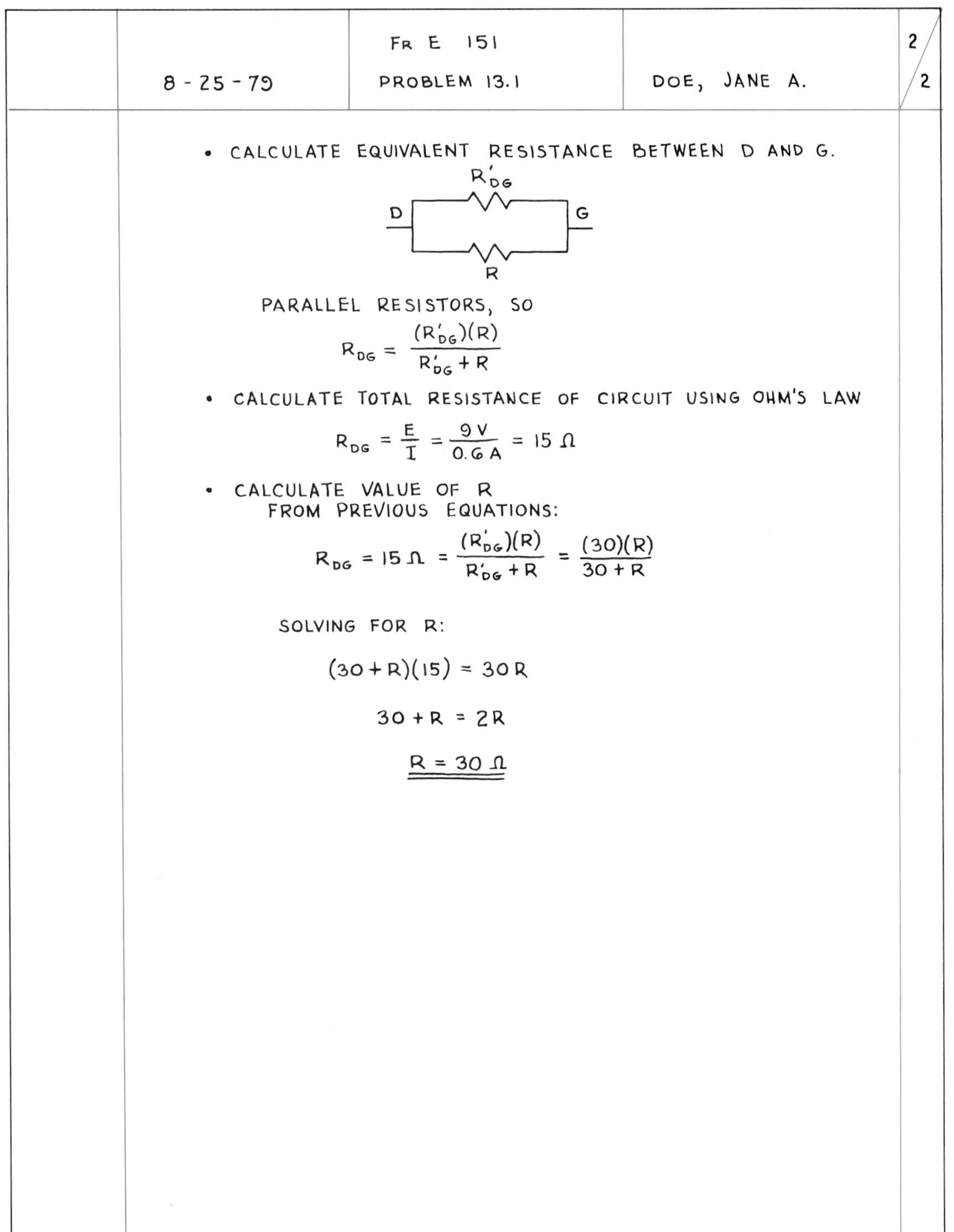

	8 - 25 - 79	FR E 151 PROBLEM 13.1	DOE, JANE A.	2/2

• CALCULATE EQUIVALENT RESISTANCE BETWEEN D AND G.

PARALLEL RESISTORS, SO

$$R_{DG} = \frac{(R'_{DG})(R)}{R'_{DG} + R}$$

• CALCULATE TOTAL RESISTANCE OF CIRCUIT USING OHM'S LAW

$$R_{DG} = \frac{E}{I} = \frac{9\ V}{0.6\ A} = 15\ \Omega$$

• CALCULATE VALUE OF R
FROM PREVIOUS EQUATIONS:

$$R_{DG} = 15\ \Omega = \frac{(R'_{DG})(R)}{R'_{DG} + R} = \frac{(30)(R)}{30 + R}$$

SOLVING FOR R:

$$(30 + R)(15) = 30R$$

$$30 + R = 2R$$

$$\underline{\underline{R = 30\ \Omega}}$$

physical setup, which may be difficult to describe adequately in words. Given data that cannot properly be placed in a diagram should be separately listed.

3. Explicitly list in sufficient detail any and all pertinent assumptions that must be made to obtain a solution or that you have arbitrarily placed on the problem. This step is vitally important for the reader's understanding of the solution and its limitations.

4. If a theory has to be derived, developed, or modified, present it next. In some cases, a properly referenced equation is sufficient. At other times, an extensive theoretical derivation may be necessary.

5. Show completely all steps taken in obtaining the solution. This is particularly important in an academic situation, because your reader, the instructor, must have the means of judging your understanding of the solution technique. Steps completed but not shown make it difficult for instructors to do so and, therefore, difficult for them to grade or critique the work.

6. Clearly indicate the final answer by underlining it with a double rule. Assign proper units according to SI rules. An answer without units (when it should have units) is meaningless.

Although this completes the layout precedure, remember that the final step of the engineering method requires that the answer be examined to determine if it is realistic. This step is often overlooked and results in incorrect solutions being presented when a student really does understand the problem but has been careless or has made an erroneous assumption.

Figures 2.3 and 2.4 illustrate satisfactory presentations of problem solutions. Study them carefully.

Problems

2.1 The cartesian components of a vector $\overline{B}$ are shown in Fig. 2.5. If $B_x = 7.2$ m and $\Delta = 35°$, find α, B_y, and B.

2.2 Refer to Fig. 2.5. If $\alpha = 51°$ and $B_y = 4.9$ km, what are the values of Δ, B_x, and B?

2.3 In Fig. 2.6, side YZ is 1.0×10^6 m. Determine the length of side XZ.

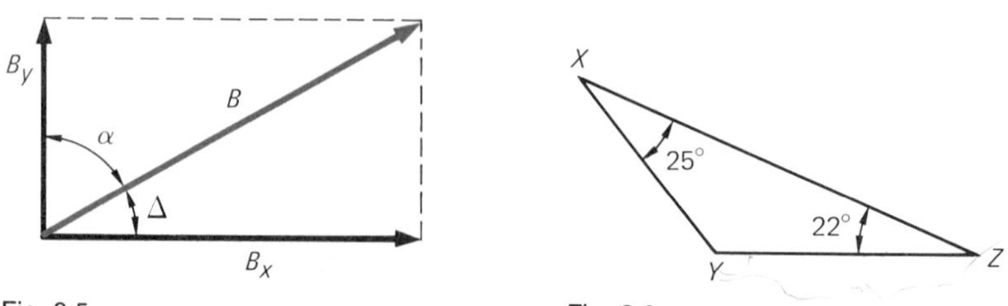

Fig. 2.5

Fig. 2.6

2.4 Calculate the length of side AB in Fig. 2.7 if side $AC = 3.6 \times 10^3$ m.

2.5 The vector $\overline{C}$ in Fig. 2.8 is the sum of vectors $\overline{A}$ and $\overline{B}$. Assume that vector $\overline{B}$ is horizontal. Given that $\alpha = 31°$, $\beta = 22°$, and the magnitude of $\overline{B} = 29$ m, find the magnitudes and directions of vectors $\overline{A}$ and $\overline{C}$.

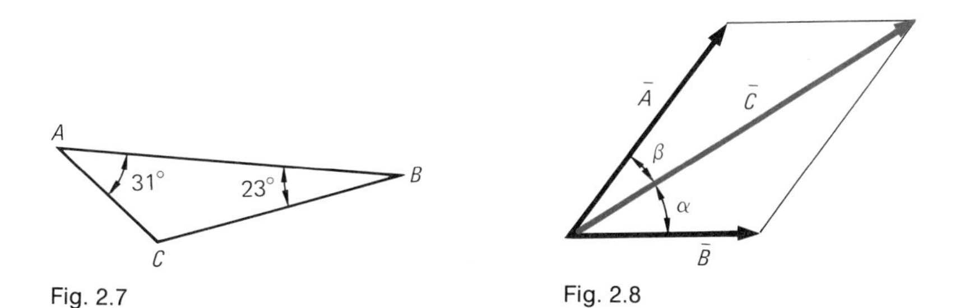

Fig. 2.7

Fig. 2.8

2.6 Vector $\overline{R}$ in Fig. 2.9 is the difference between vectors $\overline{T}$ and $\overline{S}$. If $\overline{S}$ is inclined at 25° from the vertical and the angle between $\overline{S}$ and $\overline{T}$ is 35°, calculate the magnitude and direction of vector $\overline{R}$. The magnitudes of $\overline{S}$ and $\overline{T}$ are 21 cm and 38 cm, respectively.

2.7 An aircraft has a glide ratio of 12 to 1. (Glide ratio means that the plane drops 1 m in each 12 m it travels horizontally.) A building 45 m high lies directly in the glide path to the runway. If the aircraft clears the building by 12 m, how far from the building does the aircraft touch down on the runway?

2.8 A pilot of an aircraft knows that the vehicle in landing configuration will glide 2.0×10^1 km from a height of 2.00×10^3 m. A TV transmitting tower is located in a direct line with the local runway. If the pilot glides over the tower with 3.0×10^1 m to spare and touches down on the runway at a point 6.5 km from the base of the tower, how high is the tower?

2.9 A simple roof truss design is shown in Fig. 2.10. The lower section, $VWXY$, is made from three equal length segments. UW and XZ are perpendicular to VT and TY, respectively. If $VWXY$ is 2.0×10^1 m and the height of the truss is 2.5 m, determine the lengths of XT and XZ.

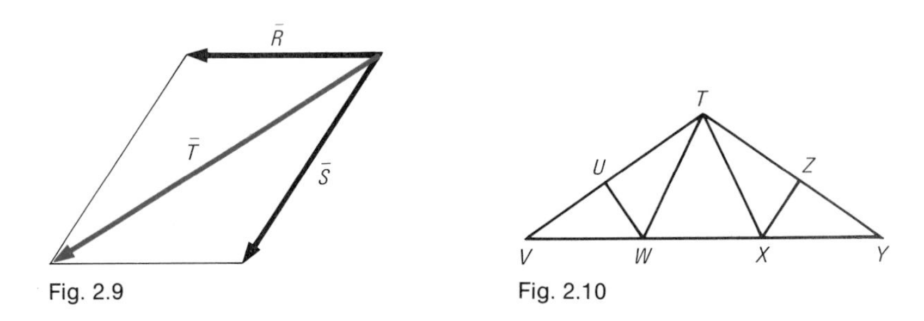

Fig. 2.9

Fig. 2.10

2.10 An engineer is required to survey a nonrectangular plot of land but is unable to measure side UT directly (see Fig. 2.11). The following data are taken: $RU = 130.0$ m, $RS = 120.0$ m, $ST = 90.0$ m, angle $RST = 115°$, and angle $RUT = 100°$. Calculate the length of side UT and the area of the plot.

Fig. 2.11

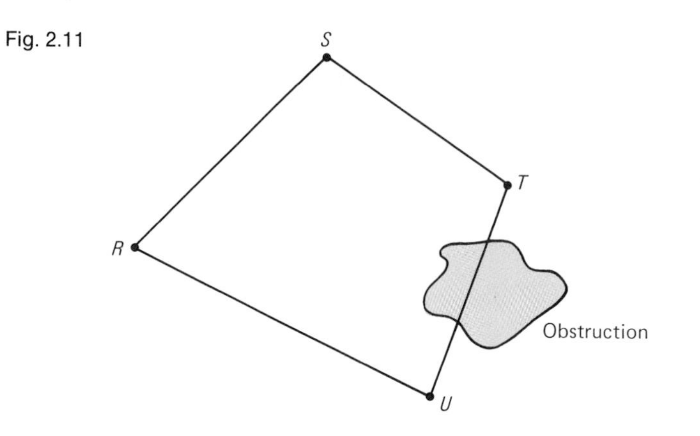

2.11 A park is being considered in a space between a small river and a highway as a rest stop for travelers (see Fig. 2.12). Boundary BC is perpendicular to the highway and boundary AD makes an angle of 75° with the highway.

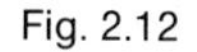
Fig. 2.12

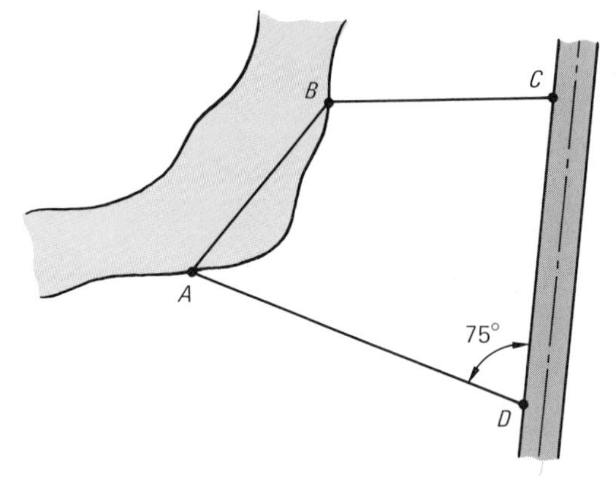

BC is measured to be 160.0 m, AD is 270.0 m, and the boundary along the highway is 190.0 m long. What is the length of side AB and the magnitude of angle ABC?

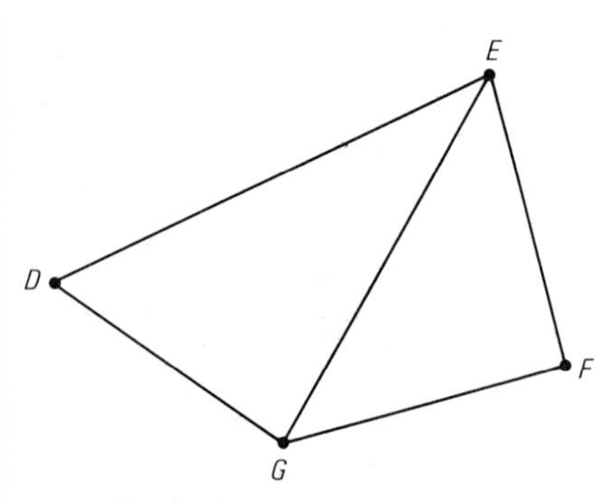

Fig. 2.13

2.12 D, E, F, and G in Fig. 2.13 are surveyed points in a land development on level terrain so that each point is visible from each other. Leg DG is physically measured as 500.0 m. The angles at three of the points are found to be: angle $GDE = 55°$, angle $DEF = 92°$, angle $FGD = 134°$. Also, angle DGE is measured at 87°. Compute the lengths of DE, EF, FG, and EG.

2.13 The height of an inaccessible mountain peak, C, in Fig. 2.14 must be estimated. Fortunately, two smaller mountains, A and B, which can be easily

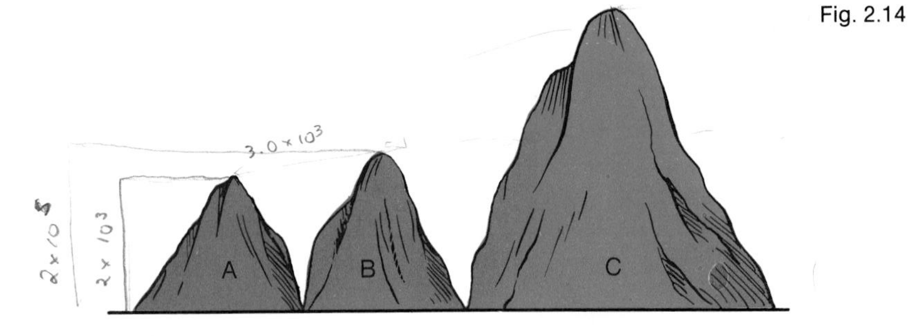

Fig. 2.14

scaled, are located near the higher peak. To make matters even simpler, the three peaks lie on a single straight line. From the top of mountain *A*, altitude 2.000×10^3 m, the elevation angle to *C* is 12.32°. The elevation of *C* from mountain *B* is 22.73°. Mountain *B* is 1.00×10^2 m higher than *A*. The straight line (slant) distance between peaks *A* and *B* is 3.000×10^3 m. Determine the unknown height of mountain *C*.

2.14 A narrow belt is used to drive a 20.00-cm-diameter pulley from a 35.00-cm-diameter pulley. The centers of the two pulleys are 2.000 m apart. How long must the belt be if the pulleys rotate in the same direction? In opposite directions?

2.15 A bicycle crank sprocket has a diameter of 15 cm and the driver sprocket on the rear wheel has a diameter of 5.0 cm. The crank and rear axle are 75 cm apart. What is the minimum chain length for this application?

2.16 A block of metal has a 90° notch cut from its lower surface. The notched part rests on a circular cylinder of diameter 2.0 cm, as shown. If the lower surface of the part is found to be 1.3 cm above the base plane, how deep is the notch? See Fig. 2.15.

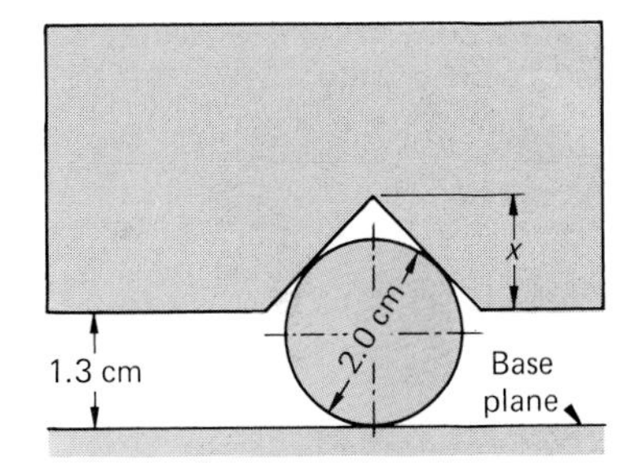

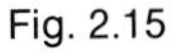
Fig. 2.15

2.17 A 1.00-cm-diameter circular gauge block is used to measure the depth of a 60° notch in a piece of tool steel. The gauge block extends a distance of 4.7 mm above the surface. How deep is the notch? See Fig. 2.16.

2.18 An aircraft moves through the air with a relative velocity of 3.00×10^2 km/h at a heading of N30°E. In a 35 km/h wind from the west:

(*a*) Calculate the *true* ground speed and heading of the aircraft.

(*b*) What heading should the pilot fly so that the *true* heading is N30°E?

2.19 In order to cross a river that is 1 km wide, with a current of 6 km/h, a novice boat skipper holds the bow of the boat perpendicular to the far river bank, intending to cross to a point directly across the river from the launch point. At what position will the boat actually contact the far bank? What direction should the boat have been headed in order to actually reach a point directly across from the launch dock? The boat is capable of making 10 km/h.

2.20 What heading must a pilot fly to compensate for a 125 km/h west wind in order to have a ground track that is due south? The aircraft cruise speed is 6.00×10^2 km/h. What is the actual ground speed?

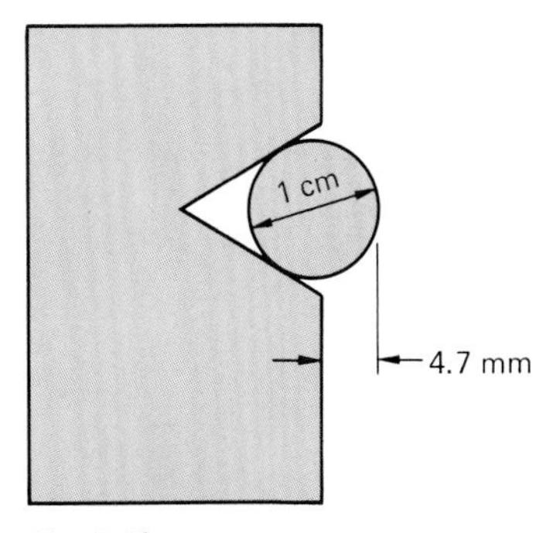

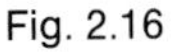
Fig. 2.16

Representation of technical information

CHAPTER 3

3.1 Introduction

This chapter provides information and guidelines that will be helpful when collecting, plotting, and interpreting technical and scientific data. Two areas in particular will be discussed in some detail: (1) graphical presentation of data and (2) graphical analysis.

Proper graphical presentation of data is necessary because calculated or experimental results are frequently collected in tabular form. Presentation in such form is generally not an optimal method of demonstrating the relationships between numerical values, since columns of numbers can sometimes be difficult to interpret. A system of graphing is thus needed. A visual impression, that is, something carefully and correctly presented in graph or chart form, is a much easier way to compare the rate of change or relative magnitude of variables.

In contrast, graphical analysis involves calculation and interpretation of data after it has been plotted. At times visual impressions are not adequate, so determination of an equation is required.

Even though it is important for engineers to be able to interpret, evaluate, and communicate different forms of data, it is virtually impossible to include in one chapter all types of graphs and charts

Fig. 3.1 Hard-copy displays of technical data. (*Beech Aircraft Corporation, Boulder Division.*)

that they may encounter. Popular-appeal, or advertising, charts such as bar charts, pie diagrams, and distribution charts, although useful to the engineer, will not be discussed here. Such types are summarized in App. C.

Numerous examples will be used throughout the chapter to illustrate methods of representation because the effectiveness of graphs depends to a large extent on the details of their construction and appearance.

3.2 Collecting and recording data

Modern science as we know it today was founded on scientific measurement. Meticulously designed experiments, carefully analyzed, have produced volumes of scientific data that have been collected, recorded, and documented. For such data to be meaningful, however, certain laboratory procedures must be followed. Formal data sheets, such as those shown in Fig. 3.2, or notebooks should be used to record all observations. Information about all instruments and experimental apparatus used should also be recorded. Sketches illustrating the physical arrangement of equipment can also be very helpful. Under no circumstances should observations be recorded elsewhere or data points erased. The data sheet is the "notebook of original entry." If there is reason for doubting the value of any entry, it may be canceled (that is, not considered) by drawing a line through it. The cancellation should be done in such a manner that the original entry is not obscured, in case you want to refer to it later.

Sometimes a measurement requires minimal accuracy, so time can be saved by making rough estimates. As a general rule, however, it is advantageous to make all measurements as accurately and reasonably as possible. Unfortunately, as different observations are made throughout any experiment, some degree of inconsistency will develop. It is a well-established fact that errors will enter into all experimental work regardless of the amount of care exercised. A more complete discussion of error is covered in Chap. 4.

It can be seen from what we have just discussed that the analysis of experimental data involves not only measurements and collection of data but also careful documentation and interpretation of results.

Experimental data once collected is normally organized into some tabular form, which is the next step in the process of analysis. Data, such as that tabulated in Table 3.1, should be carefully labeled and neatly lettered so that results are not misinterpreted.

This particular collection of data represents atmospheric pressure and temperature measurements recorded at various altitudes

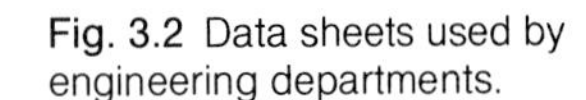

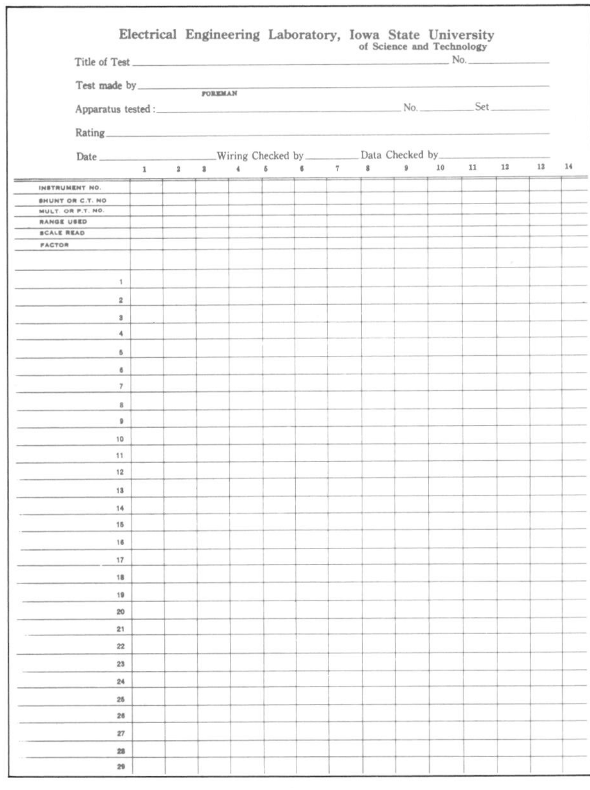

Electrical Engineering Laboratory, Iowa State University
of Science and Technology

Title of Test ______ No. ______

Test made by ______
FOREMAN

Apparatus tested: ______ No. ______ Set ______

Rating ______

Date ______ Wiring Checked by ______ Data Checked by ______

	1	2	3	4	5	6	7	8	9	10	11	12	13	14
INSTRUMENT NO.														
SHUNT OR C.T. NO														
MULT. OR P.T. NO.														
RANGE USED														
SCALE READ														
FACTOR														
1														
2														
3														
4														
5														
6														
7														
8														
9														
10														
11														
12														
13														
14														
15														
16														
17														
18														
19														
20														
21														
22														
23														
24														
25														
26														
27														
28														
29														

(a)

DEPARTMENT OF MECHANICAL ENGINEERING
IOWA STATE UNIVERSITY
OF SCIENCE AND TECHNOLOGY
AMES, IOWA

TEST OF ______ DATE ______ OBSERVERS { ______ ______

(b)

Fig. 3.2 Data sheets used by engineering departments.

Table 3.1

Height H, m	Temperature T, °C	Pressure P, kPa
0	15.0	101.3
300	12.8	97.7
600	11.1	94.2
900	8.9	90.8
1 200	6.7	87.5
1 500	5.0	84.3
1 800	2.8	81.2
2 100	1.1	78.2
2 400	−1.1	75.3
2 700	−2.8	72.4
3 000	−5.0	68.7
3 300	−7.2	66.9
3 600	−8.9	64.4
3 900	−11.1	61.9

by students during a flight in a light aircraft. The points, once tabulated, can be plotted and compared with standard atmospheric conditions thereby making possible numerous aerodynamic calculations.

Although the tabulation of data is a necessary step, you may sometimes find it difficult to visualize a relation between variables when simply viewing a column of numbers. A most important step in the sequence from collection to analysis is, therefore, the construction of appropriate graphs or charts.

3.3 General graphing procedures

The proper construction of a graph from tabulated data can be generalized into a series of steps. Each of these steps will be discussed and illustrated in considerable detail in the following subsections.

1. Select the correct type of graph paper and grid spacing.
2. Choose the proper location of the horizontal and vertical axes.
3. Determine the scale units for each axis so that the data can be appropriately displayed.
4. Graduate and calibrate the axes.
5. Identify each axis completely.
6. Plot points and use permissible symbols (that is, ones commonly used and understood).
7. Draw the curve or curves.
8. Identify each curve and add titles and other necessary notes.
9. Darken lines for good reproduction.

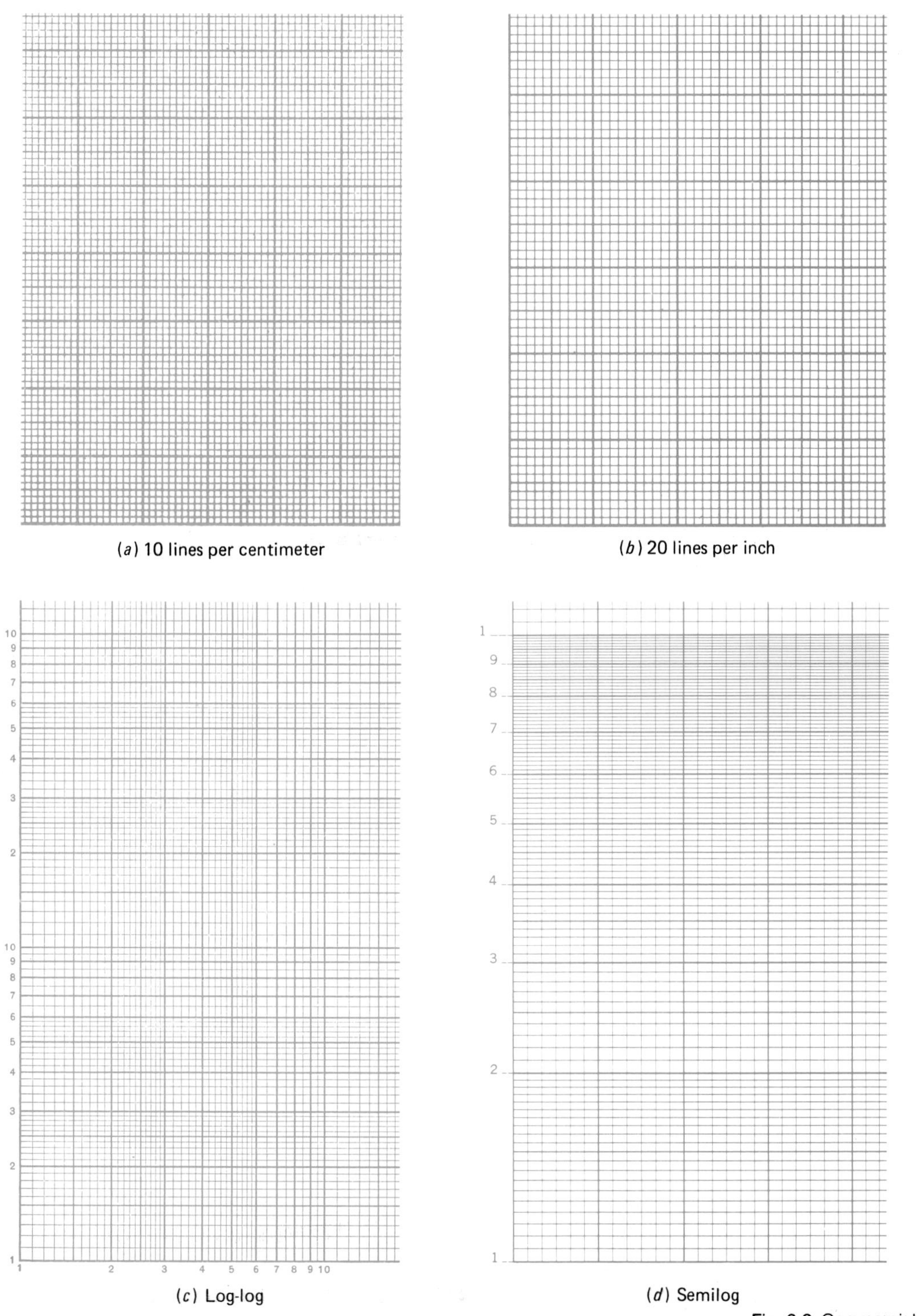

(*a*) 10 lines per centimeter

(*b*) 20 lines per inch

(*c*) Log-log

(*d*) Semilog

Fig. 3.3 Commercial graph paper.

3.3.1 Graph paper

Printed coordinate graph paper is commercially available in various sizes with a variety of grid spacing. Rectilinear ruling can be purchased in a range of lines per inch or lines per centimeter with an overall paper size of 8.5×11 in considered most typical. Figure 3.3 *a* and *b* are illustrations of both 10 lines/cm and 20 lines/in.

Closely spaced coordinate ruling is generally avoided for results that are to be printed or photoreduced. However, for accurate engineering analyses requiring some amount of interpolation, data are normally plotted on closely spaced printed coordinate paper. Graph paper is available in a variety of colors, weights, and grades. Also available is translucent paper that can be used when the reproduction system requires a material that is not opaque to light.

If the data requires the use of log-log or semilog paper, such paper can also be purchased in different formats, styles, weights, and grades. Both log-log and semilog grids are available in from one to five cycles per axis. (A later section will discuss different applications of log-log and semilog paper.) Examples of commercially available logarithmic paper are given in Fig. 3.3 *c* and *d*.

3.3.2 Axes location and breaks

The axes of a graph consist of two intersecting straight lines. The horizontal axis, normally called the x axis, is the abscissa. The vertical axis, denoted the y axis, is the ordinate. Common practice is to place the independent values along the abscissa and the dependent values along the ordinate, as illustrated in Fig. 3.4.

Many times mathematical graphs contain both positive and negative values of the variables. This necessitates the division of the

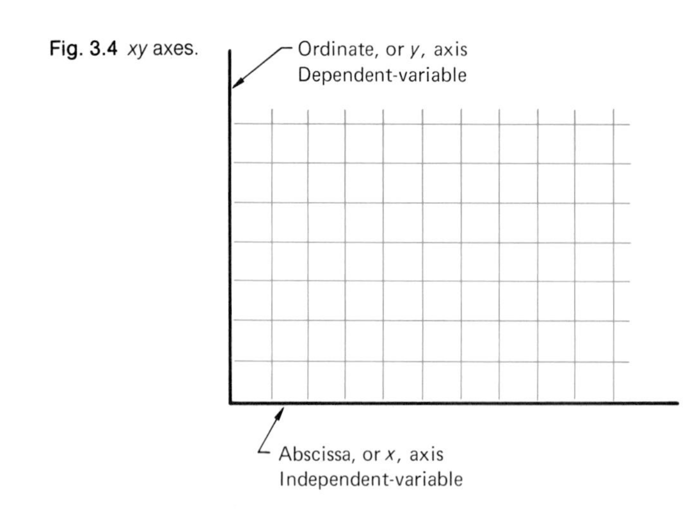

Fig. 3.4 *xy* axes.

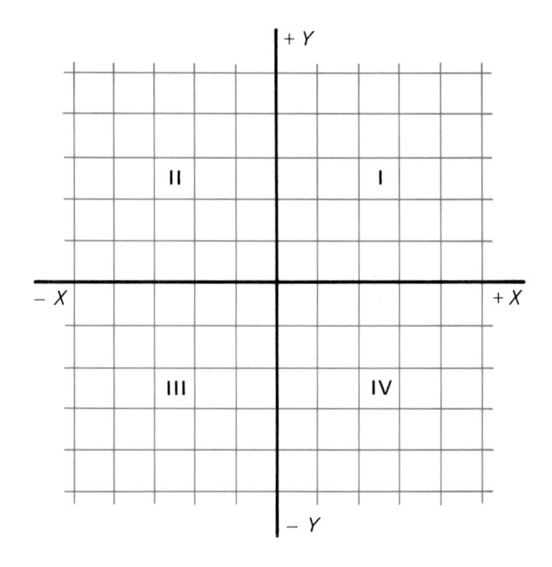

Fig. 3.5 Coordinate axes.

coordinate field into four quadrants, as shown in Fig. 3.5. Positive values increase toward the right and upward from the origin.

On any graph, a full range of values is desirable, normally beginning at zero and extending slightly beyond the largest value. To avoid crowding, the entire coordinate area should be used as completely as possible. However, certain circumstances require special consideration to avoid wasted space. For example, if values to be plotted along the axis do not range near zero, a "break" in the grid or the axis may be used, as shown in Fig. 3.6 *a* and *b*.

When judgments concerning relative amounts of change in a variable are required, the axis or grid should not be broken or the zero line omitted, with the exception of time in years, such as 1970, 1971, etc., since that designation normally has little relation to zero.

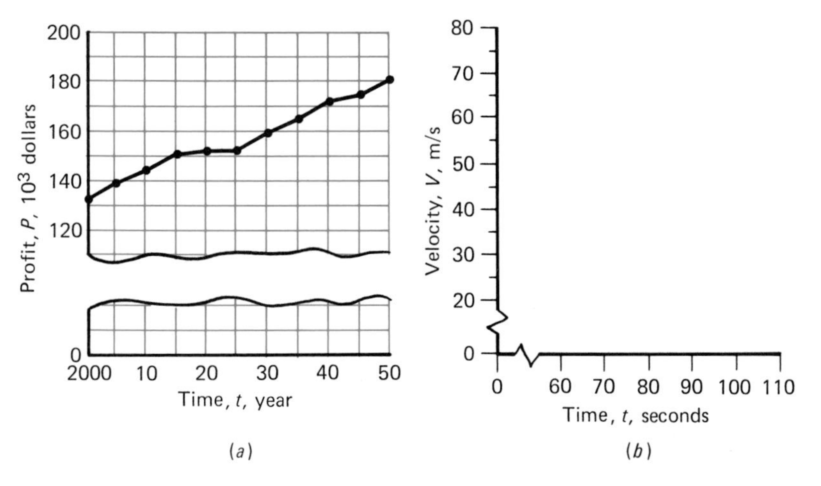

Fig. 3.6 Typical axes breaks.

Since most commercially prepared grids do not include sufficient border space for proper labeling, the axes should be placed 20 to 25 mm (approximately 1 in) inside the edge of the printed grid in order to allow ample room for graduations, calibrations, axes labels, reproduction, and binding. The edge of the grid may have to be used on log-log paper, since it is not always feasible to move the axis in. However, with careful planning, the vertical and horizontal axes can be repositioned depending on the range of the variables.

3.3.3 Scale units

The scale is a series of marks, called graduations, laid down at predetermined distances along the axis. Numerical values assigned to significant graduations are called calibrations.

A scale can be uniform, with equal spacing along the stem, as can be found on the metric, or engineer's, scales. If the scale represents a variable whose exponent is not equal to 1, or a variable that contains trigonometric or logarithmic functions, the scale is called a nonuniform, or functional, scale. Examples of both these scales are shown in Fig. 3.7.

There are occasions when commercial graph paper does not provide the proper grid spacing, size, or other necessary characteristics for a particular purpose. When preparing graphs for publication, photoreduction, presentation as slides or transparencies, or other unique uses, the ability to design and construct a scale can be a definite advantage. Any time a scale is constructed for a specific purpose, the following factors should be considered:

1. Scale length, or actual space available for the axis or stem
2. Spacing of graduations, or the number of graduations required
3. Proper calibration, or the numbers needed to represent the data
4. Titles, that is, appropriate and complete identification of each variable being plotted

The scale length necessary to represent the range of the variable properly can be controlled by the introduction of a constant, called the scale multiplier, SM. The actual value of this multiplier can be

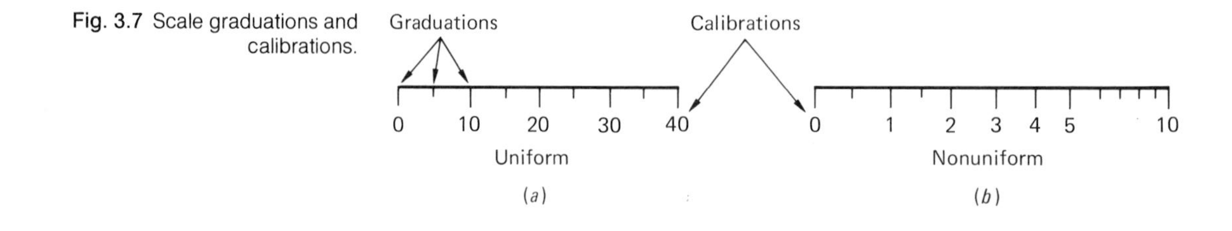

Fig. 3.7 Scale graduations and calibrations.

determined by dividing the desired scale length L by the range of limits of the variables to be represented.

The equation is

$$\text{SM} = \frac{\text{Scale length}}{\text{Range of limits}} = \frac{L}{\text{maximum limit} - \text{minimum limit}} \qquad 3.1$$

To learn how this relationship can be applied, consider the following example problem.

Example problem 3.1 A distance of 150 mm is available to lay out a time scale that varies from 0 to 50 s.

Procedure

1. Determine the scale multiplier.

$$\text{SM} = \frac{L}{\text{maximum limit} - \text{minimum limit}} = \frac{150}{50 - 0}$$

$$= 3 \text{ mm/s}$$

2. Determine the distance between any two points, such as the distance from the origin to the calibrations at 10, 20, and 40 s.

$$D = \text{SM (range)}$$

$$D_{0-10} = (3 \text{ mm/s})[(10 - 0) \text{ s}] = 30 \text{ mm}$$

$$D_{0-20} = (3 \text{ mm/s})[(20 - 0) \text{ s}] = 60 \text{ mm}$$

$$D_{0-40} = (3 \text{ mm/s})[(40 - 0) \text{ s}] = 120 \text{ mm}$$

Uniform scales are commercially available, but you can prepare a uniform scale using the procedure above. The approach is the same for the functional scales. The next example illustrates the construction of a logarithmic scale.

Example problem 3.2 Find the distance between any two points in one cycle of a logarithmic scale of length L that has a range from 1 to 10 s. For this example, let $L = 120$ mm and determine the distance from the 1-s calibration to the 2-s and 5-s calibrations.

Procedure

1. Determine the scale multiplier:

$$\text{SM} = \frac{L}{\text{maximum limit} - \text{minimum limit}} = \frac{120}{\log 10 - \log 1}$$

$$= 120 \text{ mm/s}$$

Fig. 3.8

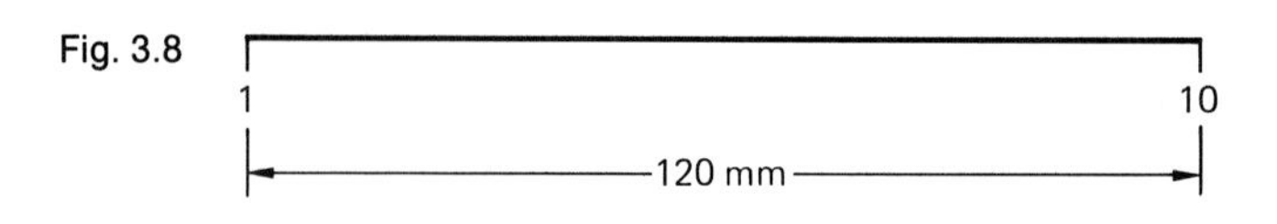

Note: Figs. 3.8 and 3.9 are drawn to half scale.

2. Determine the distances.

$$D_{1-2} = 120(\log 2 - \log 1)$$
$$= 36 \text{ mm}$$
$$D_{1-5} = 120(\log 5 - \log 1)$$
$$= 84 \text{ mm}$$

Fig. 3.9

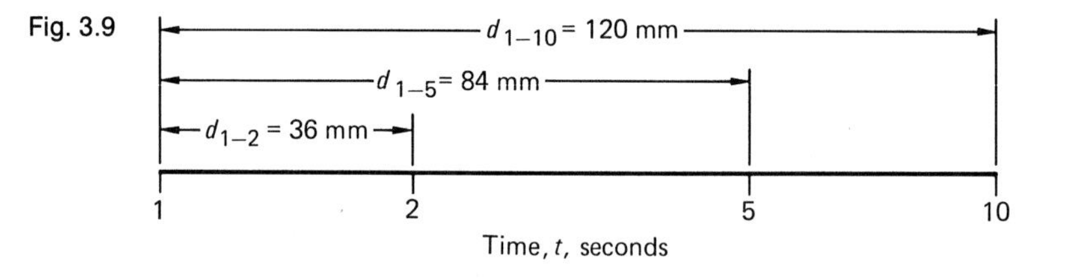

3.3.4 Scale graduations, calibrations, and designations

When plotting data, one of the most important considerations is the proper selection of scale graduations. A basic guide to use or follow is called the 1, 2, 5 rule, which can be stated as follows:

Scale graduations are to be selected so that the smallest division of the axis is a positive or negative integer power of 10 times 1, 2, or 5.

The justification and logic for this rule is clear. Graduation of an axis by this procedure makes possible interpolation of data between graduations when plotting or reading a graph. Figure 3.10 illustrates an acceptable and nonacceptable example of scale graduations.

Violations of the 1, 2, 5 rule that are acceptable involve certain units of time as a variable. Days, months, and years can be graduated and calibrated as illustrated in Fig. 3.11.

Fig. 3.10 Scale graduations.

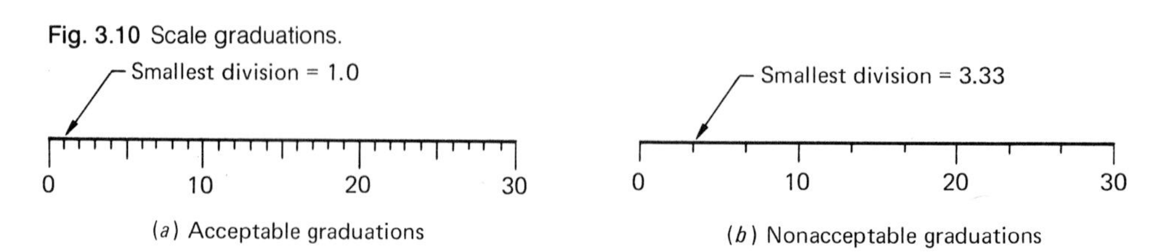

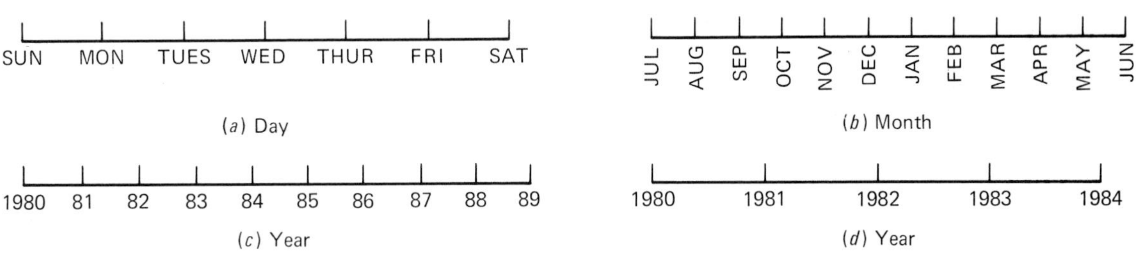

Fig. 3.11 Time as a variable.

Scale graduations normally follow a definite rule, but the number of calibrations to be included are primarily a matter of good judgment. Each application requires consideration based on the scale length and range as well as the eventual use. Figure 3.12 demonstrates how calibrations can differ on a scale with the same length and range. Both examples obey the 1, 2, 5 rule, but as you can see, too many closely spaced calibrations make the axis difficult to read.

The selection of a scale deserves attention from another point of view. If the rate of change is to be depicted accurately, then the slope of the curve should represent a true picture of the data. By contracting or expanding the axis or axes, an incorrect impression of the data could be implied. Such a procedure is to be avoided. Figure 3.13 demonstrates how the equation $y = x$ can be misleading if not properly plotted. Occasionally distortion is desirable, but it

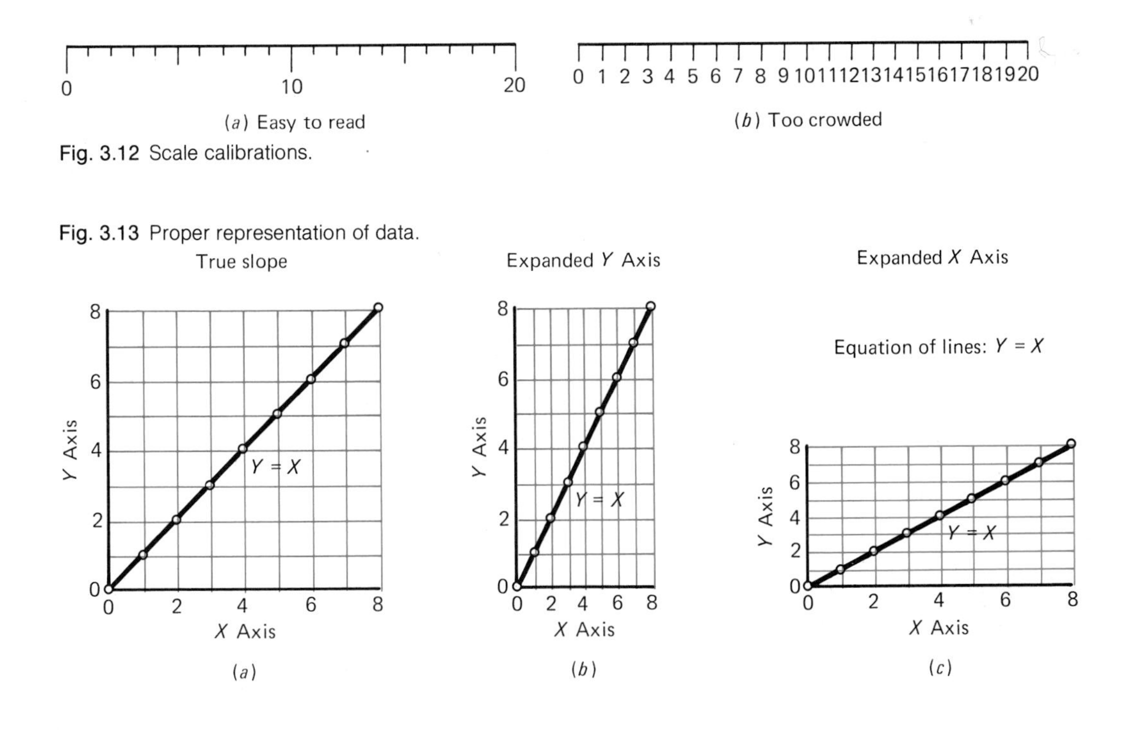

Fig. 3.12 Scale calibrations.

Fig. 3.13 Proper representation of data.

should always be carefully labeled and explained to avoid misleading conclusions.

If plotted data consist of very large or small numbers, the SI prefix names (milli-, kilo-, mega-, etc.) may be used to simplify calibrations. As a guide, if the numbers to be plotted and calibrated consist of more than three digits, it is customary to use the appropriate prefix, an example of which is illustrated in Fig. 3.14.

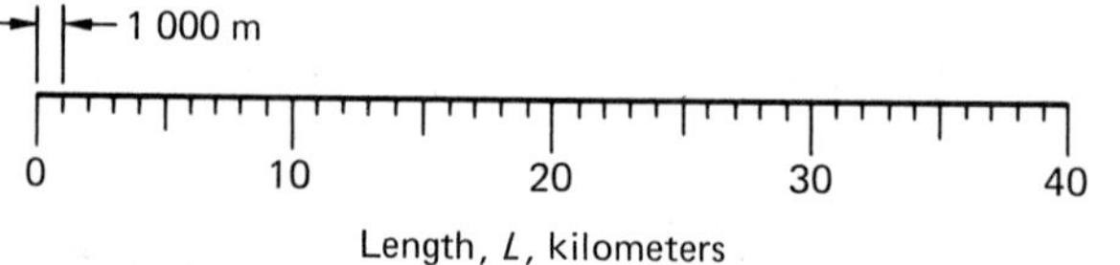

Fig. 3.14 Reading the scale.

The length scale calibrations in Fig. 3.14 contain only two digits, but the scale can be read by one's knowing that the distance between the first and second graduation (0 to 1) is a kilometer; therefore, the calibration at 10 represents 10 km.

Certain quantities, such as temperature in degrees Celsius and altitude in meters, have traditionally been tabulated without the use of prefix multipliers. Figure 3.15 depicts a procedure by which these quantities can be conveniently calibrated. Note in particular that the distance between 0 and 1 on the scale represents 1 000°C.

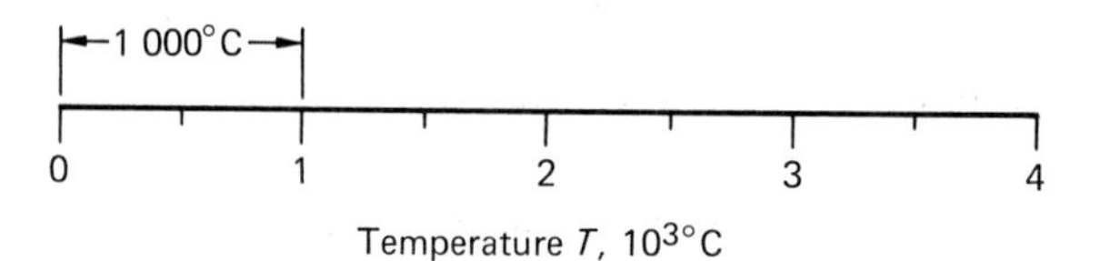

Fig. 3.15 Reading the scale.

This is another example of how the SI notation is convenient, since the prefix multipliers (micro-, milli-, kilo-, mega-, etc.) allow the calibrations to stay within the three-digit guideline.

The calibration of logarithmic scales is illustrated in Fig. 3.16. Since log-cycle designations start and end with powers of 10 (i.e., 10^{-1}, 10^{0}, 10^{1}, 10^{2}, etc.) and since commercially purchased paper is

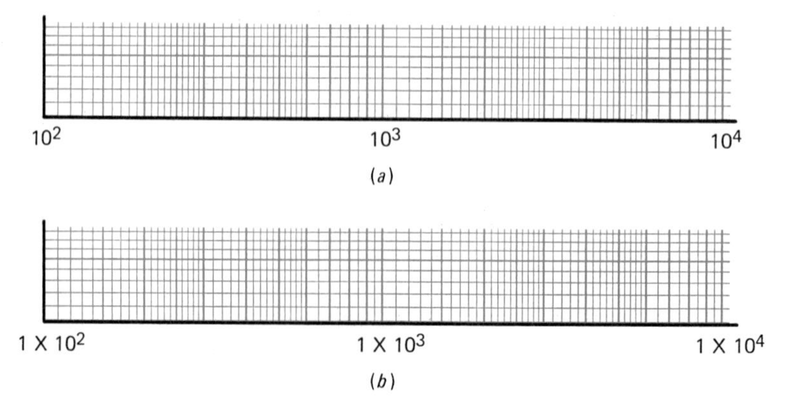

Fig. 3.16 Calibration of log scales.

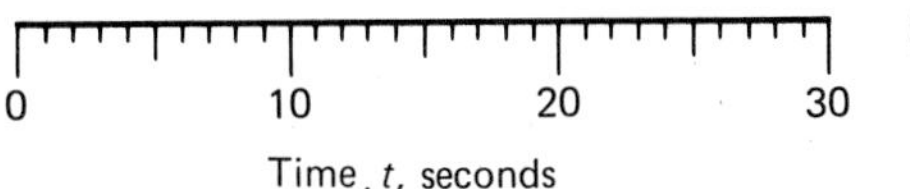

Fig. 3.17 Axis identification.

normally available with each cycle printed 1 through 10, Fig. 3.16 *a* and *b* demonstrate two preferred methods of calibration.

3.3.5 Axis labeling

Each axis should be clearly identified. At a minimum, the axis label should contain the name of the variable, its symbol, and its units. Since time is frequently used and normally plotted on the x axis, it has been selected as an illustration in Fig. 3.17.

Scale designations should be placed outside the axes, where they can be shown clearly. Labels should be lettered parallel to the axis and positioned so that they can be read from the bottom or right side of the page.

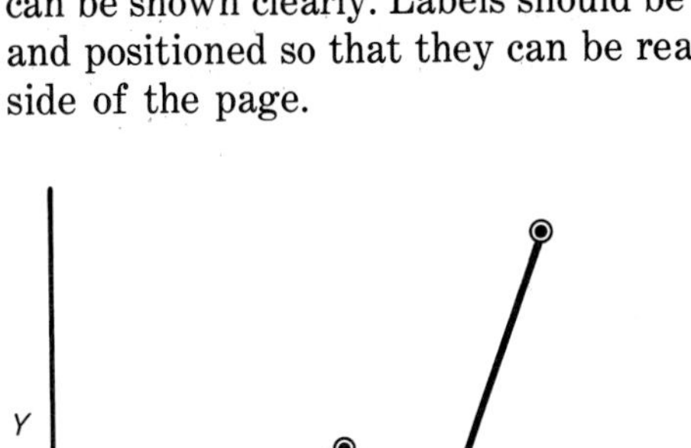

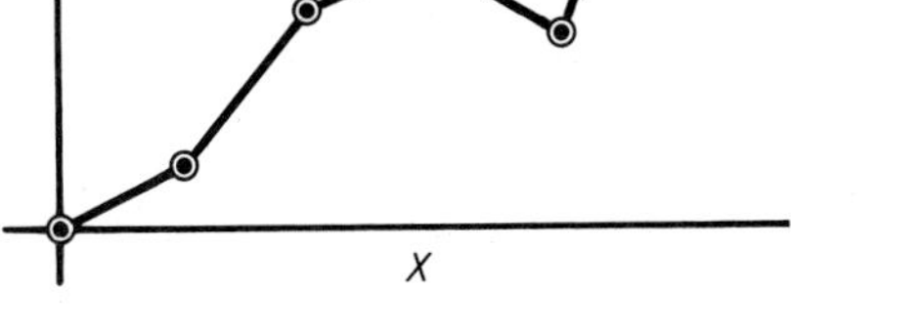

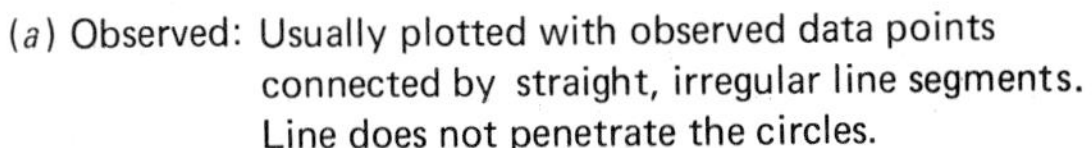

(*a*) Observed: Usually plotted with observed data points connected by straight, irregular line segments. Line does not penetrate the circles.

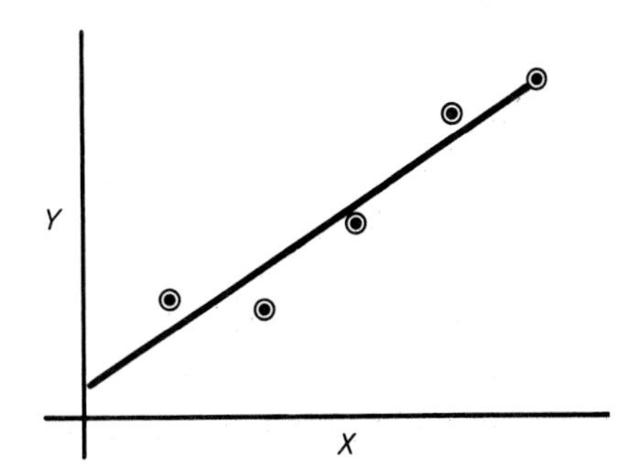

(*b*) Empirical: Reflects the author's interpretation of what occurs between known data points. Normally represented as a smooth curve or straight line fitted to data. Data points may or may not fall on curve.

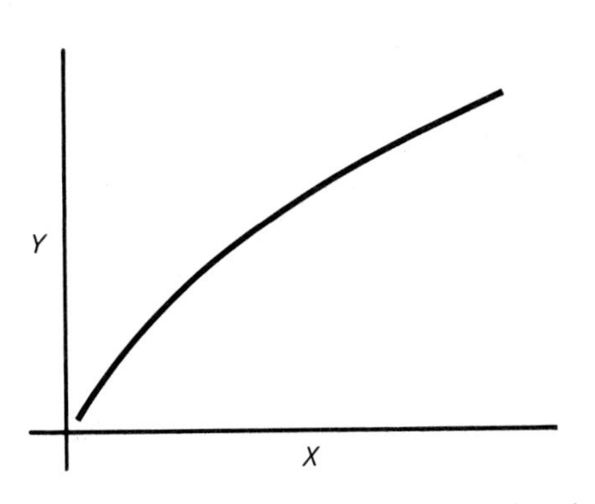

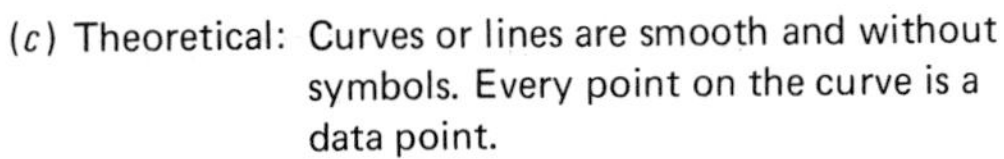

(*c*) Theoretical: Curves or lines are smooth and without symbols. Every point on the curve is a data point.

Fig. 3.18 Plotting data points.

3.3.6 Point plotting procedure

Data can normally be designated in one of three general ways, as observed, empirical, or theoretical. Observed and empirical data points are usually located by various symbols, such as a small circle or square around each data point; whereas graphs of theoretical relations are normally constructed smooth without use of symbol designation. Figure 3.18 illustrates each type.

3.3.7 Curves and symbols

On graphs prepared from observed data, resulting from laboratory experiments, points are usually designated by various symbols (see Fig. 3.19).

If more than one curve is plotted on the same grid, a combination of these symbols may be used (one type for each curve). To avoid confusion, however, it is good practice to label each curve.

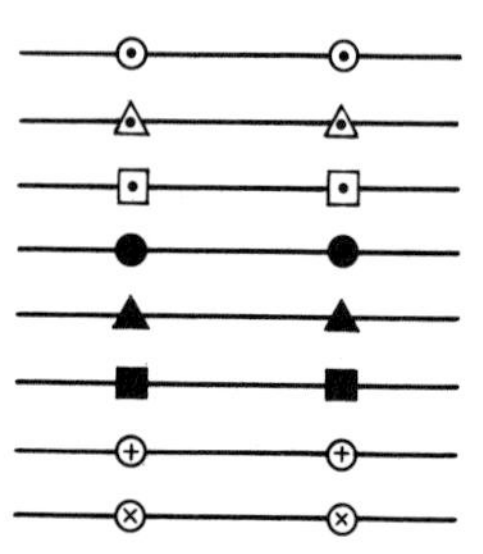

Fig. 3.19 Symbols.

When several curves are plotted on the same grid, another way in which they can be distinguished from one another is by the use of different types of lines, as illustrated in Fig. 3.20. Solid lines are normally reserved for single curves and dashed lines are commonly used for extensions. The curves should be heavier than the grid ruling.

A key, or legend, should be placed in an isolated portion of the grid, preferably enclosed in a border, to define point symbols or line types that are used for curves. Remember that the lines representing each curve should never be drawn through the symbols, so that the precise point is always identifiable. Figure 3.21 demonstrates the use of a key and the practice of breaking the line at each symbol.

3.3.8 Titles

Each graph must be identified with a complete title. The title should include a clear, concise statement of the data being represented along with items such as the name of the author, the date of the experiment, and any and all information concerning the plot including the name of the institution or company. Titles are normally enclosed in a border.

All lettering, the axes, and the curves should be sufficiently bold to stand out on the graph paper. Letters should be neat and of standard size. Figure 3.22 is an illustration of plotted experimental data incorporating many of the items discussed in the chapter.

Fig. 3.20 Line representation.

3.4 Empirical functions

An empirical function is generally described as one based on values obtained by experimentation. It is often identified as any function for which no analytic equation has previously been defined. Needless to say, a great many mathematical equations might fit such a

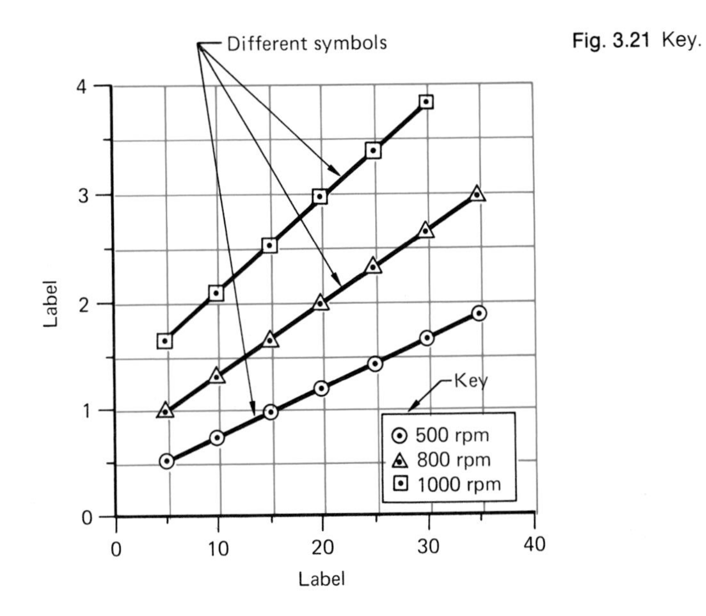

Fig. 3.21 Key.

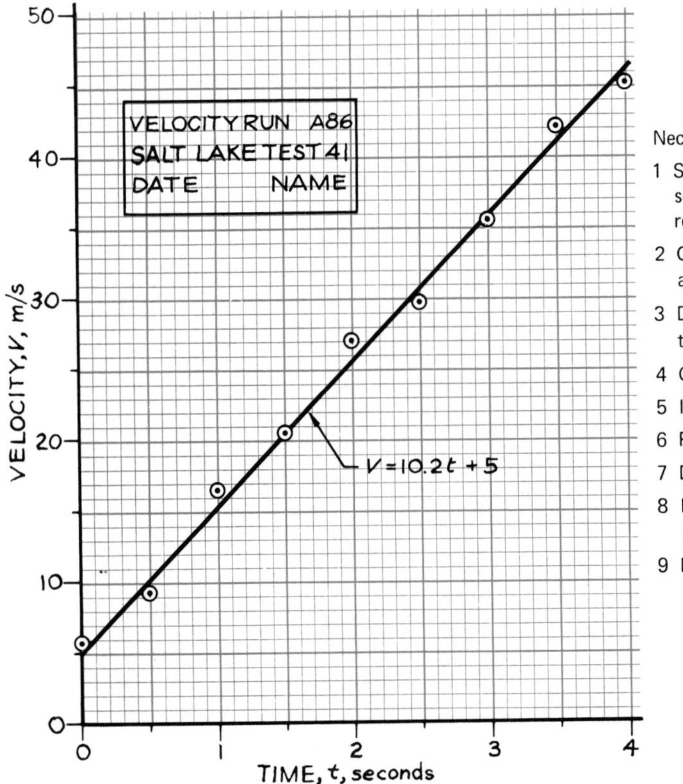

Fig. 3.22 Sample plot.

Necessary steps to follow when plotting a graph.

1. Select the type of graph paper (rectilinear, semilog, log-log, etc.) and grid spacing for best representation of the given data.
2. Choose the proper location of the horizontal and vertical axes.
3. Determine the scale units (range) for each axis to display the data appropriately.
4. Graduate and calibrate the axes (1, 2, 5 rule).
5. Identify each axis completely.
6. Plot points and use permissible symbols.
7. Draw the curve or curves.
8. Identify each curve and add title and necessary notes.
9. Darken lines for good reproduction.

function. Fortunately, it is possible to classify many empirical results into one of four general categories: (1) linear, (2) exponential, (3) power, or (4) periodic.

As the name suggests, a linear function will plot as a straight line on uniform rectangular coordinate paper. Likewise, when a curve representing experimental data is a straight line or a close approximation to a straight line, the relationship of the variables can be expressed by a linear equation.

Correspondingly, exponential equations, when plotted on semilog paper, will be linear. This can be seen by taking the log of the equation $y = be^{mx}$ ($\log y = mx \log e + \log b$). The independent variable x is plotted against the log of y.

The power equation $y = bx^m$ ($\log y = m \log x + \log b$) plots as a straight line on log-log paper, since the log of the independent variable x is plotted against the log of y.

When the data represents experimental results and a series of points are plotted to represent the relationship between the variables, it is improbable that a straight line can be constructed through every point, since some error is inevitable. If all points do not lie on the same line, an approximation scheme or averaging method may be used to arrive at the best possible fit.

3.5 Curve fitting

Different methods or techniques are available to arrive at the best "straight-line" fit. The time and expense involved increases, however, with the degree of accuracy and reliability of the method selected.

Fig. 3.23 An orientation session where engineers are discussing methods of productivity improvement with a department manager. *(Allen-Bradley.)*

Three methods commonly employed for finding the best fit are listed below:

1. Method of selected points
2. Method of averages
3. Method of least squares

Each of these techniques is progressively more accurate. The first method will be briefly described in Sec. 3.6. The most accurate method, number 3, is discussed in Chap. 14.

Number 2, the method of averages, is based on the idea that the line location is positioned to make the algebraic sum of the differences between observed and calculated values of the ordinate equal to zero.

In both methods 2 and 3, the procedure involves minimizing what are called "residuals," or the difference between an observed ordinate and the corresponding computed ordinate. The method of averages will not be applied in this book, but there are any number of reference texts available that adequately cover the concept.

3.6 Method of selected points

The method of selected points allows determination of the best fit only after the straight line has been constructed through the plotted points. Normally this is accomplished by drawing the line through a maximum number of points, with approximately an equal number falling on each side of the line.

Once the line has been constructed, two points, such as A and B, are selected *on the line* and at a reasonable distance apart. The coordinates of both points A (x_1,y_1) and B (x_2,y_2) must satisfy the equation of the line, since both are points on the line.

3.7 Empirical equations—rectangular paper

When experimental data plots as a straight line on rectangular grid paper, the equation of the line belongs to a family of curves whose basic equation is given as

$$y = mx + b \qquad 3.2$$

where m and b are constants.

To demonstrate how the method of selected points works, consider the following example.

Example problem 3.3 A racing automobile is clocked at various times t and velocities V.

Determine the equation of a straight line constructed through the points recorded in Table 3.2. Once an analytic expression has been determined, velocities at intermediate values can be computed.

Table 3.2

Time, t, s	0	5	10	15	20	25	30	35	40
Velocity, V, m/s	24	33	62	77	105	123	151	170	188

Procedure

1. Plot the data on rectangular paper. If the results form a straight line (see Fig. 3.24), the function is linear and the general equation is of the form

$$V = mt + b$$

where m and b are constants.

2. Select two points on the line, $A\ (x_1,y_1)$ and $B\ (x_2,y_2)$, and record the value of these points.

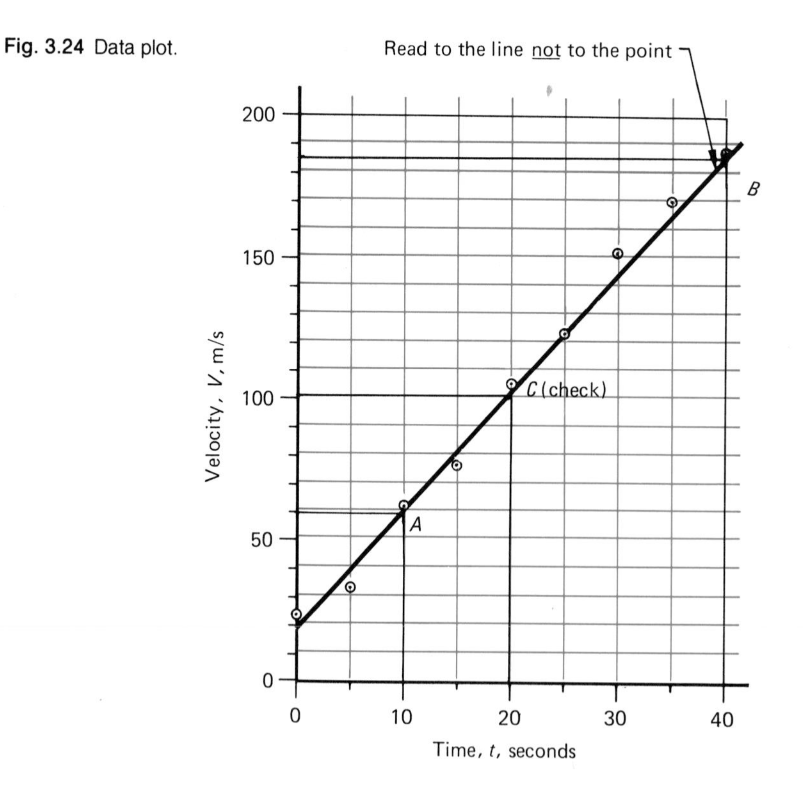

Fig. 3.24 Data plot.

A (10,60)

B (40,185)

3. Substitute the points A and B into $V = mt + b$.

$60 = m(10) + b$

$185 = m(40) + b$

4. The equations are solved simultaneously for the two unknowns.

$m = 4.2$

$b = 18.3$

5. The general equation of the line can then be written as

$V = 4.2t + 18.3$

6. Using another point, C (x_3,y_3), check for verification:

C (20,102)

$102 \approx 4.2(20) + 18.3$

$102 \approx 84 + 18.3$

It is also possible by discreet selection of points to simplify the solution. For example, if point A is selected at (0,18.3) and this coordinate is substituted into the general equation

$18.3 = m(0) + b$

the constant b is immediately known.

3.8 Empirical equations—power curves

When experimentally collected data are plotted on rectangular coordinate graph paper and the points do not form a straight line, then you must determine which family of curves the line may most closely approximate. (Knowing how to do this comes primarily from experience.) Let us consider the following familiar example.

Example problem 3.4 Suppose that a solid object is dropped from a tall building. Neglecting air friction, the values are as recorded in Table 3.3.

Table 3.3

Time, t, s	Distance, s, m
0	0
1	4.9
2	19.6
3	44.1
4	78.4
5	122.5
6	176.4

Solution To anyone who has studied fundamental physics, it is apparent that these values should correspond to the general equation for a free-falling body.

$$s = \frac{1}{2}gt^2$$

But assume for a moment that all we have is the table of values.

First it is helpful to make a freehand plot to observe the data visually. See Fig. 3.25.

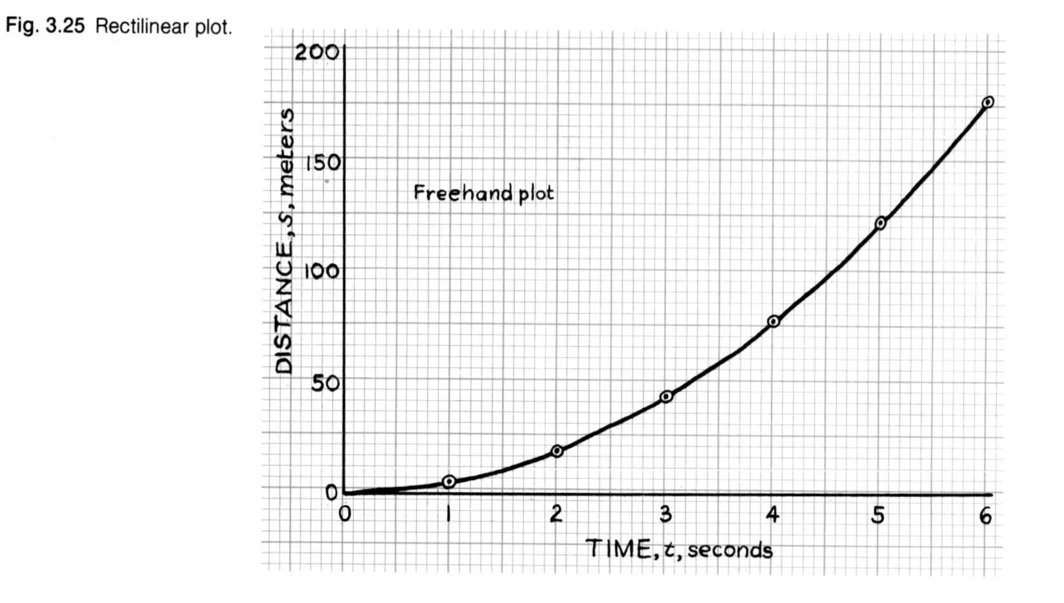

Fig. 3.25 Rectilinear plot.

From this quick plot, the data points are more easily recognized as belonging to a family of curves whose general equation can be written

$$y = bx^m \qquad 3.3$$

Remember that before the method of selected points can be applied to determine the equation of the line in this example problem, the plotted line must be straight, because two points on a curved line do not uniquely identify the line.

Mathematically, this general equation can be put in linear form by taking the logarithm of both sides,

$$\log y = m \log x + \log b$$

Table 3.4

Time, t	Distance, s	Log t	Log s
0	0		
1	4.9	0.0000	0.6902
2	19.6	0.3010	1.2923
3	44.1	0.4771	1.6444
4	78.4	0.6021	1.8943
5	122.5	0.6990	2.0881
6	176.4	0.7782	2.2465

This relationship indicates that if the log of both y and x were recorded and the results plotted, the line would be straight.

Realizing that the log of zero is undefined and plotting the remaining points (see Table 3.4), we get the graph shown in Fig. 3.26.

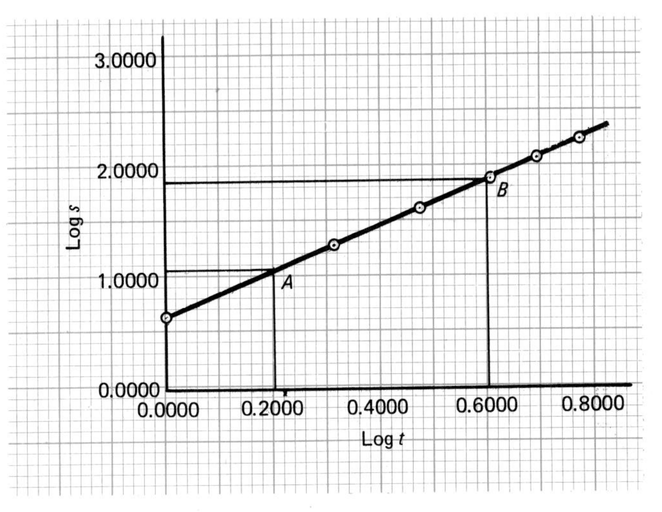

Fig. 3.26 Log-log plot on rectilinear grid.

Since the graph of log s versus log t plots as a straight line, it is possible to use the general form of the equation

$$\log y = m \log x + \log b$$

and apply the method of selected points.

When reading values for points A and B from the graph, we must remember that the logarithm of each variable has already been determined and plotted.

A (0.2000, 1.0900)

B (0.6000, 1.8900)

Points A and B can now be substituted into the general equation $\log s = m \log t + \log b$ and solved simultaneously.

$$1.8900 = m(0.6000) + \log b$$

$$1.0900 = m(0.2000) + \log b$$

$$b = 4.9$$

$$m = 2.0$$

The general equation can then be written

$$s = 4.9t^2$$

Or, $s = \frac{1}{2}gt^2$, where $g = 9.8$ m/s^2.

The only inconvenience that results from determining an equation from the originally collected data derives from the necessity of taking logarithms of each variable and then plotting the logs of these variables.

Obviously, this step is not necessary, since functional paper is commercially available with log x and log y scales already constructed. Log-log paper allows the variables themselves to be plotted directly without the need of computing the log of each value.

In the preceding example, once the general form of the equation is selected [that is, Eq. (3.3)], the data can be plotted directly on log-log paper. Since the resulting curve will be a straight line, the method of selected points can be used directly. (See Fig. 3.27.)

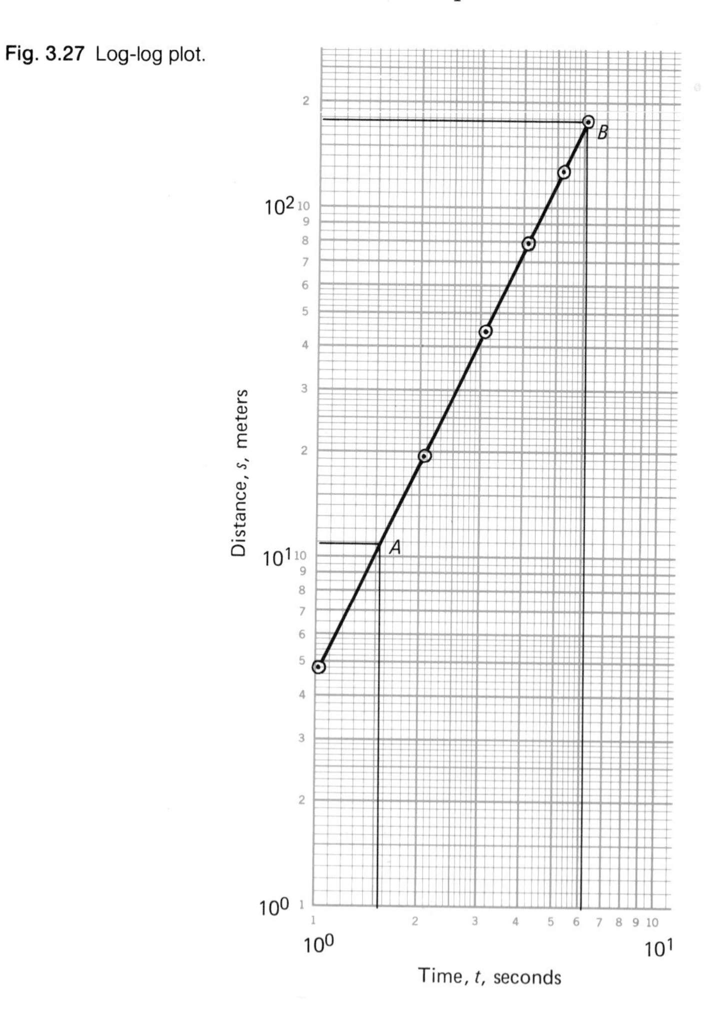

Fig. 3.27 Log-log plot.

The log form of the equation is again used:

$$\log y = m \log x + \log b$$

Select points A and B on the line:

A (1.5,10.9)

B (6,175)

Substitute the values into the general equation $\log s = m \log t + \log b$, taking careful note that the numbers are the variables and *not* the log of the variables.

$$\log 175 = m \log 6 + \log b$$

$$\log 10.9 = m \log 1.5 + \log b$$

Again solving these two equations simultaneously results in the following approximate values for the constants b and m.

$b = 4.9$

$m = 2.0$

Identical conclusions can be reached:

$$s = \frac{1}{2} g t^2$$

This time, however, one can use functional scales rather than look up the log of each number.

3.9 Empirical equations—exponential curves

If experimental data plots as a straight line on semilog paper, the curve belongs to the exponential family of curves whose most common equation has the following form:

$$y = be^{mx} \qquad 3.4$$

The values of the constants b and m can be determined using the method of selected points. Since the procedure is very similar to that for the general power equation covered in the previous section, the details will be left as an exercise for the reader.

Problems

3.1 Construct a 130-mm-velocity scale which varies from 0 to 80 m/s. Graduate, calibrate, and properly label the scale.

3.2 Construct a 150-mm-time scale with a range of 0 to 700 s. Graduate, calibrate, and properly label the scale.

3.3 Construct a nonuniform scale for the function $F(U) = U^2$. Let the scale length be 130 mm and the range vary from 0 to 5.

3.4 Construct a nonuniform scale for the function $F(P) = P^{0.25}$. Let the scale length be 130 mm and the range vary from 0 to 100.

3.5 Construct a nonuniform scale for the function $F(X) = \log X$. Let the scale length be 150 mm and the range vary from 1 to 10.

3.6 Construct a nonuniform scale for the function $F(X) = \log X$. Let the scale length be 100 mm and the range vary from 10 to 100.

3.7 Table 3.5 is data from a trial run on the Utah salt flats made by an experimental turbine-powered vehicle.

(*a*) Plot the data on rectilinear paper.
(*b*) Determine the equation for the line.
(*c*) Interpret the slope of the line.

Table 3.5

Time, *t*, s	0	10	20	30	40	50	60
Velocity, *V*, m/s	0	20	44	62	84	104	128

3.8 Table 3.6 lists values of velocity recorded on a ski jump in Colorado this past winter.

(*a*) Plot the data on rectilinear paper.
(*b*) Determine the equation for the line.
(*c*) Give the average acceleration.

Table 3.6

Time, *t*, s	0	2	4	6	8	10
Velocity, *V*, m/s	0	10.5	18.0	23.0	30.0	36.5

3.9 Table 3.7 is a collection of data for an iron-constantan thermocouple. Temperature is in degrees Celsius and the emf is in millivolts.

(*a*) Plot a graph showing the relation of emf to temperature using rectilinear paper.
(*b*) Determine the equation for the line.

Table 3.7

Temperature *t*, °C	Voltage emf mV
0	0
50	2.6
100	5.3
150	8.0
200	10.8
300	16.3
400	21.8
500	27.4
600	33.1
700	39.2
800	45.5
900	51.8

3.10 (*a*) Plot the data in Table 3.8 on semilog paper.
(*b*) Determine the equation of the curve.

Table 3.8

X	0	2	4	6	8	10	12
Y	65	29	13	6	2.5	1.2	0.5

3.11 A new production facility manufactured 20 parts the first month, but then increased production, as shown in Table 3.9.

(*a*) Plot the data on semilog paper.

(*b*) Determine the equation.

Table 3.9

Month	Jan	Feb	Mar	Apr	May	Jun	Jul	Aug
Number	20	33	54	90	148	244	402	662

3.12 The voltage across a capacitor during discharge was recorded as a function of time. (See Table 3.10.)

(*a*) Plot the data on semilog paper with time as the independent variable.

(*b*) Determine the equation of the line using the method of selected points.

Table 3.10

Voltage, V, V	120	72	45	15	6	2	0.8
Time, t, s	0	5	10	20	30	40	50

3.13 When a capacitor is to be discharged, the current flows until the voltage across the capacitor is zero. This current flow when measured as a function of time resulted in the data given in Table 3.11.

(*a*) Plot the data points on semilog paper with time as the independent variable.

(*b*) Determine the equation of the line using the method of selected points.

Table 3.11

Current, I, A	2.00	1.81	1.64	1.48	1.34	1.21	0.73
Time, t, s	0	0.1	0.2	0.3	0.4	0.5	1.0

3.14 The area of a circle can be expressed by the formula $A = \pi R^2$. If the radius varies from 0.5 to 5 cm, perform the following.

(*a*) Construct a table of radius versus area mathematically.

(*b*) Construct a second table of log R versus log A.

(*c*) Plot the values from *a* on log-log paper and determine the equation of the line.

(*d*) Plot the values from *b* on rectilinear paper and determine the equation of the line.

3.15 The volume of a sphere is $V = 4/3\pi r^3$.

(*a*) Prepare a table of volume versus radius allowing the radius to vary from 2.0 to 10.0 m in 1-m increments.

(*b*) Plot a graph on log-log paper showing the relation of volume to radius using the values from the table in part *a*.

(*c*) Verify the equation given above by the method of selected points.

3.16 A 90° triangular weir is commonly used to measure flow rate in a stream. Data on the discharge through the weir were collected and recorded in Table 3.12 as a function of time for various heights.

(*a*) Plot the data on log-log paper with height as the independent variable.

(*b*) Determine the equation of the line using the method of selected points.

Table 3.12

Discharge, Q, m^3/s	1.5	8	22	45	78	124	182	254
Height, h, m	1	2	3	4	5	6	7	8

3.17 A pitot tube is a device for measuring the velocity of flow of a fluid (see Fig. 3.28). A stagnation point occurs at point 2; by recording the height differential h, the velocity at point 1 can be calculated. Assume for this problem that the velocity at point 1 is known corresponding to the height differential h. Table 3.13 records these values.

(*a*) Plot the data on log-log paper using height as the independent variable.

(*b*) Determine the equation of the line using the method of selected points.

Table 3.13

Velocity, V, m/s	1.4	2.0	2.8	3.4	4.0	4.4
Height, h, m	0.1	0.2	0.4	0.6	0.8	1.0

Fig. 3.28

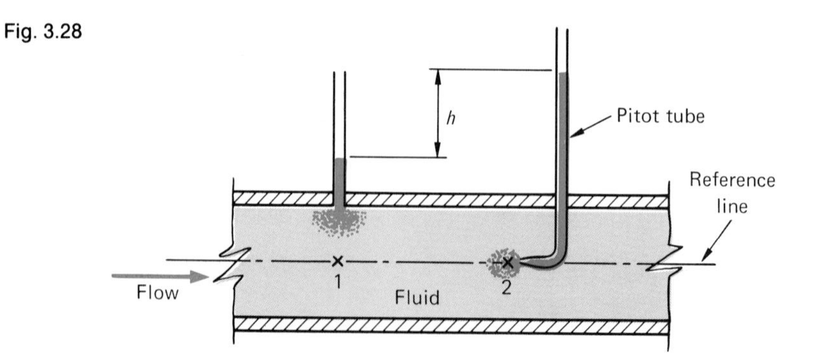

Engineering estimations and approximations

CHAPTER 4

4.1 Introduction

Much is said, and rightly so, about the great diversity among the many branches of engineering. While it is true that modern engineering has spawned a myriad of specialties, there are likewise many things that engineers have in common, one of which is the need to acquire physical measurements. The nineteenth century physicist Lord Kelvin stated that man's knowledge and understanding are not of high quality unless the information can be expressed in numbers. We all have made or heard statements such as "The water is too hot." This statement may or may not give us an indication of the temperature of the water. At a given temperature water may be too hot for taking a bath but not hot enough for making instant coffee or tea.

The truth is that pronouncements such as hot, too hot, not very hot, etc., are relative to a standard selected by the speaker and have meaning only to those who know what that standard is.

We make measurements of a vast array of physical quantities that allow addition to our knowledge and permit us to transfer this knowledge and understanding to others. Skill in making and interpreting measurements is an essential element in our practice of engineering.

Fig. 4.1 How many tons of asphalt will be used in resurfacing 65 mi of this highway? (*Honeywell, Inc.*)

4.2 Errors

To measure is to err! Any time a measurement is taken, some physical object is being compared with a standard. If we measure the distance between two points on the surface of the earth, why doesn't a repetition of the measurement produce identical results? The obvious answer is that errors occur in each attempt; only coincidence will produce exactly the same result. Errors are normally classified in two general categories: systematic and accidental (random).

4.2.1 Systematic errors

Our task is to measure the distance between two fixed points. Assume that the distance is in the range of 1 200 m and that we are experienced and competent and have equipment of high quality to do the measurement. Some of the errors that occur will always have the same sign (+ or −), and their magnitudes will depend on the number of chances of their occurrence. The errors are said to be systematic. Assume that a 25-m steel tape is to be used, one that has been compared with the standard at the U.S. Bureau of Standards in Washington, D.C. If the tape is not exactly 25.000 m long, then there will be an error each of the 48 times that we use the tape to measure out the 1 200 m. However, the error can be removed by computation. A second source of error can stem from the temperature at the time of use not being the same as that at the time when the tape was compared with the standard. Such an error can be removed if we measure the temperature of the tape and apply a mathematical correction. (The coefficient of thermal expansion for steel is 11.7×10^{-6} per kelvin.) The accuracy of such a correction depends on the accuracy of the thermometer and our ability to measure the temperature of the tape instead of the temperature of the surrounding air. Another source of systematic error can occur if the tension applied to the tape while in use is not identical with that employed during standardization. Again, scales can be used; but as before, their accuracy would be suspect. In all probability, the tape was standardized by laying it on a smooth surface and supporting it throughout. But such surfaces are seldom available in the field. The tape is suspended at times, at least partially. Nonetheless, knowing the weight of the tape, the tension that is applied, and the length of the suspended tape, we can calculate a correction and apply it.

These sources of systematic error just discussed are not all the possible sources, but they illustrate an important problem encountered in taking comparatively simple measurements. Similar problems occur in all sorts of measurements: mechanical quantities, electrical quantities, mass, sound, odors, etc. We must be aware of the presence of systematic errors, eliminate those that we can, and quantify and correct for those remaining.

4.2.2 Accidental errors

In reading Sec. 4.2.1 you may have realized that even when all the systematic errors have been detected and corrected for, the measurement is still not exact. To elaborate on this point, we will continue with the example of the task of measuring the 1 200-m distance. Several accidental errors can creep in as follows. When reading the thermometer, we must estimate the reading when the indicator falls between graduations. Moreover, it may appear that the reading is exactly on a graduation when it is actually slightly above or below the graduation. Furthermore, the thermometer may not be accurately measuring the tape temperature but be influenced instead by the temperature of the ambient air. These errors can thus produce measurements that are either too large or too small. Regarding sign and magnitude, the error is therefore random.

Errors can also result from our correcting for the sag in a suspended tape. In such a correction, it is necessary to determine the weight of the tape, its cross-sectional area, its modulus of elasticity, and the applied tension. In all such cases, the construction of the instruments used for acquiring these quantities can be a source of both systematic and accidental errors.

The major difficulty we encounter with respect to accidental errors is that although their presence may be assumed, it is impossible to predict their magnitude and sign. The refinement of the apparatus and the care in its use can reduce the magnitude of the error; indeed, many engineers have devoted their careers to this task. Likewise, awareness of the problem, knowledge about the degree of precision of the equipment, skill with measurement procedures, and proficiency in the use of statistics allow us to determine the approximate magnitude of the error remaining in measurements. This knowledge, in turn, allows us to accept the error or develop different apparatus and/or methods in our work. It is beyond the scope of this text to discuss quantifying accidental errors. However, Chap. 14 includes a brief discussion of central tendency and standard deviation, which is part of the analysis of accidental errors.

4.3 Conversions and significant digits

Significant figures, or digits, are not always handled properly. When working with modern electronic calculators, numbers are automatically carried out to eight, ten, or even more places depending on the design of the instrument, whether or not we need the information. If significant figures are understood, however, many scientific calculators can be controlled to display only the desired results.

A significant figure, or digit, is defined as any digit used in writing a number *except* those zeroes which are used only for location of the decimal point, such as 2 000, or those zeroes which do not have any nonzero digit on their left, such as 0.0015. When you read the two numbers, 2 000 and 0.0015, it is difficult to determine the number of digits that are significant. Our definition tells us that 2 000 has one significant digit and 0.0015 has two. The zeros indicate the proper placement of the decimal point. If the number 2 000 is an accurate count of a group of items, it can be reported as 2.000×10^3, thus indicating that the four digits are significant. Likewise, 1.5×10^{-3} would be a proper way to show two significant digits that would leave no doubt for the reader.

Reporting significant digits and obtaining the indicated degree of accuracy are entirely different matters. The ordinary yardstick used around the home may be graduated to $\frac{1}{16}$ in and yet not be accurate to less than $\frac{1}{8}$ in. When converted to SI units, a measurement of $6\frac{5}{16}$ in is 16.033 75 cm. However, if the yardstick is no more accurate than $\frac{1}{8}$ in (0.3 cm), then we are misleading the reader by reporting 16.033 75 cm. A more realistic value would be 16 cm.

Suppose that you determine the mass of a quantity of material to be 677 kg on a scale that is accurate to ± 0.3 percent. If you are required to convert this mass to pounds by multiplying by 2.204 6, you obtain 1 492.514 2 lb. It is clear that the decimal portion is not significant because the scale error is $\pm(0.003)(2.204\ 6)(677) = \pm 4.48$ lb. You should thus report the mass as 1 490 lb.

Conversions are but one instance where you must be careful in reporting the magnitude of numbers. As you perform arithmetic operations, it is important that you do not lose the significance of your measurement or imply accuracy that does not exist. Rules have been developed by engineering associations for which there is general agreement among the different agencies. The following rules customarily apply.

1. Multiplication and division. The product or quotient should contain the number of significant digits that are contained in the number with the fewest significant digits.
   a. $2.43 \times 17.675 = 42.950\ 25$ should be rounded to 43.0. It can be written as 4.30×10^1.
   b. $589.62 \div 1.246 = 473.210\ 273$ should be reported as 473.2.

It is generally agreed that each number which is to be multiplied or divided may be rounded to one more digit than is contained in the number with the fewest significant digits. In *a* this would produce $2.43 \times 17.68 = 42.96$, which can be rounded to 4.30×10^1.

Since most of our calculations will be performed on an electronic calculator, there is little to be gained by rounding numbers at intermediate stages of a calculation.

Make sure that you do not round off conversion factors to less than one more digit than that contained in the least significant number. If you round off further, the conversion factor may affect the accuracy of your answer.

2. Addition and subtraction. The answer should show significant digits only as far to the right as is seen in the least precise number in the calculation.

```
1 725.463
  189.2
   16.73
---------
1 931.393
```

The sum should be reported as 1 931.4 because 189.2 is the least precise number in the calculation.

The rounding rule will produce the following.

```
1 725.46
  189.2
   16.73
--------
1 931.39
```

The accuracy has not been affected, so the reported answer would be the same, 1 931.4. The rule should not cause problems in addition, but it can in arithmetic operations when the difference between numbers becomes the denominator in a division operation:

$(16.65 + 19.2) \div (772.36 - 772.34)$

Without rounding or concern for significant digits, the answer is 1 792.50. If the rule for rounding is employed, however, the expression then becomes $(16.65 + 19.2) \div (772.4 - 772.3)$ and the answer is 358.50. Which of these numbers is more nearly correct? The answer depends on the precision with which the numbers in the denominator were generated. Care must be taken in applying the rules, and the employment of common sense is absolutely necessary.

Particular care must also be taken when trigonometric functions and logarithms are part of the computation. Since most of your use of such numbers will be in conjunction with electronic calculators, you should have little problem. If you obtain the values from tables, however, make sure that you retain *at least* one more digit than is contained in the number with the least number of significant digits. Often values such as 30°, 45°, 60°, and 90° are intended to be exact, so that you must be careful in these instances. An entry like 30°00′00″ generally removes the doubts.

4.4 Accuracy and precision

Accuracy and precision are quite often used interchangeably and also incorrectly. Accuracy is a measure of the nearness of a value to the correct or true value. Precision refers to reproducibility of a measurement or its reliability. Consider again the example of measuring the 1 200-m distance discussed in Sec. 4.2.1. It is quite possible that we might measure the distance between the two fixed points four times and record 1 202.96 m, 1 203.13 m, 1 203.04 m, and 1 202.91 m. The mean of these measurements is 1 203.01 m. Our measurements stray from the mean by 0.05, 0.12, 0.03, and 0.10 m, respectively. It can reasonably be assumed that because of the closeness of the first four measurements a fifth measurement would not differ greatly. Our quantities appear to be reliable. But such results have nothing to do with the accuracy of the measurements. If the tape that we used was too short, then we obtained a measurement that is too long. Unless we account for the error in tape length, we will continue to get inaccurate, though precise, readings. We could likewise have used sloppy procedures and obtained values that were in considerable variance with each other, averaged them, and (by chance) obtained a very accurate result.

Methods of analyzing measured data are briefly discussed in Chap. 14.

Fig. 4.2 Highly accurate surface properties of metal sample are being determined by an electron microprobe. (*Ames Laboratory, U.S. Department of Energy.*)

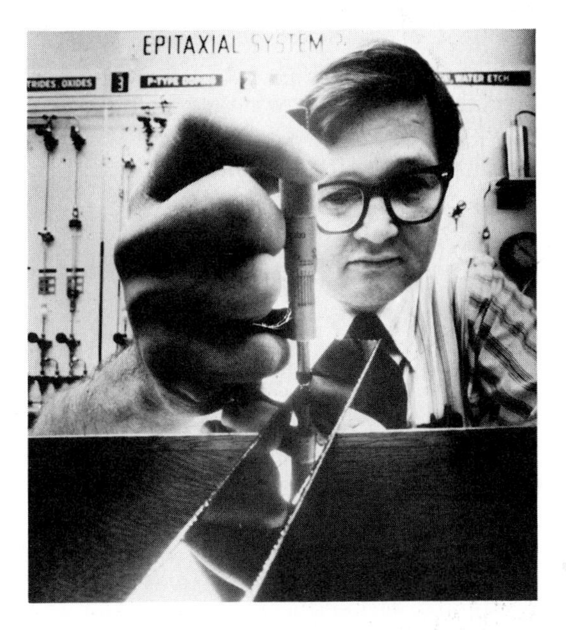

Fig. 4.3 The approximate thickness of this metal plate is being measured by the use of a micrometer. (*Westinghouse.*)

4.5 Approximations

Even though engineers try for a high level of precision, there are many times when only a close approximation is needed. For years, we used the slide rule as our computational device. We were aware that slide-rule answers had only three significant figures, so we knew when to use a machine, logarithms, or even long-hand computations to improve the precision of the answer.

When we do not have tables and references with us and must estimate a calculation with reasonable accuracy, we have to rely on our basic understanding of the problem under discussion coupled with our previous experience. If greater accuracy is needed, we will refine our first answer when we have more time and the necessary reference materials.

In the area of our highest competency, we are expected to be able to make rough estimates to provide figures that can be used for tentative decisions. Prospective investors may wish to know the cost range of two or more alternatives. They have no intention of holding us responsible for the accuracy of the estimate, even though experience later shows that it was in error by perhaps 10 or even 20 percent.

Example problem 4.1 A civil engineer is asked to meet with a city council committee to discuss their needs with respect to the disposal of solid wastes (garbage or refuse). The community, a city of 12 000 persons, must begin supplying refuse collection and disposal for its citizens for the first time. In reviewing various alternatives for disposal, a sanitary landfill is suggested. One of the councilmen is concerned about how much land is going to be needed, so he asks the engineer how many acres will be required within the next 10 years.

Discussion The engineer quickly estimates as follows:

The national average solid waste production is 2.75 kg/(capita)(d). We can assume that each citizen will thus produce 1 Mg of refuse per year:

$$(2.75 \text{ kg/d})(365 \text{ d/year})\left(\frac{1 \text{ Mg}}{1\ 000 \text{ kg}}\right) = 1 \text{ Mg/year}.$$

Experience indicates that refuse will probably be compacted to a density of 400 to 600 kg/m^3. On this basis, the per capita landfill volume will be 2 m^3 each year; and 1 acre filled 1 m deep will contain the collected refuse of 2 000 people for a year (1 acre = 4 047 m^2). Therefore, the requirement for 12 000 people will be 1 acre filled 6 m deep. However, knowledge of the geology of the particular area indicates that bedrock occurs at approximately 6 m below the ground surface. The completed landfill should therefore, have an average depth of 4 m, consequently, 1.5 acres a year or 15 acres in 10 years will be required. The patterns of the recent past indicate that some growth in population and solid waste generation should be expected. It is finally suggested that the city should plan to use about 20 acres in the next 10 years.

This calculation took only minutes and required no computational device other than pencil and paper. The engineer's experience, rapid calculations, sound basic assumptions, and sensible rounding of figures were the main requirements. And a usable estimate, designed to neither mislead nor to sell a point of view, was provided. If this project proceeds to the actual development of a sanitary landfill, the civil engineer will then gather actual data, refine the calculations, and prepare estimates upon which one would risk a professional reputation.

Example problem 4.2 Suppose that your instructor assigns the following problem. Determine the number of pieces of lumber 5 cm × 10 cm × 2.40 m that can be sawn from the tree nearest to the southeast corner of the building in which you are now meeting. How would you proceed? See Fig. 4.4 for one student's response.

Discussion The assumptions that Jim made seem to be reasonable. Although he did not allow for the width of the saw cut nor for

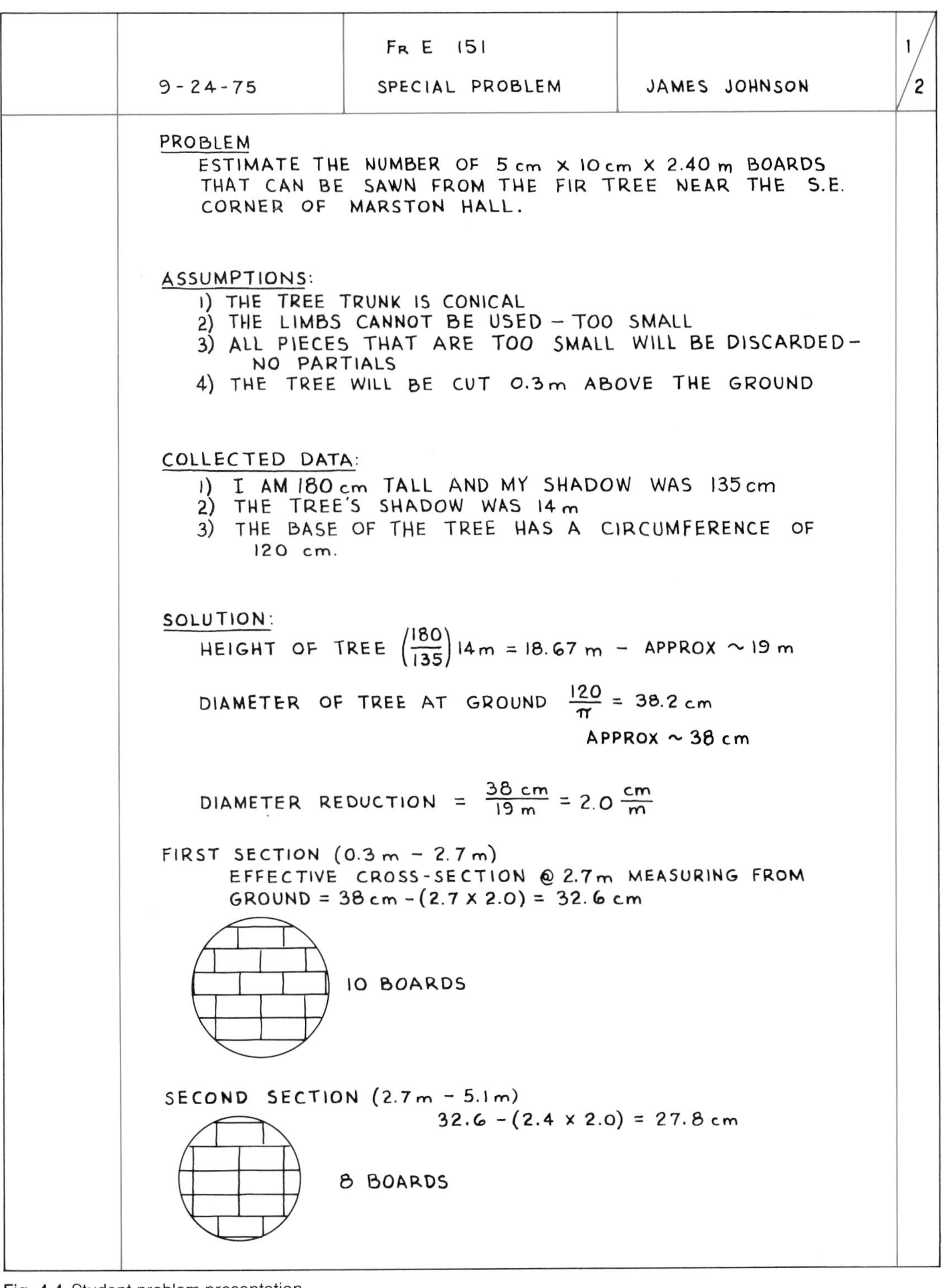

	9-24-75	FR E 151 SPECIAL PROBLEM	JAMES JOHNSON	1/2

PROBLEM

ESTIMATE THE NUMBER OF 5 cm X 10 cm X 2.40 m BOARDS THAT CAN BE SAWN FROM THE FIR TREE NEAR THE S.E. CORNER OF MARSTON HALL.

ASSUMPTIONS:

1) THE TREE TRUNK IS CONICAL
2) THE LIMBS CANNOT BE USED – TOO SMALL
3) ALL PIECES THAT ARE TOO SMALL WILL BE DISCARDED – NO PARTIALS
4) THE TREE WILL BE CUT 0.3 m ABOVE THE GROUND

COLLECTED DATA:

1) I AM 180 cm TALL AND MY SHADOW WAS 135 cm
2) THE TREE'S SHADOW WAS 14 m
3) THE BASE OF THE TREE HAS A CIRCUMFERENCE OF 120 cm.

SOLUTION:

HEIGHT OF TREE $\left(\frac{180}{135}\right) 14\text{ m} = 18.67\text{ m}$ – APPROX ~ 19 m

DIAMETER OF TREE AT GROUND $\frac{120}{\pi} = 38.2\text{ cm}$

APPROX ~ 38 cm

DIAMETER REDUCTION $= \frac{38\text{ cm}}{19\text{ m}} = 2.0\ \frac{\text{cm}}{\text{m}}$

FIRST SECTION (0.3 m – 2.7 m)

EFFECTIVE CROSS-SECTION @ 2.7 m MEASURING FROM GROUND = 38 cm – (2.7 X 2.0) = 32.6 cm

10 BOARDS

SECOND SECTION (2.7 m – 5.1 m)

32.6 – (2.4 x 2.0) = 27.8 cm

8 BOARDS

Fig. 4.4 Student problem presentation.

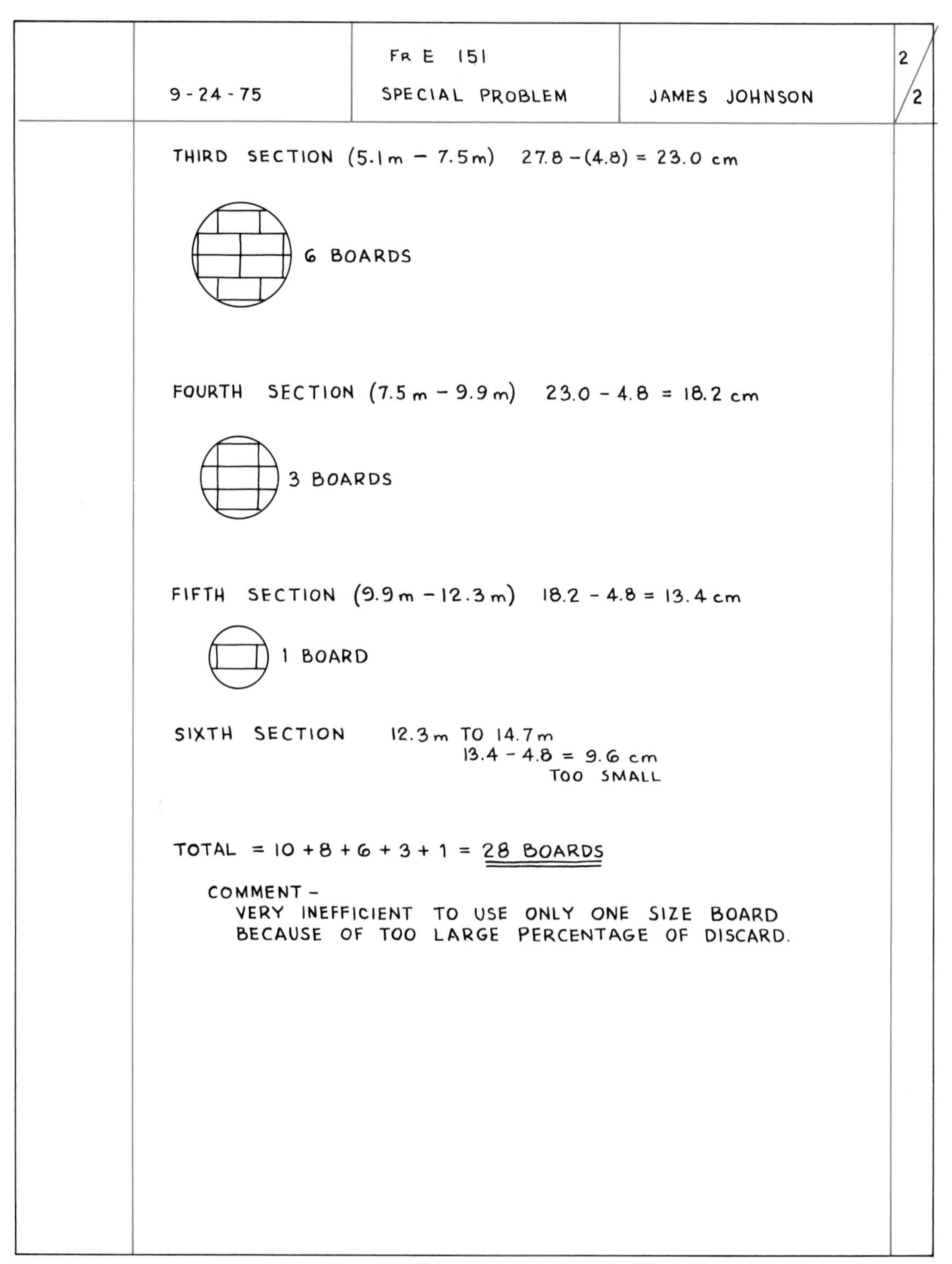

Fig. 4.4 (*Continued*)

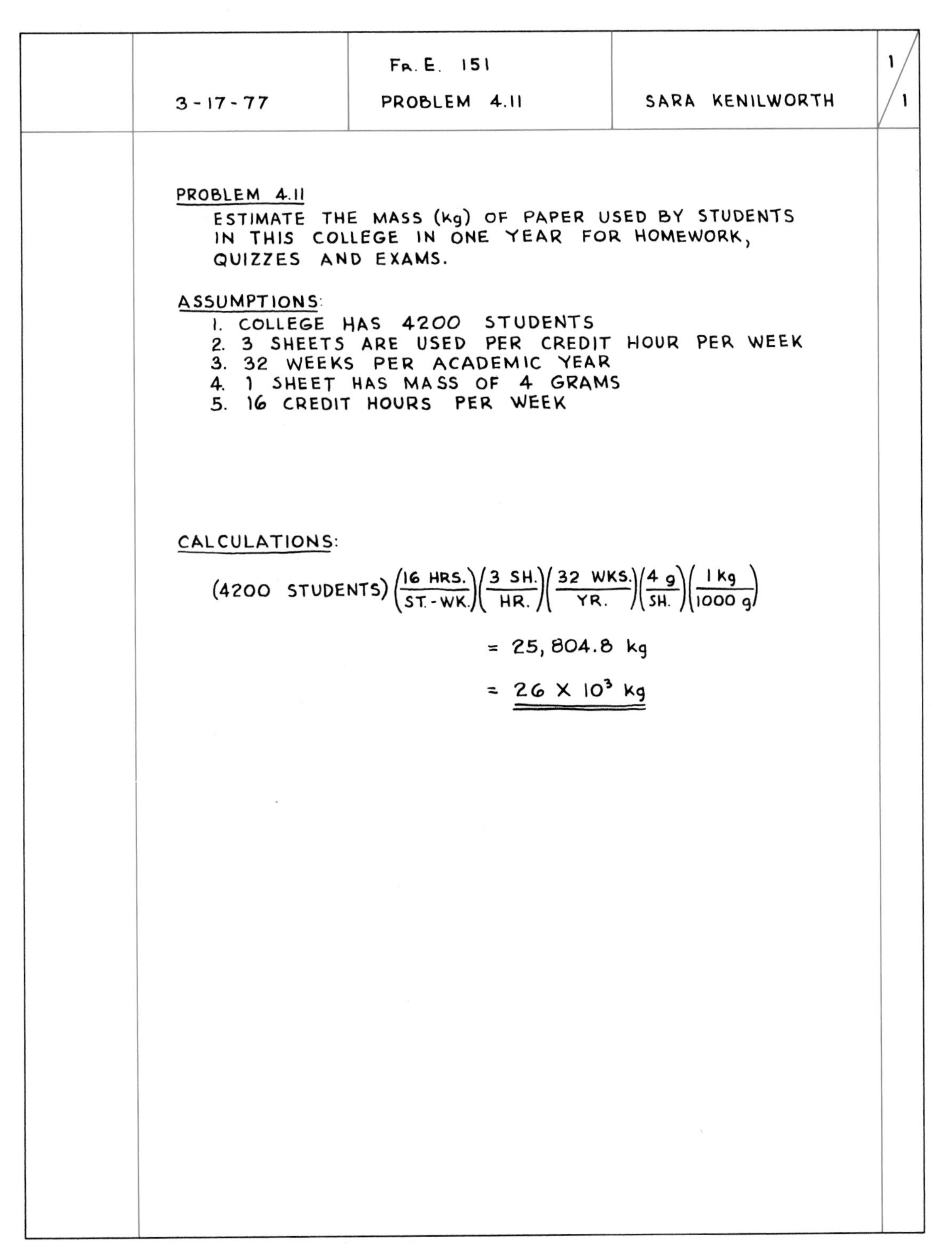

	3-17-77	FR. E. 151 PROBLEM 4.11	SARA KENILWORTH	1/1

PROBLEM 4.11

ESTIMATE THE MASS (kg) OF PAPER USED BY STUDENTS IN THIS COLLEGE IN ONE YEAR FOR HOMEWORK, QUIZZES AND EXAMS.

ASSUMPTIONS:

1. COLLEGE HAS 4200 STUDENTS
2. 3 SHEETS ARE USED PER CREDIT HOUR PER WEEK
3. 32 WEEKS PER ACADEMIC YEAR
4. 1 SHEET HAS MASS OF 4 GRAMS
5. 16 CREDIT HOURS PER WEEK

CALCULATIONS:

$$(4200 \text{ STUDENTS})\left(\frac{16 \text{ HRS.}}{\text{ST.-WK.}}\right)\left(\frac{3 \text{ SH.}}{\text{HR.}}\right)\left(\frac{32 \text{ WKS.}}{\text{YR.}}\right)\left(\frac{4 \text{ g}}{\text{SH.}}\right)\left(\frac{1 \text{ kg}}{1000 \text{ g}}\right)$$

$$= 25{,}804.8 \text{ kg}$$

$$= \underline{\underline{26 \times 10^3 \text{ kg}}}$$

Fig. 4.5 Student problem presentation.

the thickness of the bark, the omissions may very well be consistent with the degree of accuracy involved in his calculations. Most boy and girl scouts have learned to measure heights by the method he used; and all freshman engineers should be familiar with similar triangles. After determining the height and diameter of the tree, Jim applied a graphical technique for determining how many boards could be cut from each 2.4-m section of the tree. He correctly used the upper (smaller) diameter of the section. His task was then reduced to a simple counting of the boards.

Example problem 4.3 Estimate the amount of paper used by the students of this college for homework, quizzes, and examinations during one academic year. Express your answer in kilograms.

Figure 4.5 is another student's response to this problem.

Discussion The problem requires that certain data be known or assumed. The calculation is simple and straightforward once the data is known or the assumptions are made. If you are given sufficient time you can determine the correct data to replace assumptions 1, 3, 4, and 5. Assumption number 2 is open to wide disagreement, but a survey of several students on a dormitory floor could give you a reasonable number to use.

4.6 Case study

The following case study is a summary report of an actual project. The authors thank Mr. Paul Bolton, P.E., Senior Engineer, Henningson, Durham & Richardson, for granting permission to use the study.

The purpose of this case study is to examine and discuss various activities that occurred during one day in the life of an engineer. The principle character of this study will be a young man working for a private consulting firm as a licensed civil engineer.

First, a little background information. Owners of the consulting firm received a letter from officials of a nearby city asking for their interest and possible involvement in a problem currently faced by the city. The city is highly industrialized, with approximately 8 000 citizens.

The state Department of Environmental Quality (DEQ) notified the city that the effluent (treated sewage) leaving its sewage treatment plant contained excessive pollutants. (Note that the DEQ is responsible for monitoring many environmental concerns, including streams, and can file charges against cities, industries, and individuals who discharge pollutants into the natural streams of the state.) The city officials also received a series of complaints about odors coming from the sewage treatment plant; and shoppers from the surrounding area were reluctant to come to town to shop during certain times of the year. The city staff was aware of problems in the sewage collection system and had sought funds at budget hearings to correct those problems. The city council, recognizing its responsibility and liability, invited five engineering consultants to discuss separately the sewage problems with the city and to make recommendations about how the problems could be solved.

The consulting firm with which we are concerned decided to authorize an initial study of the problems and by return mail indicated to the city that a young engineer would represent the firm at an informational meeting. The meeting was scheduled, and thus begins our story.

The engineer arrived in the city early on the day of his scheduled evening meeting. He had been previously involved in meetings such as the one planned for that evening and was aware that he would be expected to answer a wide range of questions. During the day he drove about the city to refresh his memory and to develop a firm mental picture of several landmarks. He made notes on a topographic map about many features, primarily with respect to possible sites for a new treatment plant. He ascertained that adequate land for expansion was available adjacent to the existing plant, but he knew that the odor problem was an important consideration from both a physical and psychological point of view. He examined records of the treatment facilities and those of stream flow to determine how much dilution water could be expected. His study indicated two things, one good and one bad. First, the present facilities were extremely well operated and maintained; and second, they were simply not capable of properly treating the volume and type of sewage that was presently being generated in the city.

The conference that evening began in an informal manner, with introductions and explanations about why the meeting was being held. It was not a large group, but it included the mayor and city council, city staff, representatives from industry, the Chamber of Commerce, and interested citizens. Our young engineer was asked to discuss the sewage problems of the city as he saw them. It is difficult to decide how much detail to include in such a discussion without being either boring or incomplete. He decided to be brief at first and then answer questions if they arose. He explained that the existing sewage treatment plant was well designed and operated, but old. The city had grown rapidly during the past decade and several industries had been built and others enlarged. The time had come for a considerable increase in sewage treatment plant capacity; prudent planning would be to provide capacity for continued growth over the next 20 years. The engineer explained how his brief initial study had been made and was careful to indicate that a more detailed study would have to be made before reliable estimates could be established. Next, he outlined four reasonable alternatives:

1. Expansion of the sewage treatment plant at the present location, including facilities to control the odor problem. He explained at this point that all four alternatives included the construction of a sewage pumping station to elevate the sewage from a low-lying segment of the city and some new sewers to replace broken sewers that were causing basement flooding.

2. Construction of a new treatment plant $1\frac{1}{2}$ mi downstream from the existing site. He pointed out that this alternative would be the most expensive, but it had several advantages: better odor control and provision of water and sewer service to a large area of land that could then be developed for commercial, industrial, or residential purposes. The existing site could then be sold or used for other public purposes.

3. Retention of the existing plant, augmenting it with new units at the downstream site. These new units could become part of a new treatment plant at a later date.

4. Direct application of all or part of the sewage onto the land, obviating the need for any other treatment. This process is used in some places but is not particularly desirable for this city because of high land costs and the fact that land was not available nearby, hence the need for expensive sewers and high pumping costs.

Our young engineer discussed each of the alternatives in considerable detail and responded to penetrating questions by the city

Fig. 4.6 An artist's conception of the appearance of the completed water-pollution-control plant. (*Henningson, Durham & Richardson.*)

staff and the representatives of business and industry. After a period of question and answers, there seemed to be a consensus that the second alternative was the most desirable but the cost was a very large question mark. Our engineer attempted to delay making any estimate of cost on the grounds that a thorough study was necessary. Many questions about cost continued until he was unable to avoid an answer. He made his estimate as follows:

1. The consulting firm was presently supervising the construction of a sewage pump station of approximately the same size as the one he felt was needed, so he assumed that the cost would be the same.
2. On his topographic map he had sketched the location of the needed sewers and his visit with the city engineer provided him with a rough idea of the required sewer size. Sewers are normally priced on the basis of a unit cost per foot; the engineer was aware of the current costs for sewers and manholes.
3. The engineer was supervising the design of two sewage treatment plants for other cities. Preliminary cost estimates had been made for these plants, so he used those cost figures to estimate the cost of a new plant for this city.
4. Based on the amount of land required for sewage treatment plants, he estimated the land requirements and asked one of the citizens attending the meeting (a realtor) what the land would cost.
5. He then added the customary percentage fee for engineering and legal costs and arrived at an estimate of $5.5 to $6.0 million.

As expected, the size of his estimate produced mostly silence, because such a cost amounts to about $750 for each citizen. The

Fig. 4.7 The actual plant during an early stage of construction. Note the forms are in place on one of the four circular trickling filters. (*Henningson, Durham & Richardson.*)

engineer quickly explained that grants from the state and federal governments were available to cover approximately 75 percent of the cost, so that the local costs would be greatly reduced. Since he had also examined the revenue records of the city, he was able to advise them that a modest increase in the existing sewer service charge and/or a small property tax increase would produce enough money annually to retire the general obligation or revenue bonds that could be sold to pay for the construction of the needed facilities.

The engineer in this case study was required to make a large number of quick decisions and estimates, all of which were later proven good or poor after a more detailed study and close calculations were performed. His earlier experience led him to seek the needed initial data and to analyze it in sufficient detail to serve this city properly. It certainly would have been possible for him to spend more time and money in gathering his preliminary information and in performing more detailed calculations upon which to base his estimates. In this case, his firm was selected to conduct the study and prepare the construction drawings. But keep in mind that four other firms were interviewed and were not selected. Hence, their efforts and expenditures were in vain.

A footnote: This project was completed about 4 years and 8 months after the interview. The final cost was about 50 percent greater than the estimate, owing primarily to inflation and changes in regulations by the DEQ and the federal Environmental Protection Agency.

Problems

4.1 Estimate the number of drugstores presently in business in the United States. Explain the assumptions made and the source of all factual data used.

4.2 Estimate the volume of water discharged by the Mississippi River into the Gulf of Mexico each year. Explain your estimate.

4.3 Estimate the cost of collection of the solid waste (garbage) from this city for 1 year. Show all data and assumptions.

4.4 Estimate the amount of paper used by the students of your college for homework, quizzes, and exams in one academic year. Express your answer in kilograms. Improve assumptions over those on page 89.

4.5 What is the mass of textbooks purchased by the students of your college during one academic year?

4.6 How much gasoline was used by the students of your university during the past Christmas recess?

4.7 Estimate the number of hours that the engineering freshmen will spend with their academic advisors during the current quarter (semester).

4.8 Estimate the volume of water used in taking showers by the members of your class during the current quarter (semester).

4.9 Estimate the mass of the building in which your class meets.

4.10 Estimate the amount of chewing gum sold in the United States in 1 year.

4.11 How many auto tires are produced in the United States each year?

4.12 How many megagrams of railroad rails are there in this state?

4.13 How much was the total cost of natural gas for residential users in this city during January of this year?

4.14 What is the volume in cubic meters of all the classrooms in your university?

4.15 Given 15 min, how would you estimate the number of blades of grass inside the running track on campus?

PART TWO
Metric (SI)
units in
engineering

Dimensions, SI units, and conversions

CHAPTER 5

5.1 Introduction

In the past when countries were more isolated from one another than they are now, individual countries tended to develop and use their own set of measures. As the rapid increase in global communication and travel brought countries closer together and the world advanced in technology, the need for a universal system of measurement became clearly apparent. There was such a proliferation of information among all nations that a standard set of dimensions, units, and measurements was vital if this wealth of knowledge was to be of benefit to all.

We will deal in this chapter with the difference between dimensions and units and at the same time explain how there can be an orderly transition from many systems of units to one system, i.e., an international standard.

The standard currently accepted in most industrial nations (except the United States) is the International Metric System, or Système International d'Unites, abbreviated SI. The SI units are a modification and refinement of an earlier system (MKS) that designated the meter, kilogram, and second as fundamental units.

France was the first country, as early as 1840, to officially legislate adoption of the metric system and decree that its use be mandatory.

The United States almost adopted the metric system 150 years ago. In fact, the metric system was made legal in the United States in 1866, but its use was not made compulsory. In spite of many attempts since that time, full conversion to the metric system has not yet been realized in the United States, but significant steps in that direction are presently underway.

5.2 Physical quantities

Engineers are constantly concerned with the measurements of fundamental physical quantities such as length, time, temperature, force, etc. In order to specify a physical quantity fully, it is not sufficient to indicate merely a numerical value. The magnitude of physical quantities can be understood only when they are compared

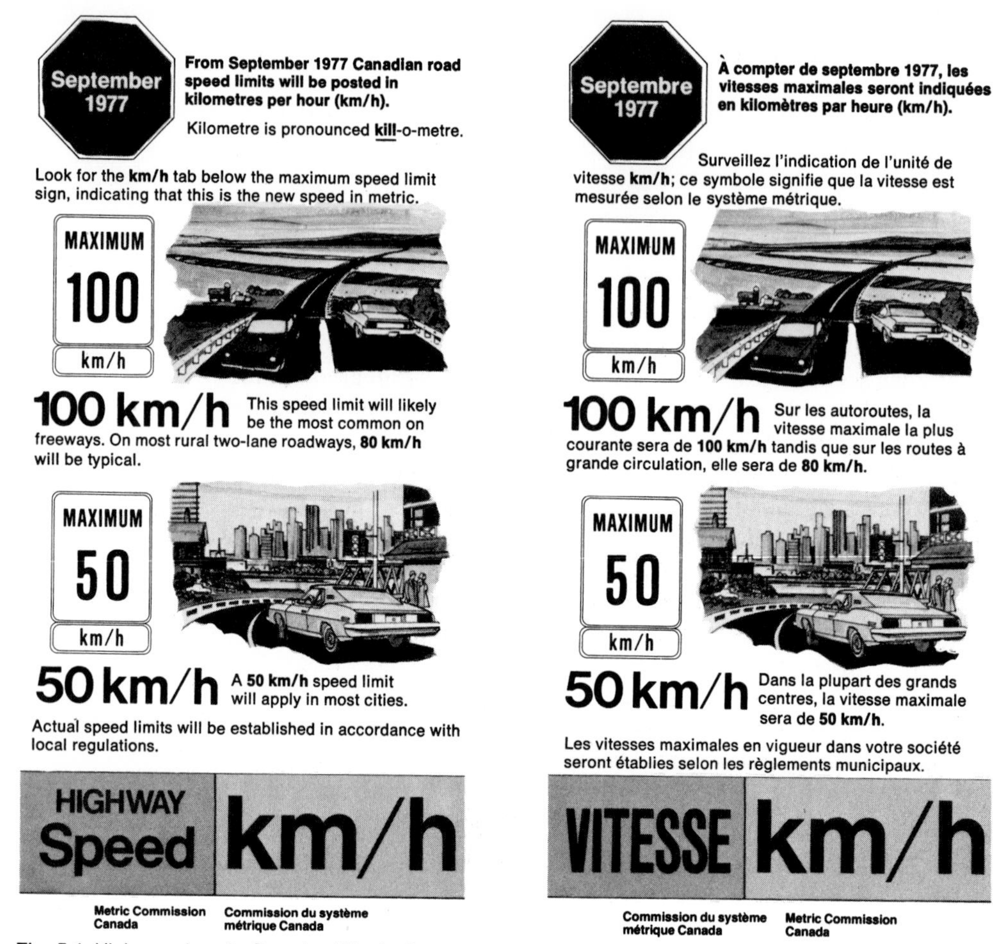

Fig. 5.1 Highway signs in Canada. (*Metric Commission of Canada.*)

with predetermined reference amounts, called units. Any measurement is, in effect, a comparison of how many units are contained within the physical quantity. For example, if we call the physical quantity Q, the numerical value V, and the unit U, then

$$Q = VU$$

For this relationship to be valid, the exact reproduction of a unit must be theoretically possible at any time. Therefore standards must be established. They are a set of fundamental unit quantities kept under normalized conditions in order to preserve their values as accurately as possible. We shall speak more about them later.

5.3 Dimensions

Dimensions are used to describe physical quantities. An important element to remember is that dimensions are independent of units. A physical quantity such as length can be represented by the dimension L, for which there are a large number of possibilities available when selecting a unit. For example, in ancient Egypt, the cubit was related to the length of the arm from the tip of the middle finger to the elbow. Measurements were thus a function of physical stature, with variation from one individual to another. Much later in Britain, the inch was specified as the distance covered by three barley corns, round and dry, laid end to end. Today we require more precision. For example, the meter is defined in terms of the wavelength of monochromatic radiation from pure Krypton-86. We can draw two important points from this discussion: (1) Physical quantities can be accurately measured and (2) each of these units (cubit, inch, and meter), although distinctly different, have in common the quality of being a length and not an area or a volume.

A technique used to distinguish between units and dimensions is to call all quantities of length simply L. In this way, each new physical quantity gives rise to a new dimension, such as T for time, F for force, M for mass, etc. (Note that there are as many dimensions as there are kinds of physical quantities.)

Moreover, dimensions can be divided into two areas—derived and fundamental. Derived dimensions are a combination of fundamental dimensions. Velocity, for example, can be defined as a fundamental dimension, V. But it is more customary as well as convenient to consider velocity as a combination of fundamental dimensions, so that it becomes a derived dimension, $V = (L)(T)^{-1}$.

It is advantageous to use as few fundamental dimensions as possible, but the selection of what is to be fundamental and what is to be derived is not fixed. In actuality, any dimension can be selected as a fundamental dimension in a particular field of engineering or science; and for reasons of convenience, it may be a derived dimension in another field.

A set of fundamental dimensions is simply a group of dimensions that can be conveniently and usefully manipulated when expressing all physical quantities of a particular field.

A dimensional system can be defined as the smallest number of fundamental dimensions which will form a consistent and complete set for a field of science. For example, three fundamental dimensions are necessary to form a complete mechanical dimensional system. These dimensions may be either length (L), time (T), and mass (M) or length (L), time (T), and force (F), depending on the specific application.

The "absolute" system (so called because dimensions within it are not affected by gravity) has as its fundamental dimensions L, T,

Table 5.1 Two Basic Dimensional Systems

Quantity	Absolute	Gravitational
Length	L	L
Time	T	T
Mass	M	$FL^{-1}T^{2}$
Force	MLT^{-2}	F
Velocity	LT^{-1}	LT^{-1}
Pressure	$ML^{-1}T^{-2}$	FL^{-2}
Momentum	MLT^{-1}	FT
Energy	$ML^{2}T^{-2}$	FL
Power	$ML^{2}T^{-3}$	FLT^{-1}
Torque	$ML^{2}T^{-2}$	FL

and M. An advantage of this system is that comparisons of masses at various locations can be made with an ordinary balance, because the local acceleration of gravity has no influence upon the results.

The "gravitational" system (one which accounts for gravity) has as its fundamental dimensions L, T, and F. It is widely used in many engineering branches because it simplifies computations when weight is a fundamental quantity in the computations. Table 5.1 illustrates two of the more basic systems in terms of dimensions. A number of dimensional systems other than those strictly used in mechanics are for systems of heat, electromagnetism, electrical dimensions, etc.

5.4 Units

Once a consistent dimension system has been selected, one can develop a unit system by choosing a specific unit for each fundamental dimension. The problem one encounters when working with units is that there can be a large number of unit systems to choose from for each complete dimension system, as we have already suggested. It is obviously desirable to limit the number of systems and combinations of systems. The SI previously alluded to is intended to serve as an international standard that will provide worldwide limitations.

There are two fundamental systems of units commonly used in mechanics today. One system used in almost every industrial country of the world is called the metric system. It is a decimal system based on the meter, kilogram, and second (MKS) as the units of length, mass, and time, respectively. The United States has used the other system, normally referred to as the British system, which has a less-well-defined unit base. It is based on the foot, pound, and second.

Numerous international conferences on weights and measures over the past 20 years have gradually modified the MKS system to the point that all countries previously using various forms of the metric system are beginning to standardize. The SI is now considered the new international system of units. The United States has adopted the system, but full use will be preceded by a long and expensive period of change. During the transition period, engineers will have to be familiar with other systems and their corresponding conversion factors.

5.5 SI Units and symbols

The International System of Units (SI), developed and maintained by the General Conference on Weights and Measures, is intended as a basis for worldwide standardization of measurements. The name and abbreviation were set forth in 1960. SI at the present time is a complete system that is being universally adopted.

This new international system is divided into three classes of units.

1. Base units
2. Supplementary units
3. Derived units

There are seven base units in the SI. The units (except the kilogram) are defined in such a way that they can be reproduced anywhere in the world.

Table 5.2 lists each base unit along with its name and proper symbol.

Each of the base units is defined below as established at an International General Conference on Weights and Measures (CGPM).[1]

1. Length: The meter is a length equal to 1 650 763.73 wavelengths in a vacuum of the radiation corresponding to the transition between the levels

Table 5.2 Base Units

Quantity	Name	Symbol
Length	meter	m
Mass	kilogram	kg
Time	second	s
Electric current	ampere	A
Thermodynamic temp	kelvin	K
Amount of substance	mole	mol
Luminous intensity	candela	cd

[1]The initials stand for Conférence Générale des Poids et Mesures.

$2p_{10}$ and $5d_5$ of the krypton-86 atom. The meter was defined by the 11th CGPM in 1960.

2. Time: The second is the duration of 9 192 631 770 periods of radiation corresponding to the transition between the two hyperfine levels of the ground state of the cesium-133 atom. The second was adopted by the 13th CGPM in 1967.

3. Mass: The standard for the unit of mass, the kilogram, is a cylinder of platinum-iridium alloy kept by the National Bureau of Weights and Measures in France. A duplicate copy is maintained in the United States. The unit of mass was adopted by the 1st and 3rd CGPM in 1889 and 1901. It is the only base unit nonreproducible in a properly equipped lab.

4. Electric current: The ampere is a constant current which if maintained in two straight parallel conductors of infinite length and of negligible circular cross section and placed one meter apart in vacuum would produce between these conductors a force equal to 2×10^{-7} newton per meter of length. The ampere was adopted by the 9th CGPM in 1948.

5. Temperature: The kelvin, unit of thermodynamic temperature, is the fraction 1/273.16 of the thermodynamic temperature of the triple point of water. The kelvin was adopted by the 13th CGPM in 1967.

6. Amount of substance: The mole is the amount of substance of a system that contains as many elementary entities as there are atoms in 0.012 kilogram of carbon-12. The mole was defined by the 14th CGPM in 1971.

7. Luminous intensity: The base unit candela is the luminous intensity, in the perpendicular direction, of a surface of 1/600 000 square meter of blackbody at the temperature of freezing platinum under pressure of 101 325 newtons per square meter. This convention was adopted by the 13th CGPM in 1967.

The units listed in Table 5.3 are called supplementary units and may be regarded as either base units or as derived units.

The unit for a plane angle is the radian, a unit that is used frequently in engineering. The steradian is not as commonly used. These units can be defined in the following way.

1. Plane angle: The radian is the plane angle between two radii of a circle that cut off on the circumference an arc equal in length to the radius.

2. Solid angle: The steradian is the solid angle which, having its vertex in the center of a sphere, cuts off an area of the sphere equal to that of a square with sides of length equal to the radius of the sphere.

Table 5.3 Supplementary Units

Quantity	Name	Symbol
Plane angle	radian	rad
Solid angle	steradian	sr

Table 5.4 Derived Units

Quantity	SI unit symbol	Name	Base units
Frequency	Hz	hertz	s^{-1}
Force	N	newton	$kg \cdot m \cdot s^{-2}$
Pressure, stress	Pa	pascal	$kg \cdot m^{-1} \cdot s^{-2}$
Energy or work	J	joule	$kg \cdot m^2 \cdot s^{-2}$
A quantity of heat	J	joule	$kg \cdot m^2 \cdot s^{-2}$
Power, radiant flux	W	watt	$kg \cdot m^2 \cdot s^{-3}$
Electric charge	C	coulomb	$A \cdot s$
Electric potential	V	volt	$kg \cdot m^2 \cdot s^{-3} \cdot A^{-1}$
Potential difference	V	volt	$kg \cdot m^2 \cdot s^{-3} \cdot A^{-1}$
Electromotive force	V	volt	$kg \cdot m^2 \cdot s^{-3} \cdot A^{-1}$
Capacitance	F	farad	$A^2 \cdot s^4 \cdot kg^{-1} \cdot m^{-2}$
Electric resistance	Ω	ohm	$kg \cdot m^2 \cdot s^{-3} \cdot A^{-2}$
Conductance	S	siemens	$kg^{-1} \cdot m^{-2} \cdot s^3 \cdot A^2$
Magnetic flux	Wb	weber	$kg \cdot m^2 \cdot s^{-2} \cdot A^{-1}$
Magnetic flux density	T	tesla	$kg \cdot s^{-2} \cdot A^{-1}$
Inductance	H	henry	$kg \cdot m^2 \cdot s^{-2} \cdot A^{-2}$
Luminous flux	lm	lumen	$cd \cdot sr$
Illuminance	lx	lux	$cd \cdot sr \cdot m^{-2}$
Activity (radionuclides)	Bq	becquerel	s^{-1}
Absorbed dose	Gy	gray	$m^2 \cdot s^{-2}$

As indicated in previous discussion, derived units are formed by combining base, supplementary, or other derived units. Symbols for them are carefully selected to avoid confusion. Those which have special names and symbols approved by CGPM are listed in Table 5.4 together with their definitions in terms of base units.

At first glance, Figure 5.2 may appear complex, even confusing. However, if you study the examples below, you will no doubt agree that a considerable amount of information is presented in a concise flowchart. To get the point of it quickly, be aware that the solid lines denote multiplication and the broken lines indicate division. The arrows pointing toward the units (circled) are significant and arrows going away have no meaning for that particular unit. Consider the pascal, for example: Two arrows point toward the circle—one solid and one broken. This means that the unit pascal is formed from the newton and meter squared, or N/m^2.

Other derived units, such as those included in Table 5.5, have no special names but are combinations of base units and units with special names.

Being a decimal system, the SI is convenient to use, because by simply affixing a prefix to the base, a quantity can be increased or decreased by factors of 10 and the numerical quantity can be kept within manageable limits. Table 5.6 lists the multiplication factors along with their prefix names and symbols.

The proper selection of prefixes will also help eliminate nonsignificant zeros and leading zeros in decimal fractions. The numerical

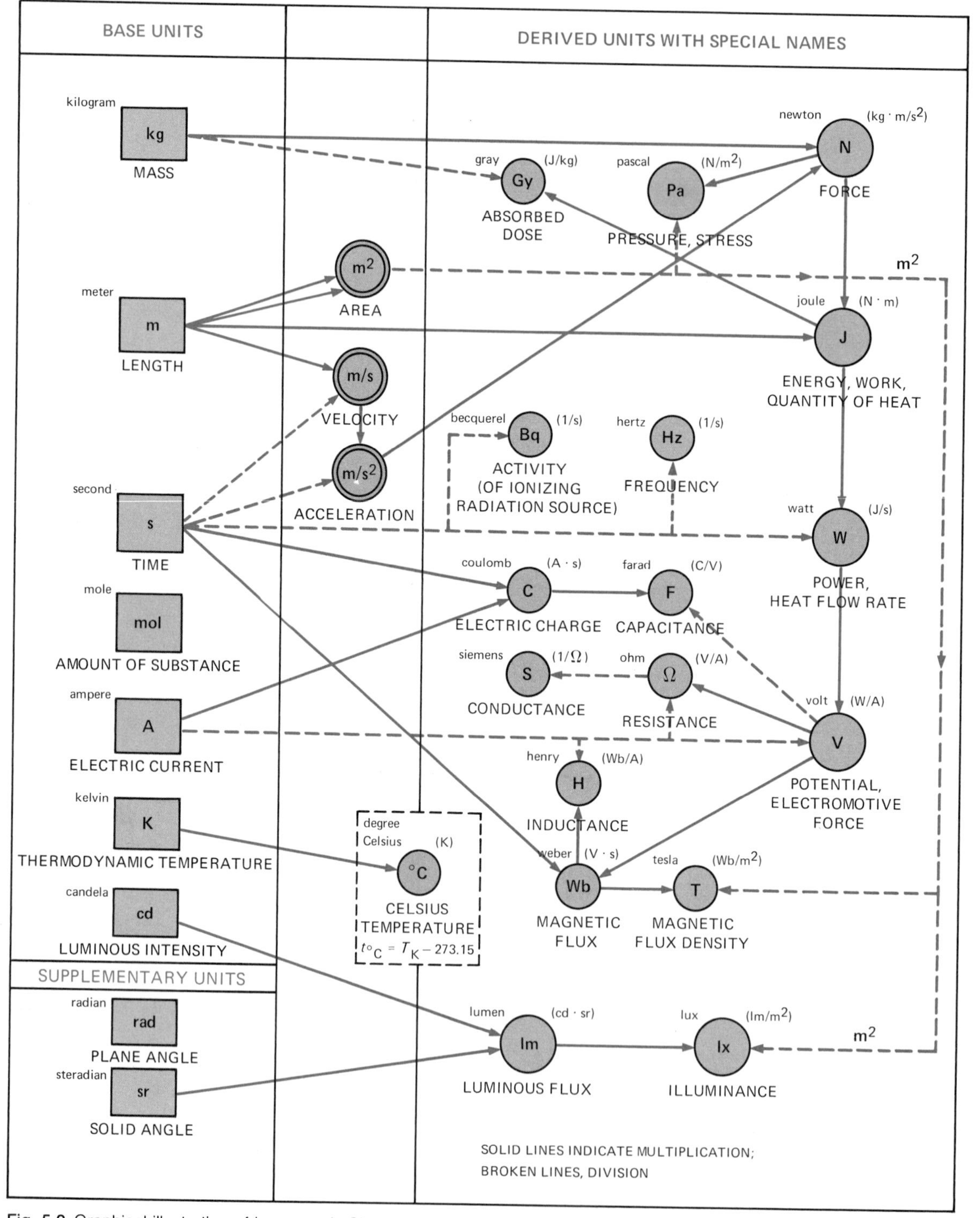

Fig. 5.2 Graphical illustration of how certain SI units are derived in a coherent fashion from base and supplementary units. (*National Bureau of Standards.*)

Table 5.5 Common Derived Units

Quantity	Units	Quantity	Units
Acceleration	$m \cdot s^{-2}$	Molar entropy	$J \cdot mol^{-1} \cdot K^{-1}$
Angular acceleration	$rad \cdot s^{-2}$	Molar heat capacity	$J \cdot mol^{-1} \cdot K^{-1}$
Angular velocity	$rad \cdot s^{-1}$	Moment of force	$N \cdot m$
Area	m^2	Permeability	$H \cdot m^{-1}$
Concentration	$mol \cdot m^{-3}$	Permittivity	$F \cdot m^{-1}$
Current density	$A \cdot m^{-2}$	Radiance	$W \cdot m^{-2} \cdot sr^{-1}$
Density, mass	$kg \cdot m^{-3}$	Radiant intensity	$W \cdot sr^{-1}$
Electric charge density	$C \cdot m^{-3}$	Specific heat capacity	$J \cdot kg^{-1} \cdot K^{-1}$
Electric field strength	$V \cdot m^{-1}$	Specific energy	$J \cdot kg^{-1}$
Electric flux density	$C \cdot m^{-2}$	Specific entropy	$J \cdot kg^{-1} \cdot K^{-1}$
Energy density	$J \cdot m^{-3}$	Specific volume	$m^3 \cdot kg^{-1}$
Entropy	$J \cdot K^{-1}$	Surface tension	$N \cdot m^{-1}$
Heat capacity	$J \cdot K^{-1}$	Thermal conductivity	$W \cdot m^{-1} \cdot K^{-1}$
Heat flux density	$W \cdot m^{-2}$	Velocity	$m \cdot s^{-1}$
Irradiance	$W \cdot m^{-2}$	Viscosity, dynamic	$Pa \cdot s$
Luminance	$cd \cdot m^{-2}$	Viscosity, kinematic	$m^2 \cdot s^{-1}$
Magnetic field strength	$A \cdot m^{-1}$	Volume	m^3
Molar energy	$J \cdot mol^{-1}$	Wavelength	m

value of any measurement should be recorded between 0.1 and 1 000. This rule is suggested because it is easier to make realistic judgments when working with numbers between 0.1 and 1 000. For example, suppose that you are asked the distance to a nearby town. It would be more understandable to respond in kilometers than meters. That is, it is easier to visualize 10 km than 10 000 m.

Moreover, the use of certain prefixes is preferred over that of others. Those representing powers of 1 000, such as kilo, mega, milli, and micro, will reduce the number you must remember. These

Table 5.6 Decimal Multiples

Multiplier	Prefix name	Symbol
10^{18}	exa	E
10^{15}	peta	P
10^{12}	tera	T
10^{9}	giga	G
10^{6}	*mega	M
10^{3}	*kilo	k
10^{2}	hecto	h
10^{1}	deka	da
10^{-1}	deci	d
10^{-2}	centi	c
10^{-3}	*milli	m
10^{-6}	*micro	μ
10^{-9}	nano	n
10^{-12}	pico	p
10^{-15}	femto	f
10^{-18}	atto	a

*Most often used.

preferred prefixes should be used, with the following three exceptions that are still common because of convention.

1. When expressing area and volume, the prefixes hecto-, deka-, deci-, and centi- may be used; for example, cubic centimeter.

2. When discussing different values of the same quantity or expressing them in a table, calculations are simpler to perform when you use the same unit multiple throughout.

3. Sometimes a particular multiple is recommended as a consistent unit even though its use violates the 0.1 to 1 000 rule. For example, many companies use the millimeter for linear dimensions even when the values lie far outside this suggested range. The cubic decimeter (commonly called liter) is also used.

Recalling the discussion of significant figures in Chap. 4, we see that the SI prefix notations can be used to a definite advantage.

Consider the previous example of 10 kilometers. When giving an estimate of distance to the nearest town there is certainly an implied approximation in the use of a round number. Suppose that we were talking about a 10 000-m Olympic track and field event. The accuracy of such a distance must certainly be greater than something between 5 000 and 15 000 m. This example is intended to illustrate the significance of the four zeros (10 000). If all four zeros are in fact significant, then the race is accurate within one meter (9 999.5 to 10 000.5). If only three zeros are significant, then the race is accurate to within 10 m (9 995 to 10 005).

There are two logical and acceptable methods available of eliminating confusion concerning zeros.

1. Use proper prefixes to denote intended significance.

Distance	Accuracy
10 000 m	5 000 to 15 000 m
10.000 km	9 999.5 to 10 000.5 m
10.00 km	9 995 to 10 005 m
10.0 km	9 950 to 10 050 m

2. Use scientific notation to indicate significance.

Distance	Accuracy
10 000 m	5 000 to 15 000 m
10.000×10^3 m	9 999.5 to 10 000.5 m
10.00×10^3 m	9 995 to 10 005 m
10.0×10^3 m	9 950 to 10 050 m

Selection of a proper prefix is customarily the logical way to handle problems of significant figures; however, there are conventions that do not lend themselves to the prefix notation. An example would be temperature in degrees Celsius, that is, 4.00×10^3°C is the conventional way to handle it, not 4.00 k°C.

5.6 Rules for using SI units

Along with the adoption of SI comes the responsibility to thoroughly understand and properly apply the new system. Obsolete practices involving English and metric units are widespread. This section provides rules that should be followed when working with SI units.

5.6.1 Unit symbols and names

1. Periods are never used after symbols unless the symbol is at the end of a sentence.

2. Unit symbols are written in lowercase letters unless the symbol derives from a proper name, in which case the first letter is capitalized.

Lowercase	Uppercase
m, kg, s, mol, cd	A, K, Hz, Pa, C

3. Symbols rather than self-styled abbreviations should always be used to represent units.

Correct	Not correct
A	amp
s	sec

4. An s is never added to the symbol to denote plural.

5. A space is always left between the numerical value and the unit symbol.

Correct	Not correct
43.7 km	43.7km
0.25 Pa	0.25Pa

Exception: No space should be left between numerical values and the symbols for degree, minute, and second of angles, and degree Celsius.

6. There should be no space between the prefix and the unit symbols.

Correct	Not correct
mm, MΩ	k m, μ F

7. When writing unit names, all letters are lowercase except at the beginning of a sentence, even if the unit name is derived from a proper name.

8. Plurals are used as required when writing unit names. For example, henries is plural for henry, etc. The following exceptions are recommended.

Singular	Plural
lux	lux
hertz	hertz
siemens	siemens

With these exceptions, unit names form their plurals in the usual manner.

9. No hyphen or space should be left between a prefix and the unit name. There are three cases where the final vowel in the prefix is omitted, but these are the only exceptions: megohm, kilohm, and hectare.

10. The symbol should be used in preference to the unit name because unit symbols are standardized. An exception to this is made when a number is written in words preceding the unit, e.g., we would write ten meters, not ten m. The same is true the other way, e.g., 10 m, not 10 meters.

5.6.2 Multiplication and division

1. When writing unit names as a product, always use a space (preferred) or a hyphen.

Correct usage
newton meter or newton-meter

An exception to this rule is made for watthours.

2. When expressing a quotient using unit names, always use the word "per" and not a solidus (/). The solidus, or slash mark, is reserved for use with symbols.

Correct	Not correct
meter per second	meter/second

3. When writing a unit name that requires a power, use a modifier, e.g., squared or cubed, after the unit name. For area or volume, the modifier can be placed before the unit name.

Correct	Correct
meter per second squared	square millimeter

4. When expressing products using unit symbols, the center dot is preferred.

Correct
N · m for newton meter

5. When denoting a quotient by unit symbols, any of the following methods are accepted form.

Correct

$$\mathrm{m/s} \text{ or } \mathrm{m \cdot s^{-1}} \text{ or } \frac{\mathrm{m}}{\mathrm{s}}$$

In more complicated cases, negative powers or parentheses should be considered; use $\mathrm{m/s^2}$ or $\mathrm{m \cdot s^{-2}}$ but not m/s/s for acceleration, use $\mathrm{kg \cdot m^2/(s^3 \cdot A)}$, or $\mathrm{kg \cdot m^2 \cdot s^{-3} \cdot A^{-1}}$, but not $\mathrm{kg \cdot m^2/s^3/A}$ for electric potential.

5.6.3 Numbers

1. To denote a decimal point, use a period on the line. When expressing numbers less than 1, a zero should be written before the decimal marker.

Example
15.6
0.93

2. Since a comma is used in many countries to denote a decimal point, its use is to be avoided in grouping digits. Where it is desired to avoid this confusion, recommended practice calls for separating the digits into groups of three, counting from the decimal to the left or right, and using a small space to separate the groups.

Correct and recommended procedure
6.513 824 76 851 7 434 0.187 62

5.6.4 Calculating with SI units

Before we look at some suggested procedures that will simplify calculations in SI, the following positive characteristics of the system should be reviewed.

Only one unit is used to represent each physical quantity, e.g., the meter for length, the second for time, etc. The SI metric units are coherent; that is, each new derived unit is a product or quotient of the fundamental and supplementary units without any numerical factors. Since coherency is a strength of the SI system, it would be worthwhile to demonstrate this characteristic by using two examples. Consider the use of the newton as the unit of force instead of pound-force or kilogram-force. It is defined by Newton's second law, $F = ma$. It is the force that imparts an acceleration of 1 m/s^2 to a mass of 1 kg. Thus,

$$1\ \mathrm{N} = (1\ \mathrm{kg})(1\ \mathrm{m/s^2})$$

Consider also the joule, a unit that replaces the Btu, calorie, foot-pound-force, electronvolt, and horsepower-hour to stand for any form of energy. It is defined as the amount of work done when an applied force of one newton acts through a distance of one meter in the direction of the force. Thus,

$$1\ \mathrm{J} = (1\ \mathrm{N})(1\ \mathrm{m})$$

To maintain the coherency of units, however, time must be expressed in seconds rather than minutes or hours, since the second is the base unit. Once coherency is violated, then a conversion factor must be included and the advantage of the system is diminished.

But there are certain units *outside* SI which may be used even though they diminish the system's coherence. These exceptions are listed in Table 5.7.

Calculations using SI can be simplified if you

1. Remember that all fundamental relationships such as the following still apply, since they are independent of units.

$$F = ma \qquad \mathrm{KE} = \tfrac{1}{2}mv^2 \qquad E = RI$$

2. Recognize how to manipulate units and gain a proficiency in doing so. Since watt = J/s = N · m/s, you should realize that N · m/s = $(\mathrm{N/m^2})(\mathrm{m^3/s})$ = [(pressure)(volume flow rate)].

Table 5.7 Other Units—Not SI

Quantity	Name	Symbol
Time	minute	min
	hour	h
	day	d
	week	week
Plane angle	degree	°
Volume	liter	L
Temperature	degree Celsius	°C

3. Understand the advantage of occasionally adjusting all variables to base units. Replace N with $kg \cdot m/s^2$, Pa with $kg \cdot m^{-1} \cdot s^{-2}$, etc.

4. Develop a proficiency with exponential notation of numbers to be used in conjunction with unit prefixes.

$1\ mm^3 = (10^{-3}\ m)^3 = 10^{-9}\ m^3$

$1\ ns^{-1} = (10^{-9}\ s)^{-1} = 10^9\ s^{-1}$

5.7 Special characteristics

A term that should be avoided when using SI is "weight." Frequently we hear statements such as, "The man weighs 100 kg." A better statement would be, "The man has a mass of 100 kg." To clarify any confusion, let's look at some basic definitions.

First, the term mass should be used to indicate only a quantity of matter. Mass, as we know, is measured in kilograms against an international standard.

Force, as defined previously, is measured in newtons. It denotes an acceleration of 1 m/s^2 to a mass of 1 kg.

The acceleration of gravity varies at different points on the surface of the earth as well as with distance from the earth's surface. The accepted standard value of gravitational acceleration is 9.806 650 m/s^2.

Gravity is instrumental in measuring mass with a balance or scale. If you use a beam balance to compare an unknown quantity against a standard mass, the effect of gravity on the two masses cancels out. If you use a spring scale, mass is measured indirectly, since the instrument responds to the local force of gravity. Such a scale can be calibrated in mass units and be reasonably accurate when used where the variation in the acceleration of gravity is not significant.

The following example problem clarifies the confusion that exists in the use of the term weight to mean either force or mass. In everyday use, the term weight nearly always means mass; thus, when a person's weight is discussed, the quantity referred to is mass.

Example problem 5.1 A "weight" of 100.0 kg[1] is suspended by a rope (see Fig. 5.3). Calculate the tension in the rope in newtons when the mass is lifted vertically with an acceleration of 15.0 m/s.

Fact: The local gravitational constant is 9.81 m/s^2.

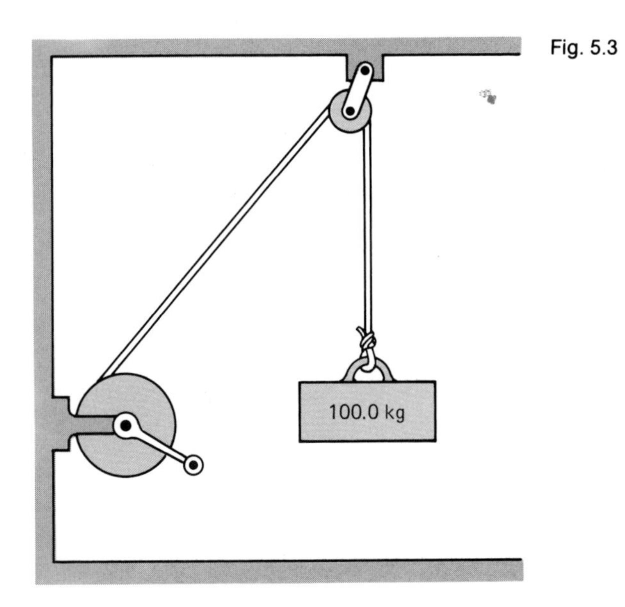

Fig. 5.3

Solution Tension in rope cord when mass is at rest:

$$F_1 = mg$$

$$= \frac{(100.0 \text{ kg})(9.81 \text{ m})}{\text{s}^2}$$

$$= 0.981 \text{ kN}$$

To raise with an acceleration of 15.0 m/s^2:

$$F_2 = ma$$

$$= \frac{(100.0 \text{ kg})(15.0 \text{ m})}{\text{s}^2}$$

$$= 1.50 \text{ kN}$$

$$\text{Total} = F_1 + F_2$$

$$= 2.48 \text{ kN}$$

[1]The unit itself indicates mass. If the units were 100.0 N, then the quantity would contain the gravitational constant appropriate at that elevation.

5.8 Conversion of units

Although the SI system will eventually be the international standard, there are many other systems in use today. It would be fair to say that nearly the entire work force of graduate engineers has been schooled using terminology such as slugs, pound mass, pound force, etc., and a very high percentage of the total United States population is more familiar with degrees Fahrenheit than degrees Celsius.

For this reason, as well as for the fact that total conversion will take time, you must be able to convert from old systems to SI.

Four typical systems of mechanical units presently being used in the United States are listed in Table 5.8. The table does not provide a complete list of all possible quantities; it is presented to demonstrate the different terminology that is associated with each unique system.

If a physical quantity is expressed in any system other than SI, it is a simple matter to convert the units from one system to another. To do this, the basic unit conversion must be known and a logical unit analysis must be followed.

Mistakes can be minimized if you always remember that conversion factors are nothing but equivalent measurements in different systems, that is,

$$1.0 \text{ in} = 25.4 \text{ mm}$$

$$\text{or } 1 = \frac{25.4 \text{ mm}}{1.0 \text{ in}}$$

Example Problem 5.2 demonstrates a systematic procedure to use when performing a unit conversion. The construction of a series

Table 5.8 Mechanical Units

Quantity	Absolute system		Gravitational system	
	MKS	CGS	Type I	Type II
Length	m	cm	ft	ft
Mass	kg	g	slug	lbm
Time	s	s	s	s
Force	N	dyne	lb	lbf
Velocity	$m \cdot s^{-1}$	$cm \cdot s^{-1}$	$ft \cdot s^{-1}$	$ft \cdot s^{-1}$
Acceleration	$m \cdot s^{-2}$	$cm \cdot s^{-2}$	$ft \cdot s^{-2}$	$ft \cdot s^{-2}$
Torque	$N \cdot m$	$dyne \cdot cm$	$lb \cdot ft$	$lbf \cdot ft$
Moment of Inertia	$kg \cdot m^2$	$g \cdot cm^2$	$slug \cdot ft^2$	$lbm \cdot ft^2$
Pressure	$N \cdot m^{-2}$	$dyne \cdot cm^{-2}$	$lb \cdot ft^{-2}$	$lbf \cdot ft^{-2}$
Energy	J	erg	$ft \cdot lb$	$ft \cdot lbf$
Power	W	$erg \cdot s^{-1}$	$ft \cdot lb \cdot s^{-1}$	$ft \cdot lbf \cdot s^{-1}$
Momentum	$kg \cdot m \cdot s^{-1}$	$g \cdot cm \cdot s^{-1}$	$slug \cdot ft \cdot s^{-1}$	$lbm \cdot ft \cdot s^{-1}$
Impulse	$N \cdot s$	$dyne \cdot s$	$lb \cdot s$	$lbf \cdot s$

Fig. 5.4 Military systems are designed in accordance with SI standards. (*General Dynamics, Pomona Division.*)

of horizontal and vertical lines separating the individual quantities will aid the thought process and help ensure a correct unit analysis. In other words, the units to be eliminated will cancel out, leaving the desired SI results.

The final answer should be checked to make sure it is reasonable. For example, the results of converting from inches to millimeters should be approximately 25 times larger than the original number.

Example problem 5.2 Convert 6.7 in to millimeters.
Conversion factor: 1 in = 25.4 mm.

Solution

$$6.7\ \text{in} = \frac{6.7\ \text{in}}{1}\left|\frac{25.4\ \text{mm}}{1\ \text{in}}\right. = 1.7 \times 10^2\ \text{mm}$$

Example problem 5.3 Convert 85.0 lbm/ft³ to kilograms per cubic meter.
Conversion factor: 1 ft = 0.304 8 m
1 lbm = 0.453 6 kg

Solution

$$85.0\ \text{lbm/ft}^3 = \frac{85.0\ \text{lbm}}{\text{ft}^3}\left|\frac{(1)^3\ \text{ft}^3}{(0.304\ 8)^3\ \text{m}^3}\right|\frac{0.453\ 6\ \text{kg}}{\text{lbm}}$$

$$= 1.36 \times 10^3\ \text{kg/m}^3$$

The problem of unit conversion becomes more complex if an equation has a constant with hidden dimensions. It is necessary to work carefully through the equation converting the constant K_1 to a new constant K_2 consistent with the equation units.

Consider the following example problem given with English units.

Example problem 5.4 The velocity of sound in air (c) can be expressed as

$$c = 49.02\sqrt{T}$$

where c is in feet per second
T is in degrees Rankine
Express c in meters per second when T is in kelvins.

Procedure

1. First, the given equation must have consistent units, i.e., it must have the same units on both sides. Squaring both sides we see that

$$c^2 \text{ ft}^2/\text{s}^2 = (49.02)^2 T\,°\text{R}$$

It is obvious that the constant 49.02^2 has units in order to maintain unit consistency.

$$c^2 \text{ ft}^2/\text{s}^2 = \frac{(49.02)^2 \text{ ft}^2}{\text{s}^2\ °\text{R}} \left| \frac{T\,°\text{R}}{} \right.$$

2. The next step would be to convert the constant 49.02 (K_1) to a new constant (K_2) that will allow us to calculate c in meters per second given T in kelvins.

$$\frac{(49.02)^2 \text{ ft}^2}{\text{s}^2\,°\text{R}} = \frac{(49.02)^2 \text{ ft}^2}{\text{s}^2\,°\text{R}} \left| \frac{(0.304\ 8)^2 \text{ m}^2}{1 \text{ ft}^2} \right| \frac{9\,°\text{R}}{5 \text{ K}} = \frac{401.84 \text{ m}^2}{\text{s}^2 \text{ K}}$$

3. Substitute this new constant 401.84 (K_2) back into the original equation

$$c^2 = 401.84T$$

$$c = 20.05\sqrt{T}$$

where c is in meters per second
T is in kelvins.

Problems

5.1 List the proper names of the seven base units and proper symbol identification for each.

5.2 Express each quantity in its proper ''dimension,'' using the absolute system (such as MLT^{-2} for force) and also indicate the proper SI base units.
(*a*) Force (*b*) Velocity (*c*) Pressure (*d*) Momentum
(*e*) Energy (*f*) Power (*g*) Torque

5.3 The correct use of rules is very important in SI units. The following list is not correct. Make the appropriate corrections.
(*a*) .756 (*b*) 62,541 (*c*) 273 Kelvin (*d*) hr

(*e*) 72 j (*f*) 65.8 N. (*g*) 15C (*h*) 9.3mm
(*i*) 15.5pa (*j*) 35 amp (*k*) 15 A's (*l*) 39 p F

5.4 When multiplication and division of units are to be accomplished, certain rules apply. The list below is not correct. Make the appropriate changes.
(*a*) Newtons × meters (*b*) meter/sec. (*c*) cubic seconds
(*d*) N × m (*e*) meter/s (*f*) m/s/s

5.5 Rework the following paragraph, with respect to the correct way of presenting units of measure in the SI. "A 1000.0Kg mass was to be raised at the rate of .59 meter/sec. It was decided that an electric motor using 20 amps would be required but that the motor should shut-down if its temperature were to increase over 1°Kelvin/hr."

5.6 If a 100.0-kg mass is to be lifted by a rope, what force would be required under the following conditions?
(*a*) At sea level, where $g = 9.807 \text{ m/s}^2$
(*b*) On the top of a mountain, where $g = 9.75 \text{ m/s}^2$
(*c*) On the surface of the moon, where $g = 1.63 \text{ m/s}^2$

5.7 A mass of 50.0 kg is suspended by a rope. Let $g = 9.81 \text{ m/s}^2$. Calculate the tension in the rope in newtons, when
(*a*) The mass is lifted vertically with an acceleration of 10.0 m/s^2
(*b*) The mass is lowered vertically with an acceleration of 6.5 m/s^2

5.8 Convert the following to SI, using proper significant figures.
(*a*) 10.75 in (*b*) 2.50×10^3 Btu (*c*) 48.0°F
(*d*) $6.5 \text{ ft}^3\text{/min}$ (*e*) 2.0×10^4 gal/d (*f*) 340.0 hp
(*g*) 55.0 mi/h (*h*) 743.9 oz (*i*) $4.38 \times 10^2 \text{ lbf/ft}^2$

5.9 If 2.00×10^3 lbm is raised vertically through a distance of 20.0 ft, what is the work done in newton-meters? Use an acceleration of gravity of 9.81 m/s^2. Work = FL.

5.10 In Prob. 5.9, what is the power, in kilowatts, if the mass is raised in 1 min and 10 s? Power = work/time and has units of J/s = N · m/s.

5.11 The velocity of an automobile is measured at 54.7 ft/s in a 30 mi/h zone. Is the driver speeding? If so, how much, expressed in meters per second.

5.12 A police officer stops a motorist for traveling 120 km/h in a 55 mi/h zone. If the fine is $5 for each mile per hour over the limit, what is the penalty for speeding? (Round to the nearest mile per hour.)

5.13 Express standard atmospheric pressure of 14.696 lbf/in^2 in pascals.

5.14 Determine the volume and mass of water contained in a cylindrical water tower that is 20.0 ft high and 4.0×10^1 ft in diameter. Water at 50°F has a density of 1.94 slugs/ft^3. Express your answer in SI units.

5.15 The density of standard air at 14.696 lbf/in^2 and 70°F is 2.328×10^{-3} slugs/ft^3. What is the mass in kilograms of air in a typical classroom that measures 20.0 × 30.0 × 10.0 ft.?

5.16 A pump delivers water at the rate of 1.7×10^4 gal in a 24-h period. What is the equivalent mass flow rate in kilograms per second if the density of water is 1.94 slugs/ft^3?

5.17 A hollow copper ball 2.25 in in diameter (outside) has a wall thickness of 0.250 in.

(*a*) Compute the outside surface area in square centimeters.
(*b*) Compute the mass if the density of copper is 5.5×10^2 lbm/ft^3.
(*c*) Will the ball float if placed in water with a density of 62.4 lbm/ft^3?

5.18 If pressure P is given in pound-force per square foot, specific volume v in cubic feet per pound mass, and temperature T in degrees Rankine, the perfect gas equation for air can be written

$$P = \frac{53.3\,T}{v}$$

Determine the new constant when pressure is in pascals, specific volume is in cubic meters per kilogram, and temperature is in kelvins.

PART THREE

Computers and calculators in engineering

Introduction to computing systems

CHAPTER 6

6.1 Introduction

With over 150 000 computers in the United States and more than 1.5 million persons involved in what has become a multibillion dollar computer business, it is hard to believe that as recently as 30 years ago the computer business was essentially nonexistent. Computers are vital in most phases of business and industry today, with rapid developments in their use taking place in transportation, health care, business management, education, and national defense. The development of the minicomputer and microprocessor has brought the computer into the home, automobile, and shopping center. The impact of computers on individuals has been and will continue to be direct, sweeping, and extreme.

This unparalleled advance in computing capability puts special demands on you as an engineer. You must learn how computers work, what they can do, and what they are used for. You will also have to surmise what their effect on society might be. Otherwise, you may find it difficult to compete in the engineering profession. When used properly, computers can greatly extend your effectiveness. Not only do they save you time in doing routine computations

Fig. 6.1 Computers are found in most areas of industry today. Here a computer system controls a manufacturing operation. (*Allen-Bradley*.)

but their speed enables you to examine many more alternatives before making decisions. Even at a cost of several hundred dollars per hour for time on a large computing system, a few minutes of computation can yield substantial results. As a comparison, it has been estimated that for a moderately sized computer system, one machine-second is equivalent to one person-month for arithmetic computations.

A computer may be defined broadly as a device capable of accepting information or data, processing it (that is, manipulating it), and providing usable results from it. Because engineers use the computer as a means of computing, they must understand the computer's capabilities and have access to it by knowing how to program data into it. To program is to prepare a detailed sequence of operating instructions for a particular problem. Most engineering problems are mathematical in nature, so engineers must have a complete knowledge of the problem parameters and necessary mathematical analyses before they can determine which type of computing machine would be most efficient and economical. (It may not be feasible to program a computer solution if a rough estimate of the answer is all that is desired at the time.)

The five basic steps for logical problem analysis and solution (the engineering method) given in Chapter 2 are repeated here to show how and where the computer can assist the problem solver.

1. Recognize and understand the problem
2. Accumulate facts
3. Make necessary assumptions and select the appropriate theory or principle
4. Solve the problem
5. Verify and check

The computer enters the solution process at step 4. Selection of the appropriate computational device depends on the complexity and accuracy requirements of the mathematical model developed in step 3. The computer can assist at step 5 by performing calculations and comparing results from alternate methods of solution. In order for the computer to provide the needed computational assistance, it must be supplied a definite procedure.

The procedure prescribed for solving a specified problem by means of well-defined rules or processes is called an algorithm. The most difficult part of writing an algorithm is that of stating precisely and fully each step necessary. (It must be remembered that a computer cannot supply missing steps because it is incapable of creative thinking.) To appreciate the challenge of writing an algorithm, let's consider writing one for calculating the square root of a number. We can verify the correctness of our algorithm by applying it to find the square root of 42 to three significant figures.

Steps in the algorithm	Example
1. Divide number by 2 and use it as first trial root.	$\frac{42}{2} = 21 =$ trial root
2. Add trial root to the fraction formed from the original number divided by the trial root and divide result by 2.	$\frac{21 + \frac{42}{21}}{2} = \frac{23}{2} = 11.5$
3. Square result	$11.5^2 = 132.25$
4. Compare with original number. (If result is not within prescribed accuracy, let the computed trial root from step 2 be the new trial root and repeat steps 2, 3, and 4.)	$132.25 > 42$ Trial root $= 11.5$
5. Continue the process until desired accuracy is obtained.	

The reader should continue to execute the algorithm, verifying that the successive trial roots are 7.576, 6.560, and 6.481 – the latter of which, when squared, yields 41.99. The square-root algorithm thus yields 6.48 as the $\sqrt{42}$ to three significant figures.

The remainder of the chapter is devoted to a discussion of various types of computers, their component parts, and their applications. A brief history of computing equipment will begin our discussion.

Fig. 6.2 A modern-day, large-scale computing system located on a major university campus being monitored by the computer operator.

The ensuing chapter on preparation for programming will illustrate the necessary procedures for preparing a problem for computer solution.

6.2 Evolution of computing equipment

The first counting devices were undoubtedly based on the 10 fingers of the hands. Pebbles were used to count higher values. Around 1000 B.C., the first form of a calculating instrument, the abacus, appeared. It is a rectangular frame with beads strung on parallel wires. It can be used to add and subtract rapidly by manually sliding and thereby grouping the beads.

In 1642, Blaise Pascal, a French scientist, constructed the first mechanical calculator generally considered to be the forerunner of the modern desk calculator. Pascal's device replaced the beads and wires with toothed wheels with 10 cogs (teeth) per wheel, each cog representing the number 1. The wheels were placed side by side to enable the carrying operation to take place. When one wheel completed a revolution, a ratchet caused the adjacent wheel on the left to move one notch, thereby effecting a carrying, or borrowing, operation as in longhand arithmetic. As with the abacus, the Pascal device could perform multiplication and division only by successive additions and subtractions.

Up to this point, computing devices were digital in nature; that is, they dealt only with data in the form of discrete numbers (digits). Next appeared the first forms of another method of computing, the analog. Analog computation is a means of representing numbers by a continuous range of physical quantities such as lengths, rotations, voltages, or currents. For example, a tempera-

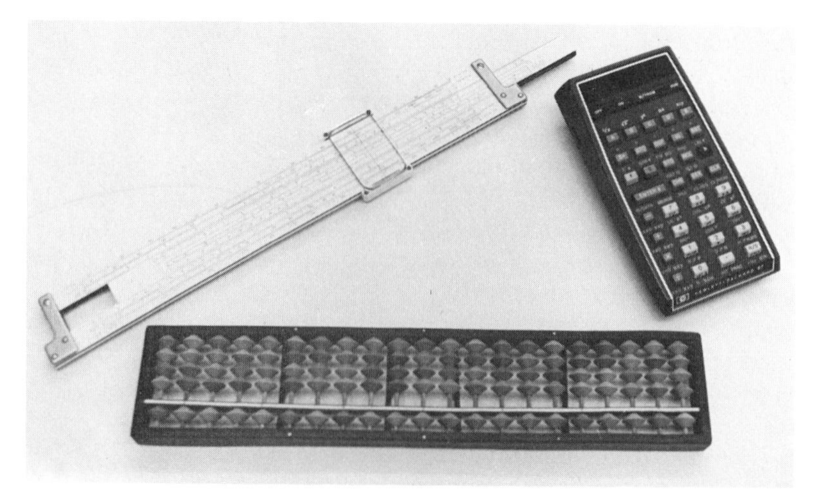

Fig. 6.3 Three of the most-important hand-held calculating devices ever used by engineers. The electronic calculator has evolved within the past decade.

ture may be represented by a voltage, which is its analog. It is possible then to add two numbers by adding the voltages which represent the numbers.

Consider the measurement of the passage of one minute of time. The passage of the minute can be measured *digitally* by counting 60 seconds, one at a time; *analogically*, by observing one revolution of the second hand on a clock. An analog is thus a physical variable that remains similar to another variable in that the proportional relationships are the same over some specified range.

The first form of analog computation, the slide rule, used length to represent numbers. In 1614, John Napier described the natural logarithms, which allowed a person to multiply and divide by adding exponents (logarithms). Henry Briggs and Napier jointly published a table of logarithms using 10 as a base. One year later Edmund Gunter constructed the logarithmic line, which enabled the user to add and subtract lengths (representing logarithms of numbers) with a pair of dividers. About 1630, William Oughtred invented the sliding logarithmic scales, thereby helping the user add or subtract lengths rapidly (thus multiplying or dividing numbers). This device is considered the forerunner of the slide rules we know today. By 1650, two types of computers were in common use; the digital, in the form of the abacus and Pascal's machine; and the analog, in the form of the slide rule.

Baron von Leibniz, a German mathematician, provided the next major digital mechanical computing device. His machine, built in 1671, could perform multiplication directly through the use of gears, instead of as a series of additions, which was required by Pascal's machine. Unfortunately, the poor machining techniques used in manufacturing Leibniz's calculator limited its reliability, so its capability was not realized for more than 50 years. Mechanical calculators, which were manufactured as recently as the 1950s, incorporated many of the features of Leibniz's calculating machine.

The first attempt at producing a computer with all the elements of a modern digital computer [namely, memory, control (sequencing of the operations), and calculating unit] was made in 1833 by Charles Babbage, an English mathematician. Babbage worked 20 years on the machine, but manufacturing limitations prevented its successful completion. Babbage attempted to incorporate one important idea that is basic to modern computers: the ability of the computer to modify the course of a calculation according to intermediate results. In the study of computer programming this concept is called branching. Because he developed this idea, Babbage is often referred to as the grandfather of the computer.

The first large-scale digital computer using modern principles was the Harvard Mark I, completed in 1944. Powered electrically, this computer used many of Babbage's techniques. The Mark I was extremely slow by today's standards. For example, division of two 10-digit numbers took about a minute compared with what would be just a few microseconds today.

The first electronic digital computer (powered by electron-charge carriers, that is), the Electronic Numerical Integrator and Computer (ENIAC), was completed in 1946. This machine contained over 18 000 vacuum, or electron, tubes, required 130 kW of power, and occupied an area of 140 m^2. The significant contribution of this machine was a speed increase of 10 000 over that of the electromechanical computers.

As we have seen, it was a short time between the construction of the first large computer and today's widespread use of sophisticated high-speed equipment. New technology in the past 30 years has far surpassed the computing achievements in the 3 000-year period from the abacus to ENIAC.

6.3 Digital computers

In the discussion of the evolution of computing devices, the terms calculator and computer were used. The basic difference between the automatic digital computer and desk calculator is that the computer will perform a series of computations without human intervention once the sequence of operations is begun. The sequence of instructions which tells a computer how to perform a set of computations is called a program. A program is simply an algorithm

Fig. 6.4 One of the fastest growing portions of the computer industry is that of minicomputers. The versatility of these machines allows for a multitude of applications in industry.

expressed in the language of a particular computer. Within the program is usually a set of criteria that causes the computer to vary the sequence of operations depending on the outcome of a computation. For example, if an intermediate result is positive (greater than zero), the program will instruct the computer to follow a certain sequence of instructions. If the result is negative the computer will be instructed to follow a different sequence of instructions. It must be emphasized again that all steps of the computation must be supplied to the machine, since it cannot plan for itself. The following discussion describes the equipment that constitutes a digital computer system.

The five major operational components of a digital computer are illustrated in Fig. 6.5. The arrows indicate flow of information. The components within the dotted outline are called the central processing unit (CPU).

6.3.1 Input/output components

The input/output (I/O) equipment is a major portion of a digital computer system. Computer equipment other than the CPU and storage is called peripheral equipment. Peripheral equipment works in conjunction with a computer but is not a physical part of the computer itself. A card or paper-tape reader, magnetic tape handler, and line printer (all of which are discussed later) are all peripheral equipment. The I/O equipment is the primary portion of the peripheral equipment that provides the computer system with external communication. I/O systems are designed to move information between peripheral devices and main storage, into which data can be inserted, in which it can be retained, and from which it can be retrieved. The CPU and storage systems can process data at almost incomprehensible rates. But before doing so, the necessary

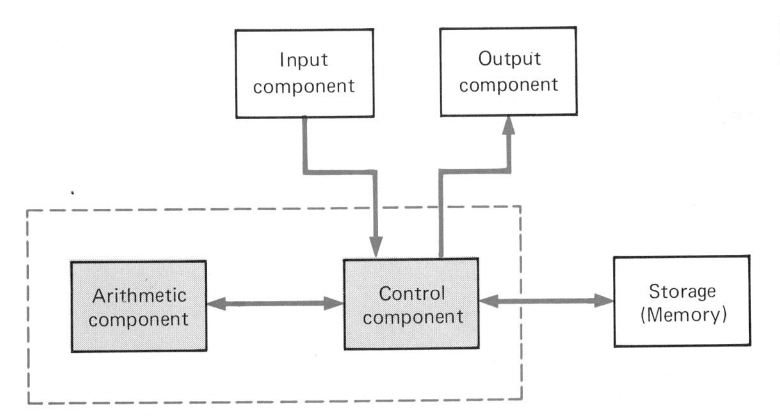

Fig. 6.5 Operational components of a digital computer.

instructions and data must be communicated to the machine. The programmer prepares instructions in a source language (similar to ordinary mathematics). For example, to compute a value of distance d from the equation $d = \frac{1}{2}at^2$ (algorithm), where a is acceleration and t is time, the program statement is written in one source language, called FORTRAN, as follows.

```
D = A * T * * 2/2.0
```

Note the similarity of the FORTRAN statement with the mathematical equation. The program is put into the machine by means of an input device and is then compiled. Compilation is the process of converting the source program into an object program and checking for source language correctness. After the program has been executed, the results are communicated back to the programmer through an output unit. The I/O function is simply a means of communicating with a high-speed electronic moron.

Transferring information between relatively slow I/O devices and high-speed CPU and storage components is the basis of I/O system design. Since I/O components are both mechanical (paper-feed devices on printers) and electrical (magnetic-tape readers), a large difference in operating speeds is inherent in many systems because of the mechanical components. The development of the time-sharing concept has led to more efficient use of I/O components in conjunction with high-speed CPUs. Time sharing means that several I/O channels have access to a single CPU and main storage system. Figure 6.7 illustrates the time-sharing concept schematically. Time sharing is accomplished with the use of a com-

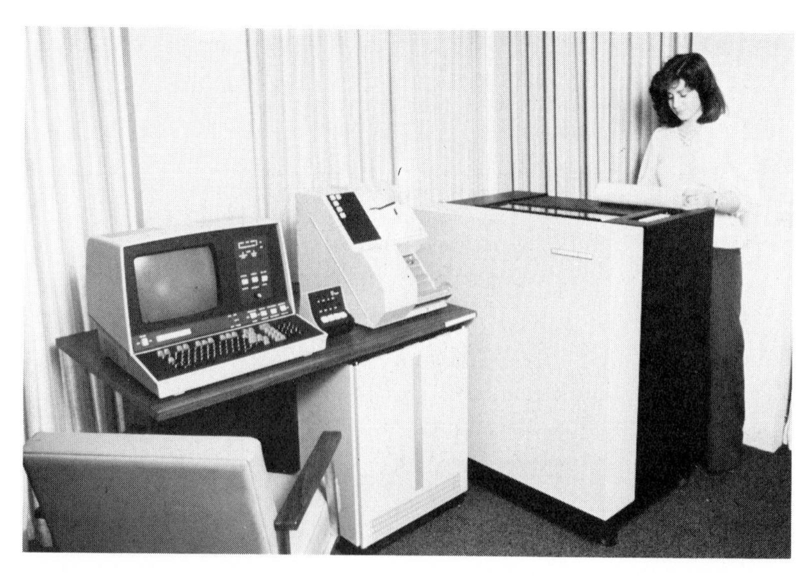

Fig. 6.6 A remote terminal for a computing system shows some of the types of input-output equipment available. (*Control Data Corporation.*)

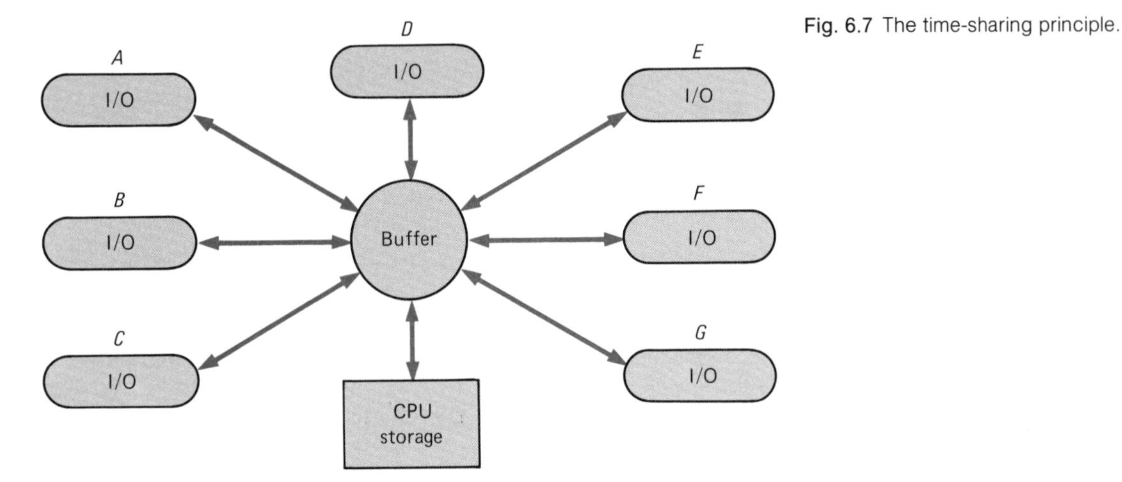

Fig. 6.7 The time-sharing principle.

munication buffer, or storage device, that compensates for the tremendous difference in the flow of data between I/O devices and a CPU by holding the information and feeding it at the rate that the I/O device operates.

Requests for processing are placed through the appropriate I/O device. The requests are in turn stored in the buffer for processing by the CPU and storage component. The computer can scan all the I/O terminals in negligible time. If one user has a request that may take several minutes to generate on the I/O terminal, this does not tie up the CPU and storage system, since it can process requests, in the order presented to the buffer, from any of the other I/O terminals. The high-speed CPU and storage systems are therefore never tied up waiting for information from the slower I/O components. If systems are synchronized properly, the user will probably not realize that the computer is being used for other requests simultaneously.

Fig. 6.8 A remote entry terminal that allows, through time sharing, several hookups to a central computing system simultaneously. (*Beech Aircraft Corporation, Boulder Division.*)

Some of the more common I/O devices are punched cards, punched paper tapes, printed pages, magnetic tapes, optically readable characters on cards or tapes, and visual displays. Voice-recognition systems are in limited use today, but with the possibility that they may become more common in the next decade or so.

The punched card is the most widely used input device. Originally developed by Hollerith in 1880, the punched card contains 80 columns, each column representing a single character. Punched cards take a long time to prepare (one to five characters per second) but can be punched without connecting to the computer. Once the cards are punched, checked, and arranged, a card reader can process several hundred cards per minute. Punched paper tape is essentially the same as punched cards, in that both display a pattern of holes or cuts to represent data.

The most common output medium is the printed page. Printed pages can be generated by a line printer that operates by impact printing (by means of devices that actually strike the paper to produce print), producing one line at a time with a maximum rate today of over 1000 lines per minute.

Magnetic tape is a relatively high-speed I/O medium. It can provide highly reliable data rates of up to 680 000 characters per second. Information is stored on the tape in parallel channels along the length of the tape by magnetized spots called bits. A tape may contain 7 or more channels. Character densities of 1 600 per inch have been achieved. The tape may be retained indefinitely or erased and reused. Another form of magnetic I/O media is the magnetic ink character. Common in the banking business, the

Fig. 6.9 An automatic card reader is one of the most common input devices for large computing systems.

Fig. 6.10 A line printer can produce copy at a rate of over 1 000 lines per minute.

magnetic ink character is visually readable as well as magnetically readable. The numbers on the bottom of a paycheck are examples of magnetic ink characters.

Optically readable characters are a popular means of representing data that is prepared for the computer manually. Instead of all the information being written out, a character is denoted by marking a space on a card. The card is then read by a device that electrically senses the marks and the information is stored. Optical character reading (OCR) machines have been developed that will read up to 14 000 characters per second. A typical example of optically readable information is the mark sense answer sheet used for examinations.

Fig. 6.11 A bank of magnetic-tape drives used for input and output of information for a large computing system.

Visual output has recently been seen to be growing in use in education, business, and industry. The cathode ray tube (CRT) is the central element of a visual component. When used in the health field, a simple coding procedure from a terminal will bring onto the display screen a patient's medical history. In the area of computer graphics, a drawing can be completed and corrected by use of a light pen on a CRT. Once the drawing has been corrected to the satisfaction of the designer, a print can be made from the visual display by additional peripheral equipment. Visual output displays are being used extensively in the field of education as an effective teaching tool (see Fig. 6.12).

The smaller-scale computers and programmable calculators usually have a single digital display for output and a keyboard for input. Some models have the additional capability of magnetic card I/O. The important fact to remember about keyboard and digital output (which may need to be written down) is the human intervention. While serving a useful purpose and providing an economical method of solution to certain problems, the programmable pocket calculator is restricted in capability by relatively inefficient and slow I/O components. See Fig. 6.13.

Fig. 6.12 An interactive terminal is very useful as a teaching medium and design aid. The user communicates with the system through the keyboard or by touching the screen. (*Control Data Corporation.*)

Fig. 6.13 A programmable calculator that can work many problems in most environments. Operating on batteries, this machine is a versatile tool for today's engineers.

6.3.2 Storage and CPU components

The storage (memory), control unit, and arithmetic unit of a computing system consist of solid-state electronic devices connected in thousands of circuits. Today, large computer systems using transistors contain more than 100 000 circuits, with the use of larger systems foreseeable in the future. A basic consideration for measuring the performance of a computer is the speed of the operating circuits. Today's speeds are measured in nanoseconds (1 ns $= 10^{-9}$ s). Even though the speed of the circuitry has increased, the physical size has decreased manyfold as a result of advances in electronic hardware technology. Moreover, the cost of operation has decreased while reliability has increased. For example, transistors and other solid-state semiconductors use less power and occupy less space than did their vacuum-tube predecessors.

One might ask if there is a limit to the improvement in computer circuitry. A limit to the ability of a computer to perform operations rapidly is the speed of light for the propagation of computer information along circuit wires. There is a delay of 6 ns for every meter of interconnecting wire that a signal must travel. Other limitations are response time in flip-flop devices (electrical units that assume an on or off state depending on an input signal) and packing density for storage devices. For example, a semiconductor (large-scale integrated circuit) memory in a modern computer has on the order of 10^3 binary digits (bits) per cubic centimeter. The human brain, the "most packed" computer known, has a packing density of 10^{10} bits/cm^3.

The memory of a large computer is not actually concentrated in one place; it is scattered throughout the machine because of the many functions required of it. One of the basic storage devices is the operations register, which is used to store and decode the operation code for the next instruction to be carried out by the computer. Actual calculations, such as addition, multiplication, and logic analyses, are performed in and at the direction of these registers. The next category of storage is the inner memory, a set of storage registers identified by addresses, which enables the control component either to write into or read from a particular register. Additional memory, called auxiliary memory, is added to most large computers to handle excess data above the capacity of the inner memory.

To describe the forms and capabilities of the various storage devices some definitions are needed.

1. Access time. The time required for data to be called from storage and delivered to the necessary register; or the time required to transfer data from one point in the memory to another.

Fig. 6.14 A bank of disk drives serves as auxiliary memory for a large computing system.

2. Random access. The process of obtaining data from, or placing data into, storage when the time required for such access does not depend on the location of the data most recently obtained or placed into storage.

3. Nondestructive readout. A situation in which data is retrieved from a storage location and the contents of that location are unaltered.

Operations registers operate at a very high speed and thus require high-speed access to the inner memory. For example, 50 ns is common for addition, so inner memory must have a correspondingly fast access time. The faster the arithmetic speed, the larger and faster must be the inner storage.

Perhaps the most common storage units today are the magnetic-core units. A magnetic core is a tiny doughnut-shaped ring of ferromagnetic material about 1 mm in diameter. Cores are placed like beads on sets of wires. A single core element is depicted in Fig. 6.15. The element can be magnetized only by sending one-half the magnetizing current through each of the perpendicular wires. Each core can thus represent an item of data by being either magnetized or not magnetized corresponding to the binary digits (bits) of 1 or 0. The sense wire can detect the magnetic state of the core. Magnetic-core memories have an access time of about 10^{-6}s (1 μs).

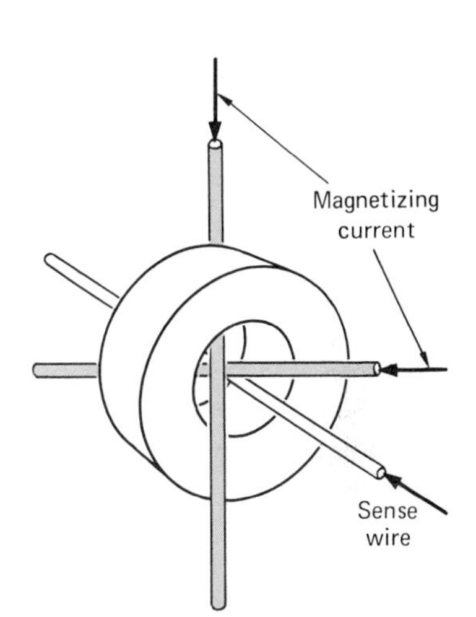

Fig. 6.15 Schematic of a magnetic-core element.

Other inner storage memories used are magnetic thin film and integrated semiconductor circuits. The semiconductor memory has somewhat faster access times than do the magnetic cores but has not proved as reliable as the core memory.

Auxiliary storage devices are usually magnetic tape or punched cards. These are much slower than core memory but provide almost unlimited storage for a fraction of the cost of the high-speed memories.

Research to develop faster, less-expensive memories continues. One device, called a magnetic bubble memory, has proved feasible. Packing densities of 10^5 bits/cm^3 are possible with access rates of $3(10^6)$ bits/s. New types of memories (for example, holographic memories) are being explored that will have a capacity of more than 10^8 bits with random access time of 1 μs.

6.4 Analog computers

In many engineering applications (for example, an aircraft control system), the problem variables change with time and the engineer is interested in determining the time behavior of these variables. The mathematical model for the system is usually a set of differential equations, and their solution requires calculus. The digital computer can perform the required operations for solution using discrete numerical values. But as problems become more complex, digital solutions require more time and thus become more costly. Moreover there may be errors in the solution if it becomes necessary to use simplifying assumptions to decrease the computer time required for a solution. The engineer must decide whether the results justify the cost of computing or whether other methods should be used to obtain the desired results.

An experimental approach (attempting to predict real-system behavior by laboratory methods) is one alternative for investigation. Design could progress on the basis of experimental results from a prototype. However, the prohibitive cost of a prototype of a space vehicle or nuclear reactor, for example, obviously demonstrates the difficulty of direct experimentation. A scale model for experimentation is also a possibility. For example, a model of an aircraft can be tested extensively in a wind tunnel to assist in solving design problems. But as with prototype models, scale models can be expensive and very difficult to alter in an attempt to improve performance.

The scale model simulates the original device and if properly designed is a physical analog. It may be possible to simulate a physical system if the mathematical equations describing the physical

Fig. 6.16 An analog computer can simulate a physical system, thus enabling the engineer to predict performance without having to build a prototype.

system can be written. The equations are programmed into an electronic analog computer and the real system is simulated electronically. Behavior of the real system, not its physical characteristics, is simulated.

In an electronic analog computer, the variables are represented by electrical voltages. It is not necessary to solve the equations describing the problem, nor even prescribe a method of solution, since the analog computer does this. As an example, consider the flight of a rocket fired vertically upward. The equations of motion for the rocket are written and then a circuit is wired into the analog computer so that voltages in the circuit behave as do the variables in the problem. As the computer operates, the voltages change proportionally to the actual variables and a continuous plot of such quantities as height, velocity, and acceleration can be recorded. Note that changes in the initial data (velocity, acceleration, etc.) or end conditions (velocity, altitude, etc.) can be made quite easily by altering the voltages representing these conditions. This is more difficult to do on a prototype or scale model.

One limitation of analog computer systems is degree of accuracy. For instance, consider the different ways the digital and analog computers perform a multiplication. A desk calculator will yield an answer accurate to the number of significant places in the display, usually 10 or more. However, a slide rule (analog computer) is accurate to three significant figures because of the need to measure a linear distance. A 30-m-long slide rule may yield five significant figures, but one can see the impracticality of such a device. Analog computers are usually accurate to within a few percent, which may be quite acceptable in many engineering applications.

6.5 Hybrid computers

In comparing analog and digital computers, the inherent characteristics can be summarized as follows.

Digital computer
- Discrete numbers
- Sequential operations
- High accuracy

Analog computer
- Variables are continuous quantities
- Simultaneous operations (high-speed solution)
- Limited accuracy

To take advantage of digital computer accuracy and analog computer speed, the hybrid computer has evolved. The development of the hybrid computer requires conversion devices to change analog signals into digital data, and vice versa. Figure 6.17 is a schematic of a hybrid computer for control of a space vehicle.

For complex problems, hybrid computers have proven to be faster and more economical than either digital or analog computers. It is expected that computers of the future will incorporate more and more analog and digital features to a point where the two kinds may no longer be separately identifiable.

6.6 Minicomputers and microcomputers

The general trend in the computing industry up to the late 1960s had been to develop large systems to centralize and concentrate computing power. The reason was that the larger the system, the less the cost of each executed operation. This is the same argument used to compare transportation costs by stating that the cost per mile per person is less using buses than individual cars. The public has not yet accepted mass transportation in lieu of the convenience of the automobile, just as they have not accepted the totally centralized computer facilities. Although many agree that the smaller decentralized computers are inefficient, the recent rapid growth of the minicomputer and microcomputer industry indicates that convenience is more of a consideration than efficiency.

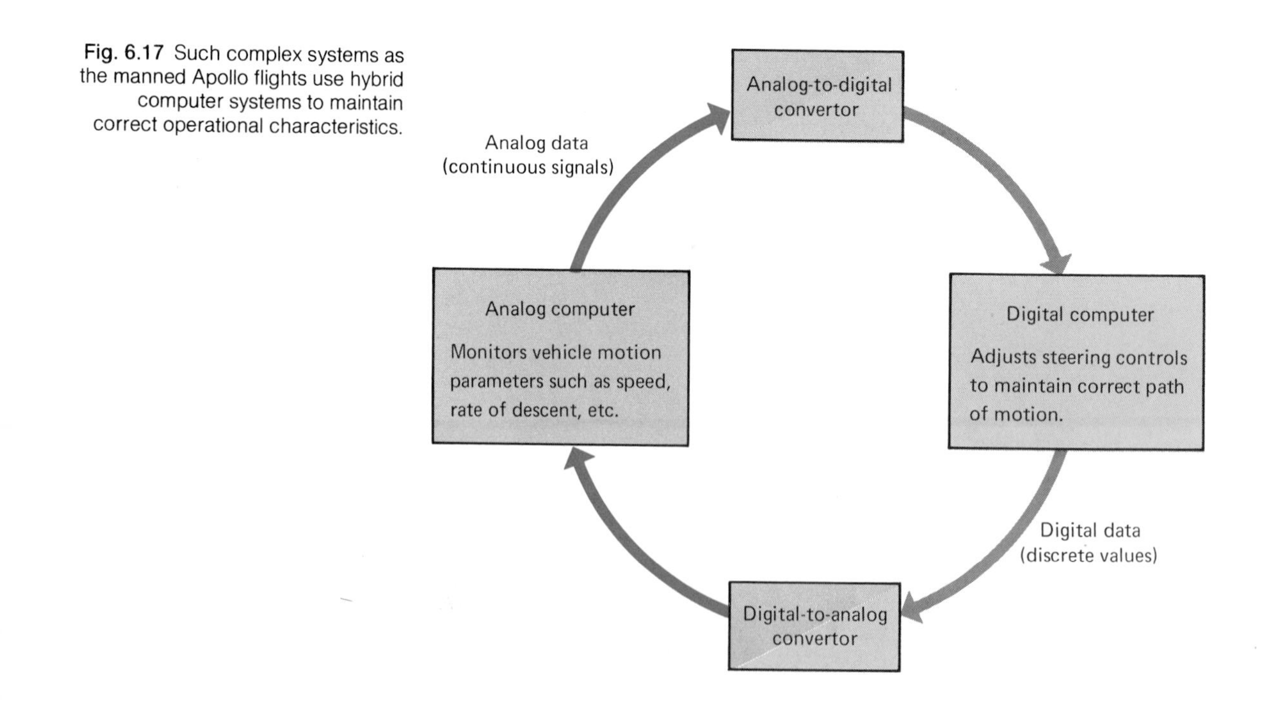

Fig. 6.17 Such complex systems as the manned Apollo flights use hybrid computer systems to maintain correct operational characteristics.

Some brief statements about the characteristics that identify the digital computer classifications are necessary at this point. Generally maxi (large) computers are complete systems that cost more than $20 000, use at least 32-bit memory word size, and have arithmetic speed for addition measured in nanoseconds. Large-scale computers are usually installed with all the software (programs) necessary for immediate operation.

Minicomputers are generally classified as computers in the $3 000 to $20 000 price range that have a 16-bit memory word length and an addition speed of about 1 μs and are generally small enough to be classified as desk-top computers. As with large-scale computers, the minis are free-standing general-purpose computers with the capability of increased storage. See Fig. 6.18.

The microcomputer consists of at least one microprocessor (about the size of a razor-blade box) and additional integrated circuitry. The basic microprocessor chip can be purchased for less than $20. However, costs for support software and other hardware items for operation can push the cost up to $1 000 or more. The main advantages of the microcomputers are low cost and the capability of customizing to specific functions. By its nature the microcomputer is special purpose; however, many different functions can be performed by one microcomputer by interchanging or adding microprocessors to the system. Microcomputer sales are expected to be in billions of dollars in the 1980s.

6.7 A look ahead

The rapid development of computers during the past 30 years gives an indication of things to come. Society today is oriented to the

Fig. 6.18 An engineer works with a stand-alone minicomputer system that supplies paper-tape results or results graphed by a plotter. (*Beech Aircraft Corporation, Boulder Division.*)

Fig. 6.19 The relative size of a calculator chip belies its capacity. The microcomputer board shown below contains the peripheral equipment required to make a microcomputer with the chip (right) as the basic component.

future and computers have contributed to this orientation. Science fiction of a few years ago may become reality by the year 2000. It behooves the individual to gain an understanding of the potential of the computer and how society may be affected by new developments. The following possibilities exist for the future. In fact, some of them are actualities today.

1. Computer control of traffic flow. As an example, the city of Toronto, Canada, is operating in the sixth phase of a project to improve the city's surface transit system. Major goals include the coordination of buses and their routes by computer to improve rider convenience. The position of

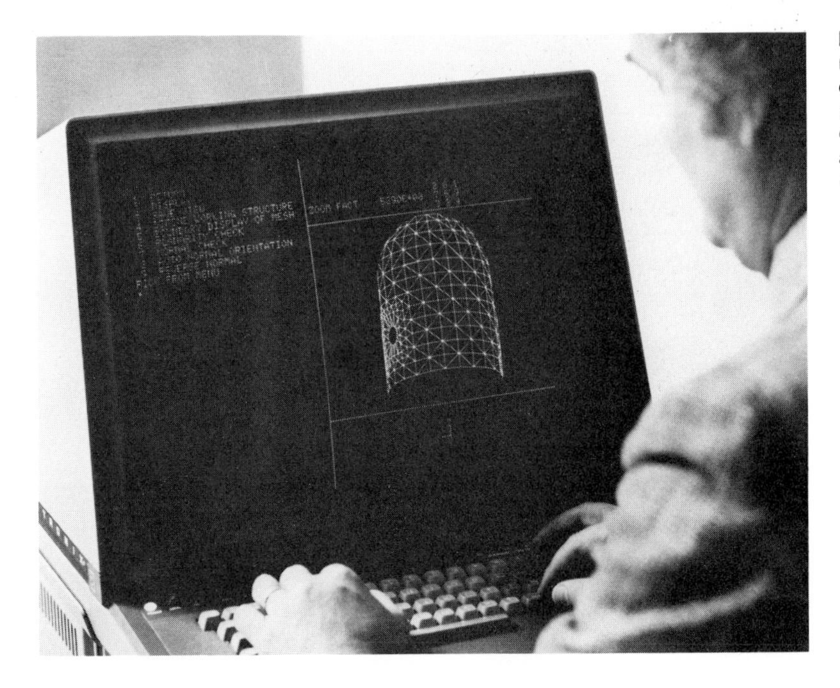

Fig. 6.20 Interactive terminals can reduce an engineer's design and evaluation time and thereby improve productivity. This engineer is evaluating structural alternatives for a proposed design. (*Control Data Corporation.*)

a bus and the number of passengers are instantaneously known at the control centers. A future possibility is display signs at bus stops providing continuously updated information about time of arrival and available seating. Extending this idea farther into the future is not difficult in the area of general automobile travel. Message boards controlled by microcomputers in some places now flash traffic and road conditions to the driver and suggest alternate routes. Also foreseeable is an automobile autopilot controlled by a minicomputer.

2. A cashless society. Several examples of this are already popular today. The self-service banks dispense cash and even accept utility bill payments all at the request of a plastic card which sends a centralized minicomputer into action. Consider this possibility. A commuter in a hurry to catch a train runs to the turnstile, inserts a coded plastic card, and is admitted to the train. The turnstile is connected to a minicomputer which automatically debits the commuter's account. Many more examples of the trend toward a cashless society can be cited.

3. Computer education. Students in several schools are able to learn what careers they may be best suited for by answering preprogrammed questions from a centralized computer system. An obvious extension of this program is to have the computer display and terminal located within the home where educational lessons can be studied at the convenience of the student. Again a centralized computer will evaluate the student's progress instantaneously.

4. Computer-aided health services (see Fig. 6.21). In the near future, all hospital patients will be monitored by computer. In addition, doctors will

Fig. 6.21 Monitoring medical patients with computer systems can reduce manyfold the time required for reliable diagnoses by hospital personnel. (*Hewlett-Packard.*)

have computer terminals available for patient medical histories and reliable diagnoses of patient symptoms as soon as the symptoms are presented to a computer.

5. Personal computers. Rapidly decreasing costs for microprocessor and microcomputer products soon will make custom-designed computational equipment accessible to anyone for virtually any use. A personal computer may be used to store messages, recipes, and the like for instant recall when needed. Devices such as the automatic television program control that allows you to "program" your set for the entire evening and sit back and watch as the stations are switched automatically at the proper time for your requested viewing will appear more and more in the future. Microcomputers will help us conserve energy by controlling thermostats through monitoring of

weather conditions on the outside; and by controlling automobile carburetors to provide the optimum air-fuel ratio.

You can add many more items to this list of possible applications and perhaps you will share in the development of some of these products.

They are only a few of the many possibilities that exist. A great deal of automation will take place in industry and business with the increased development of minicomputers and microcomputers. This is certain to create a reduction in the work force that will be partly compensated by shorter work weeks and earlier retirements. Computers will help make life more livable only if their limitations and applications are fully understood.

Problems

6.1 Write a brief summary of the contribution to the development of the modern computer by each of the following.

(*a*) Edmund Gunter
(*b*) Howard Aiken
(*c*) Herman Hollerith
(*d*) John Von Neumann
(*e*) J. P. Eckert

6.2 Write a short report on an application of each of the following classes of computers in the engineering field of your choice.

(*a*) Large-scale digital computer
(*b*) Minicomputer
(*c*) Microcomputer
(*d*) Electronic analog computer

6.3 Write an effective algorithm for the following

(*a*) Cooking a hard-boiled egg
(*b*) Multiplying two integer numbers of two digits each
(*c*) Finding the roots of the equation $Ax^2 + Bx = C$.
(*d*) Computing a worker's net pay knowing the hourly rate and hours worked and assuming appropriate deductions.

6.4 One of the classic algorithms was developed about 250 B.C. by Eratosthenes. Called the Sieve of Eratosthenes, the algorithm can find all the prime numbers between 1 and some specified integer *N*. Find documentation of the algorithm and use it to determine the prime numbers

(*a*) Between 1 and 18
(*b*) Between 26 and 54

6.5 Following the engineering method, develop a procedure to determine whether any three given lengths can constitute a right triangle.

6.6 Using the engineering method of problem solution, determine the solutions to the following equations to four significant figures.

(*a*) $x = \sin x + 0.5$
(*b*) $x \log x - 1 = 0$

6.7 From a manufacturer or supplier, obtain an owner's manual or other descriptive information for one of the following devices. Make a brief oral report to the class emphasizing the functions and operation of the microprocessor or microcomputer contained in the device.

(*a*) Programmable microwave oven
(*b*) Television hockey, table tennis, or similar games
(*c*) Automatic bank tellers
(*d*) Electronic timers for televisions
(*e*) Auto fuel-injection control system
(*f*) Temperature controllers for home or industrial furnaces

Preparation for computer solutions

CHAPTER 7

7.1 Introduction

In using the engineering method of problem solution, you oftentimes reach a point at which a decision must be made about what computational device will be used. In many cases, you simply use a calculator to process the numerical values, or you leave the answer in algebraic form. Problems of a sophisticated nature or ones involving lengthy, repetitive calculations should be considered for computer solution, but the decision to do so should not be made hastily. Because the digital computer performs calculations very quickly, a large, expensive volume of unwanted, incorrect computations may be generated unless a well-thought-out program is developed.

In Chap. 2 the engineering method of problem solving was described in terms of the following five-step procedure:

1. Recognize and understand the problem
2. Accumulate facts
3. Make necessary assumptions and select appropriate theory or principle

Fig. 7.1 A modern digital computing system. (*Honeywell, Inc.*)

4. Solve the problem
5. Verify and check

The mechanics of computation fits into this procedure at step 4. Before deciding on the method of computation and the equipment to be used, you must do several things in order to compute correct answers. First, you must define the problem and then outline all the knowns and unknowns. You arrive at step 3 where you develop the mathematical statements that must be solved or calculated.

At this point, you must first determine how to perform the calculations necessary to solve the problem. That is, you must proceed as though there is a problem within a problem. You must also decide what type of computational equipment to use. Let's suppose for purposes of discussion that the computation is complex and iterative in nature and that something more than a simple calculator is needed. The decision to write a computer program is not made without due consideration of the available alternatives and without proper weighing of the advantages and disadvantages. So to proceed, assume that you have determined that the best way to solve the problem is by use of a computer or programmable calculator.

Once this decision has been made, a three-phase procedure is recommended for obtaining a successful solution.

4*a*. Construct a flowchart for the solution procedure

4*b*. Write the program in a computer language appropriate for the machine you will use

4*c*. Follow the correct procedures for getting your program and data into the computer for processing and for retrieving and interpreting the output (results)

Computers and calculators perform only operations that they are directed to do. Directions are given to the computer via the program, which is a series of steps, a routine, that must be sequentially followed to solve the problem. Devising the plan is called programming. The one who prepares the sequence of instructions, without necessarily converting them into the detailed codes is called a programmer. In the steps, just given, 4*a* can be completed without regard to the machine to be used, but steps 4*b* and 4*c* must take into account the computer and peripheral equipment. We will devote the remainder of this chapter therefore to step 4*a*, the construction of flowcharts. The details of the programming language and operation of the computer equipment are beyond the scope of this text. For further information about programming, consult one of many texts that are available.

7.2 Flowcharting

As mentioned briefly in Chap. 6, the portion of a computer program concerned with sequencing the steps to the problem solution is called an algorithm. The algorithm does not include details on how to input or output data, perform the arithmetic operations, or make decisions based on computations. It instead states clearly the order in which the solution steps are to be executed. If all the solution steps are executed but in an incorrect order, you will get incorrect results. To illustrate the specificity of the commands, let us consider the steps required in taking a shower.

1. Remove clothing
2. Turn on water
3. Adjust water temperature
4. Step into shower
5. Thoroughly wet the body

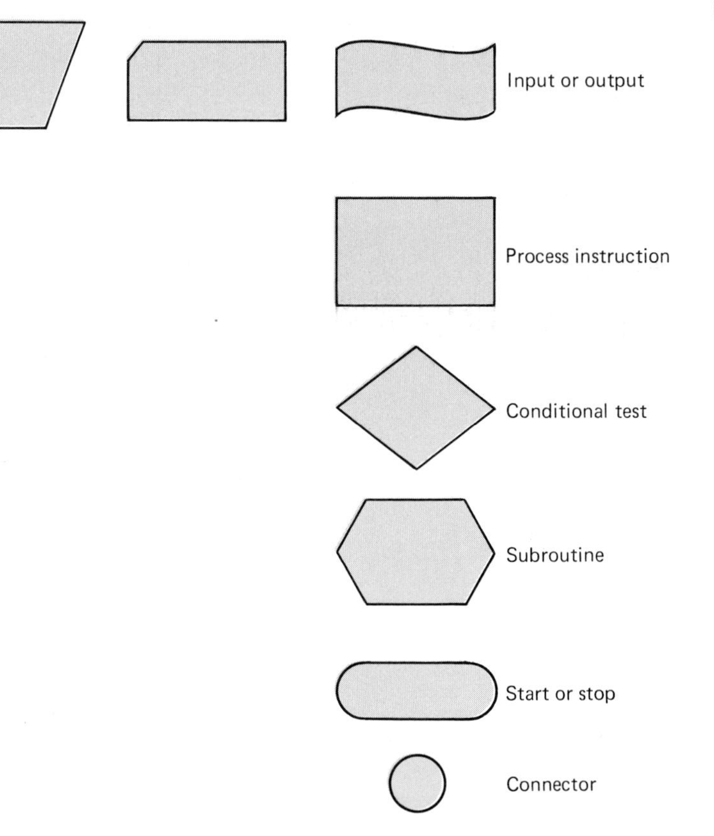

Fig. 7.2 Flowchart symbols.

6. Apply soap
7. Rinse
8. Turn off the water
9. Step out of shower
10. Dry

More detailed instructions could be given (and will be as the flowchart is constructed later), but this is a satisfactory algorithm. You would obviously obtain an unsatisfactory result if you performed the steps in the order 5, 2, 3, 9, 4, 8, 1, 10, 7, 6, or any other combination besides 1 through 10.

A clearer presentation of the steps required for solving a problem can be made by constructing a flowchart. The instructions are enclosed in boxes and various symbols and arrows are used to indicate the sequence (flow) of solution steps. The symbols used in the flowchart are shaped to indicate a particular action required of the computer. Some common symbols are shown in Fig. 7.2.

Figure 7.3 shows the flowchart for the process of taking a shower.

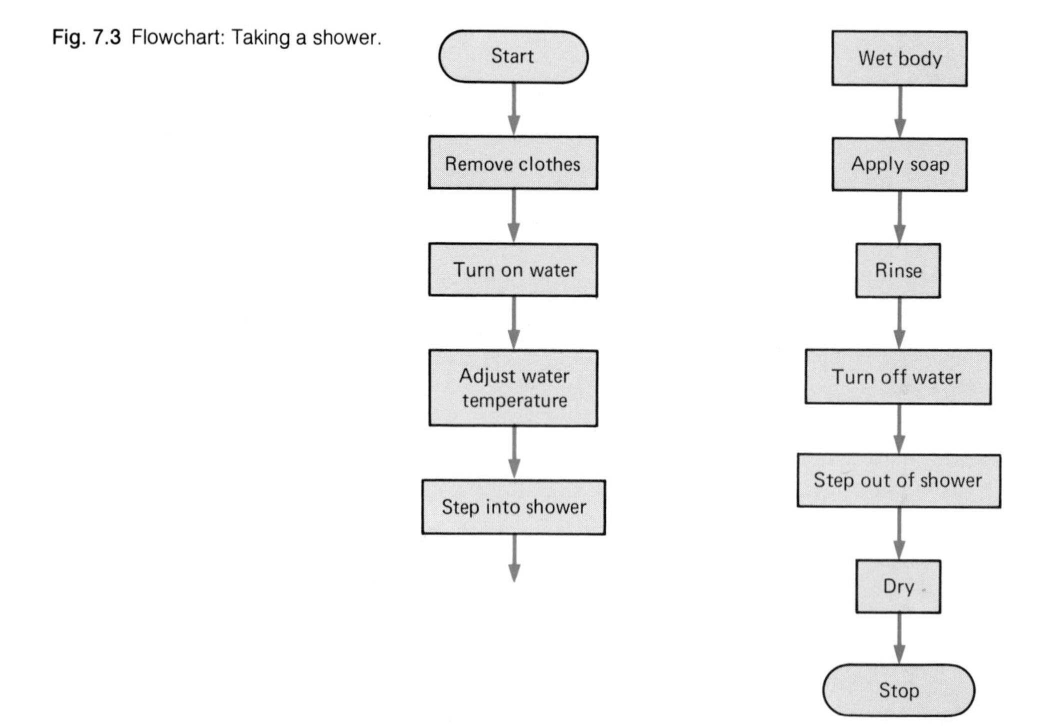

Fig. 7.3 Flowchart: Taking a shower.

7.3 Conditional testing

The process of taking a shower may be refined by specifying a comfortable water temperature and a cleanliness factor. A modified flowchart is shown in Fig. 7.4. Note the addition of the two decisions, or conditional tests. In this case, a "no" answer to the test requires adjusting a variable and cycling back to a previous step in the flowchart. There is no way to determine how many cycles (loops) will be required before a comfortable water temperature is attained. This condition can cause the computer to run endlessly, unless the programmer specifies a limit on the number of loops or places a time limit on the program. Figure 7.5 shows one method of limiting the number of loops, that of using a counter and stopping on a predetermined value of the counter.

By following the arrows in Fig. 7.5, you can see that the looping will continue until a comfortable temperature is attained *or* until five adjustments have been made. After five adjustments have been made, the flowchart directs that the next instruction, "step into shower," be undertaken.

The notation $N = N + 1$ in Fig. 7.5 although algebraically incorrect simply means that the current value of N is to be replaced by a new value equal to $N + 1$. Often this is indicated in a flowchart as $N \rightarrow N + 1$. This step is the key to counting with the computer. You can also count by 2s, 5s, 10s, etc.

Example problem 7.1 Construct a flowchart for calculating the sum of the squares of the even integers from 2 through 20.

Procedure This problem will need no data input. The required initial and final values of the integers, call them x, can be generated by the computer at the beginning of the program. The result, call it SUM, should be kept in a storage location so that it can be output at the completion of the program. This storage location must be initially zero. A correct flowchart is shown in Fig. 7.6.

You should study this flowchart carefully, because several examples of computer programming concepts are illustrated. First each variable is initialized: that is, it is given a numerical value the first time it is encountered in the program. Both the initial value ($x = 2$) and final value ($N = 20$) are specified. A variable, called SUM, is designated initially as zero. The process instructions $y = x^2$ and $\text{SUM} = \text{SUM} + y$ are the mathematical computations necessary to carry out the problem solution. A conditional test, encountered next, will eventually stop the operations when $x = N$ or, in this case, when $x = 20$. The loop contains the counting mechanism (incrementing the integer x by 2). Finally, the instruction to print the solution concludes the sequence of steps. This example would serve

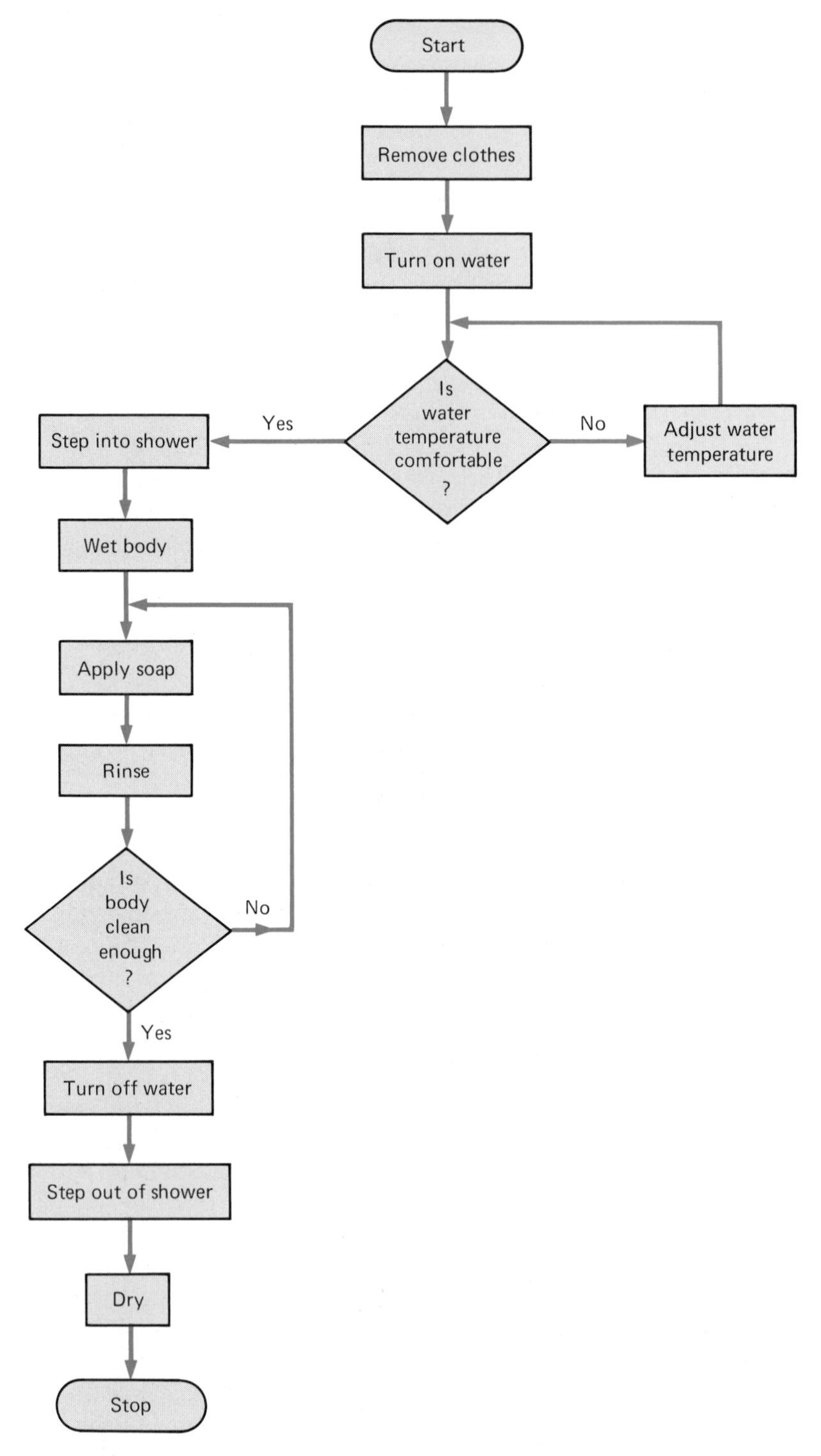

Fig. 7.4 Flowchart: Showering.

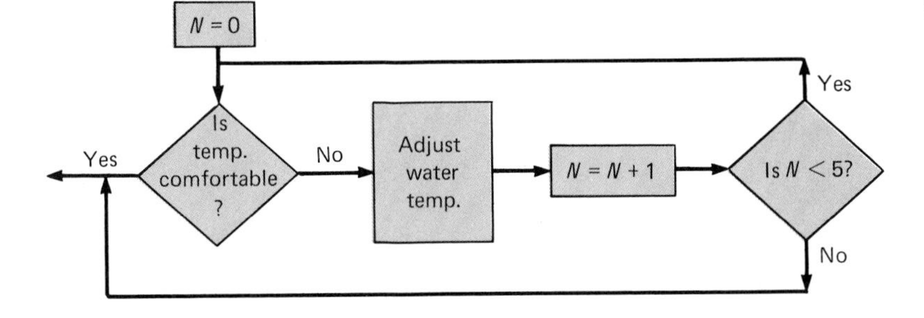

Fig. 7.5 Flowchart: Conditional test.

well as a first program for you to write if you are just beginning to learn a computer language.

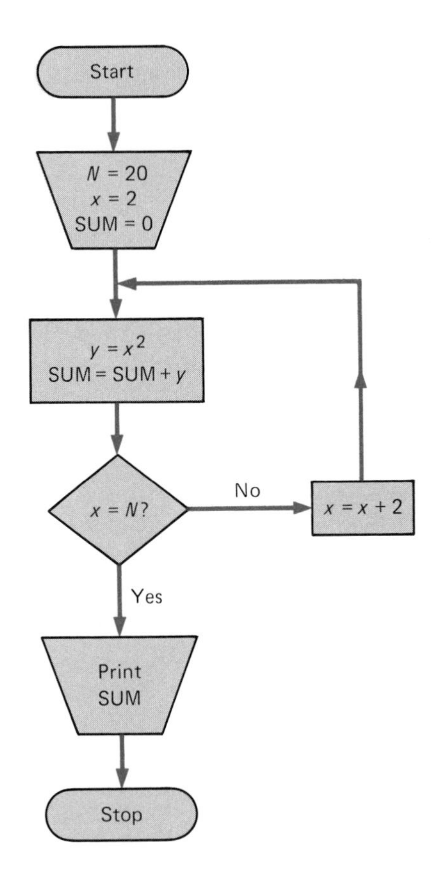

Fig. 7.6 Flowchart: Summing even integers.

Example problem 7.2 Mr. Jones operates a lounge and allows people to play a game of chance, the rules of which are as follows. Two dice are tossed; if the sum of the dice is less than 7, A must give a coin to B; if the sum of the dice is greater than 7, B must give a coin to *A;* if the sum of the dice is 7, *A* and *B* each give a coin to Jones. The two players begin with an equal number of coins, all of the same denomination. Construct a flowchart.

Procedure Although the game is quite simple to play, charting the logic requires several steps, since all possibilities for the outcome of each roll of dice must be stipulated. A flowchart illustrating the game is shown in Fig. 7.7.

If this game were programmed for computer analysis, the flowchart would have to be reworked and variables defined. The input required would be the number of coins that *A* and *B* have initially.

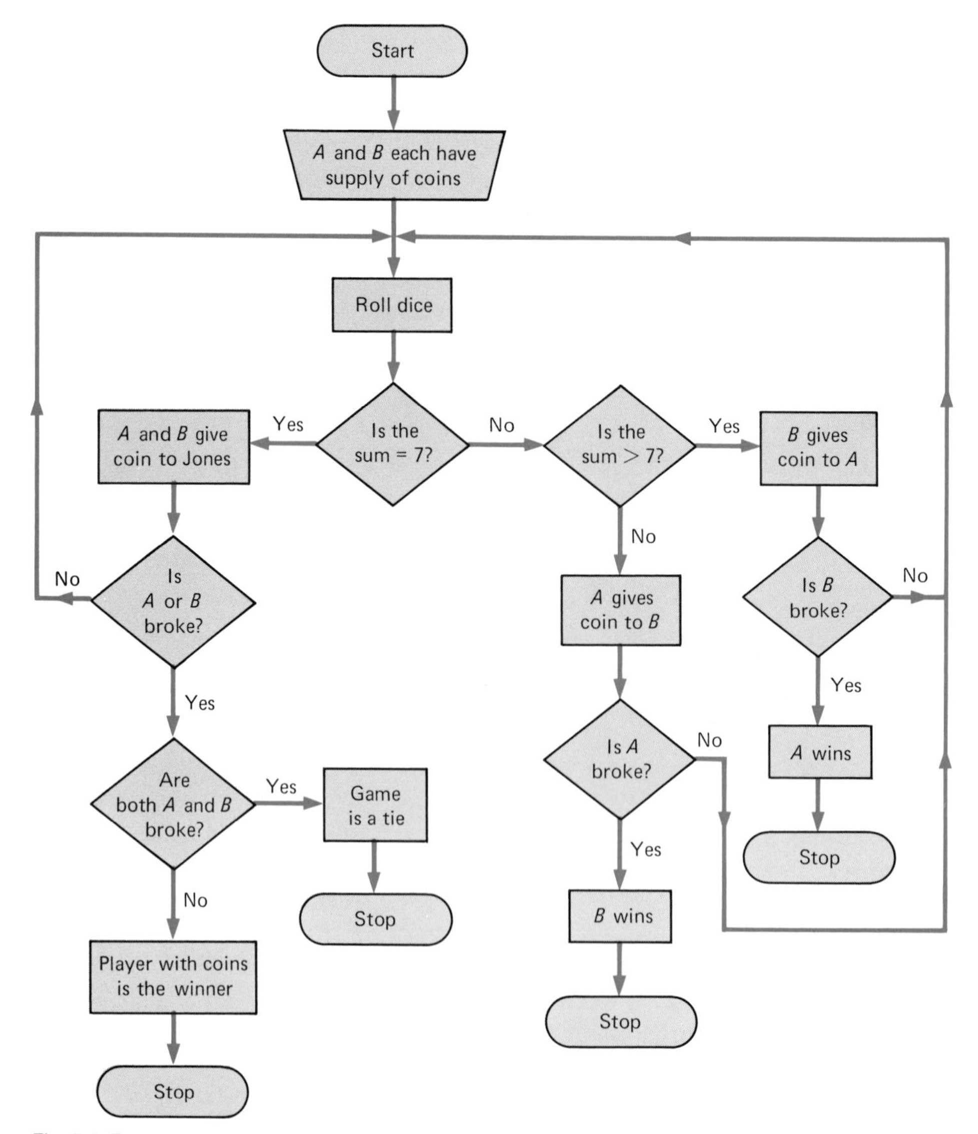

Fig. 7.7 Flowchart: Coin game

The sum of the numbers rolled with the dice would have to be specified for each turn. The output should include an identification of the winner and the number of coins possessed by the winner and Jones.

Example problem 7.3 The sine of an angle can be approximately calculated from the following series expansion.

$$\sin x = \sum_{i=1}^{N} (-1)^{i+1}\left[\frac{x^{2i-1}}{(2i-1)!}\right]$$

$$= x - \frac{x^3}{3!} + \frac{x^5}{5!} - \frac{x^7}{7!} + \cdots + (-1)^{i+1}\left[\frac{x^{2i-1}}{(2i-1)!}\right]$$

where x is the angle, in radians. The degree of accuracy is determined by the number of terms of the series that are summed for a given value of x. Prepare a flowchart to calculate the sine of an angle of P degrees and cease the summation when the last term calculated has a magnitude less than 10^{-7}. Of course, an exact answer for the sine of the angle would require the summation of an infinite number of terms. Assume that the computer used does not have the capability of computing a factorial directly, so that the flowchart must indicate the procedure necessary to compute a factorial.

Procedure: One possible flowchart is shown in Fig. 7.8. We will discuss several features of this flowchart, after which you should evaluate on paper the first three or four terms of the series expansion to make sure that you understand the problem.

1. The magnitude of the quantity controlling the number of terms summed is called ERR.
2. i denotes the summation variable and is initialized as 1.
3. M represents $(2i - 1)!$ and is initialized as 1 (its value in the first term of the series).
4. SUM is the accumulated value of the series as each term is added to the previous total.
5. The angle is input in degrees and then immediately converted to radians by multiplying by $\pi/180$.
6. W is assigned as the quantity x^{2i-1}.
7. Z represents the quantity $(-1)^{i+1}$, which will be either $+1$ or -1 depending on the current value of i. It must be $+1$ for the first term of the series and alternate signs after that.
8. TERM is the value of each term beginning with $(+1)(x^{2(1)-1}/[2(1) - 1]!)$, or simply x.
9. SUM is equal to $0 + x$ the first time through the procedure.

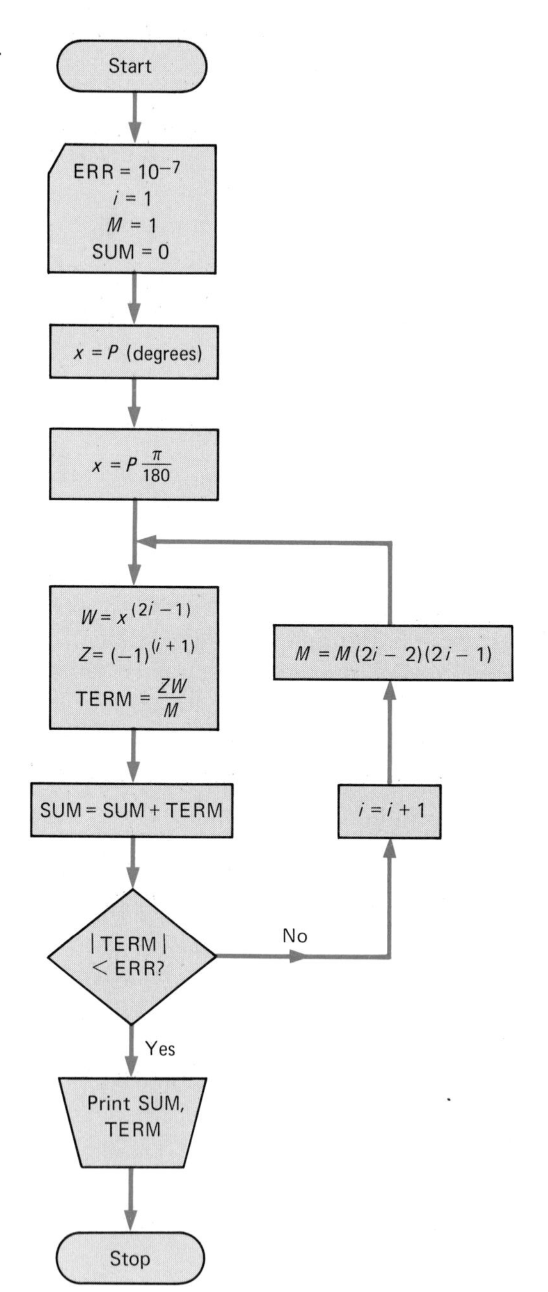

Fig. 7.8 Flowchart: Sine series.

10. The absolute value of TERM is now checked against the control value ERR to see if computations should cease. Note that the absolute value must be used because of the alternating signs. If a "no" answer is received to this conditional test the appropriate incrementing of the variable must be undertaken.

11. i is increased by 1. This time through, i becomes 2 since it was initialized as 1.

12. M becomes $(1)(2 \times 2 - 2)(2 \times 2 - 1) = 1 \cdot 2 \cdot 3 = 3!$

13. Following the flowchart directions, we now return to the evaluation of W, Z, and TERM with the new value of i.

$$W = x^3$$

$$Z = -1$$

$$\text{TERM} = -\frac{x^3}{3!}$$

14. SUM then becomes $x - x^3/3!$. TERM is again tested against ERR.

15. The process repeats until the magnitude of the last term is less than 10^{-7} and the machine is instructed to report the value of $\sin x$ and the value of the last term to make certain the standard of accuracy has been attained.

You should check several terms for a given value of x and then repeat the process for different angles. You will note that as the size of the angle varies, the number of terms required to achieve the standard of accuracy also varies. For example, for a value of $P = 5°$, the third term of the series is about $4(10^{-8})$, much less than the 10^{-7} requirement. But for $P = 80°$, the third term is approximately $4.4(10^{-2})$. When $P = 80°$, seven terms of the series are required before the magnitude of the last term becomes less than 10^{-7}.

The preceding examples should give you some insight into the construction of a flowchart as a prelude to writing a computer program for your particular computer.

The mechanisms for calculating, testing, incrementing, looping, etc., vary with the computational device and the programming language. The flowchart, however, should be valid for all computer systems, because it graphically portrays the steps that must be completed to solve the problem.

Problems

7.1 Draw a flowchart to solve for P when $P = RQ + K + R^K$.

7.2 Draw a flowchart to solve for M when $M = 3B + BE/F$.

7.3 Draw a flowchart to calculate the sum of the areas of circles whose radii are 2, 3, 4, and 5 cm.

7.4 Draw a flowchart to calculate $\cos X$ when

$$\cos X = 1 - \frac{X^2}{2!} + \frac{X^4}{4!} + \cdots$$

correct to six decimal places. X = angle, in radians, and N is a positive integer. HINT: The general term of the series is

$$(-1)^{N+1}\frac{X^{2N-2}}{(2N-2)!}$$

7.5 Draw a flowchart to simulate the required courses in your curriculum.

7.6 Draw a flowchart to calculate the sum of all the numbers between 0 and 99 whose square roots are integers.

7.7 Draw a flowchart to calculate the value of y when

$$y = \sum_{x=2}^{100} \frac{e^x}{\ln x}$$

7.8 Draw a flowchart to calculate the compound interest accumulated by P dollars at i percent annual interest:

(*a*) Compounded annually

(*b*) Compounded daily

HINT: [Sum = $P(1 + i)^n$, where i is the interest rate per period and n is the number of periods.]

7.9 Draw a flowchart to calculate the temperature in kelvins when the Fahrenheit temperature is known.

7.10 A group of engineers have an office pool for football games of their alma mater. The winner is the one that ends the season with the fewest demerits. Demerits are awarded after each game on the following basis: three demerits for each one point that one errs in guessing the correct difference in the final scores of the two teams; and one demerit for each point one is incorrect in the final score of each team. [Example: One engineer predicts the home team to win 14 to 7 but the final score is home team 10, visitors 21. Demerits = $(11 + 7)3 + (14 - 10) + (21 - 7) = 72$ demerits.] Draw a flowchart describing this pool.

7.11 Draw a flowchart to calculate the sum of the squares of the even integers between 0 and 15.

7.12 Draw a flowchart to depict the process of buying a car from a used-car lot.

7.13 Draw a flowchart that shows the process of a blackjack (21) dealer. Assume that the dealer has already completed dealing to the players, has one card himself and is ready to complete the deal to himself. The rules are that he must draw a card if he has less than 16 points and cannot draw if he has 16 or more points.

7.14 Draw a flowchart to arrange a list of 20 numbers in descending order.

7.15 Draw a flowchart that selects from a list of numbers only numbers whose square root is an integer.

7.16 Draw a flowchart to calculate the present worth of a sinking fund. The formula is

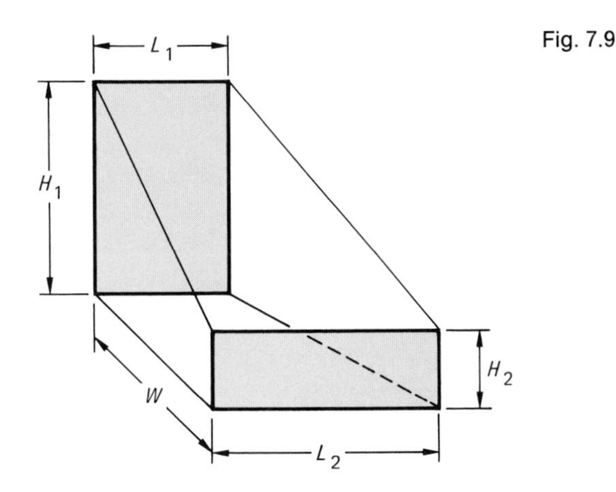

Fig. 7.9

$$\text{Present worth} = A\frac{(1+i)^n - 1}{i(1+i)^n}$$

See Chap. 13 for the definition of the variables.

7.17 Given two points with coordinates (x_1,y_1) and (x_2,y_2), draw a flowchart to calculate the length of the line connecting the two points.

7.18 Do the same as in Prob. 7.17 except design the flowchart to calculate the direction of the line. Specify the output as an angle. Assume that the line goes from (x_1,y_1) toward (x_2,y_2).

7.19 Draw a flowchart to convert a Roman numeral to an Arabic number.

7.20 The prismoidal formula is

$$U = (A_1 + 4A_m + A_2)\frac{W}{6}$$

where A_1 = area of one end
A_2 = area of other end
A_m = area at midpoint

Construct a flowchart to calculate the volume of a solid figure such as the one shown in Fig. 7.9.

7.21 Draw a flowchart to solve the quadratic equation $Ax^2 + Bx + C = 0$ for x. Account for all possible solutions.

7.22 The area A of a triangle can be found by $A^2 = s(s-a)(s-b)(s-c)$, where a, b, c = sides of triangle and $s = \frac{1}{2}(a + b + c)$. Draw a flowchart to calculate the area of a triangle by this formula.

7.23 Draw a flowchart to determine if three given lengths can form a triangle.

7.24 Suppose I have chosen a number between two limits, say 0 and 100. Your task is to guess the number. All I will report is that your guess is too high, too low, or correct. Draw a flowchart of your process if each successive guess is midway between your last guess and the current upper or lower extreme value. (If your first guess is 40 and I say, "too low," your next guess is 70 and I say, "too high," your third guess should be 55.) Use only integers.

PART FOUR

Applied engineering concepts

Mechanics

CHAPTER 8

8.1 Introduction

Mechanics is a branch of engineering science that deals with various effects of forces on bodies. Forces can cause a body to move, to be deformed, or when properly balanced, to remain stationary. The study of rigid bodies at rest or moving with uniform velocity is called *statics*. The study of bodies experiencing irregular motion is called *dynamics*. *Mechanics of materials* is involved with the change in size and shape of bodies. The concern of this chapter will be limited to an introduction to statics. During the next few terms most of you will have full courses devoted to each of the topics noted above.

8.2 Scalar and vector quantities

The study of engineering mechanics involves a number of physical quantities such as acceleration, force, mass, time, velocity, and volume, some of which can be expressed as scalars, and others as vectors. A *scalar* quantity has only magnitude, while a *vector* quantity has direction as well as magnitude. Mass and time are scalar quantities, and force is a vector quantity.

Fig. 8.1 Design analyses of equipment such as this must anticipate environmental extremes. (*FMC Corporation.*)

Fig. 8.2

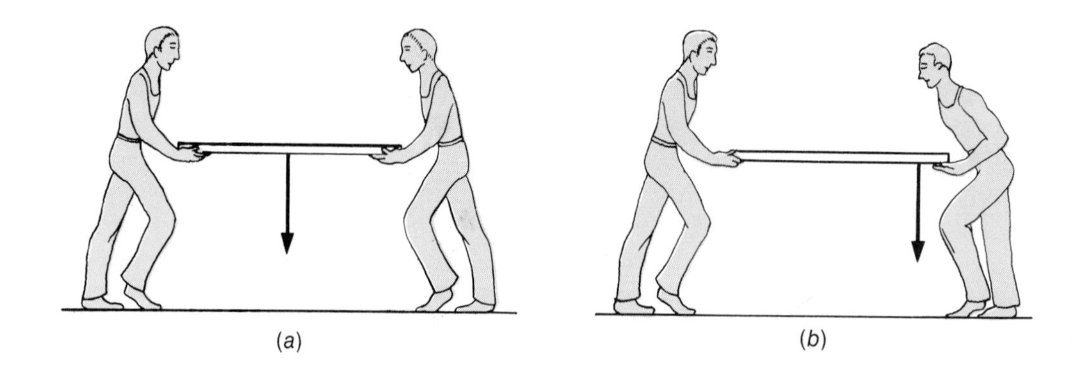

The effect of a force can be demonstrated by considering the difference in the two sketches in Fig. 8.2. In Fig. 8.2*a* it can be seen that the people share the force equally. In Fig. 8.2*b*, the one on the right obviously must support the greater portion of the force. Since a force has magnitude, direction, and location, if any of these characteristics is changed, there will be a change in the force supported by each person.

8.2.1 Forces

When one body acts upon another body, it tends to change the motion of the body acted upon. This action is called a force. Newton's third law states that if body P acts upon body Q with a force of a given magnitude and direction, body Q will *react* upon P with a force of equal magnitude and opposite direction. Therefore, to describe a force you must give its magnitude, its direction, and the location of at least one point along its line of action. Figure 8.3 shows two common ways of depicting a force.

The reference line for angle θ may be chosen arbitrarily. Figure 8.3*a* shows the customary way of measuring the angle: counterclockwise from a positive horizontal line. The direction could have been shown as the angle with respect to a vertical line; or the angle could have been measured from the horizontal on the left. If two points along the line of action of the force are known, say A and B in Fig. 8.3*b*, then the location and direction of the force could be identified as from A to B, or AB. Incidentally, care must be used

Fig. 8.3

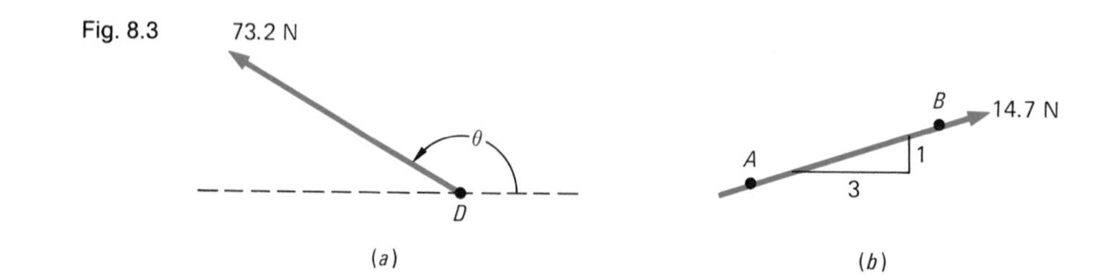

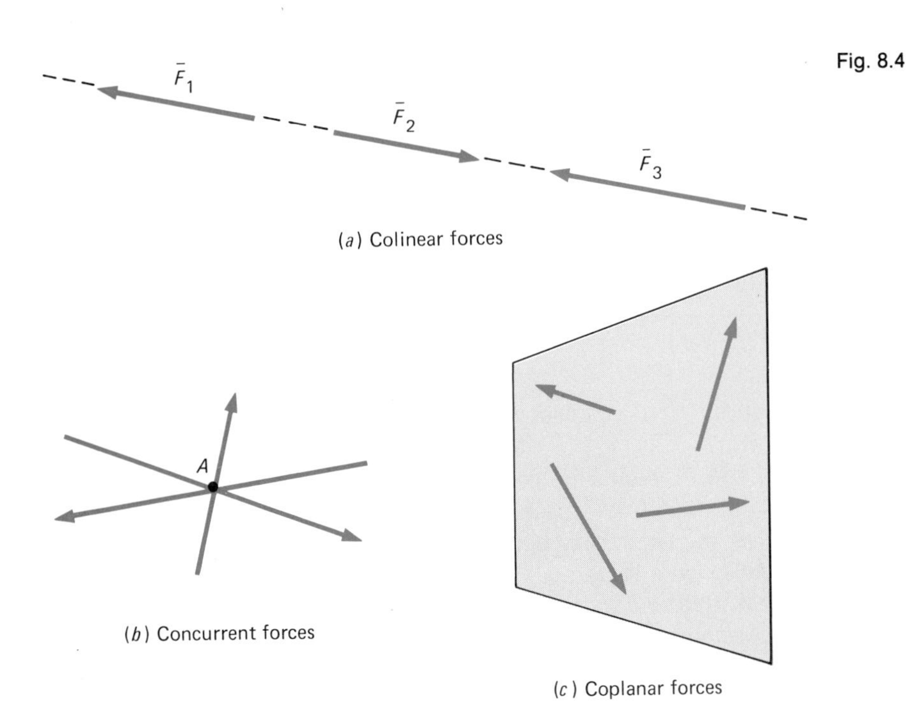

Fig. 8.4

(a) Colinear forces

(b) Concurrent forces

(c) Coplanar forces

here, because BA is, by convention, exactly opposite in direction to AB.

Forces acting along the same line are *colinear*. Forces that pass through the same point are said to be *concurrent*. Forces in the same plane are *coplanar*. Figure 8.4 illustrates these force systems. In this text, the forces will be understood to be only coplanar: they may or may not be concurrent or colinear.

8.2.2 Resolution of forces

If a force is known, then its magnitude, direction, and point of application have been determined. Quite often it is convenient in calculating the effect of a force to divide the force into components. The most commonly used system for doing this is by use of the familiar xy axes, or cartesian coordinate system.

If force $\overline{F}$ has a magnitude of 700.00 N and acts through point A at an angle of 30° to the horizontal, as shown in Fig. 8.5, $\overline{F}$ can be replaced by two components F_x and F_y as follows:

$$F_x = F\cos\theta = 700.00(0.866\,03) = 606.22 \text{ N}$$

$$F_y = F\sin\theta = 700.00(0.500\,00) = 350.00 \text{ N}$$

(If these calculations are not obvious to you, you should refer to

Fig. 8.5

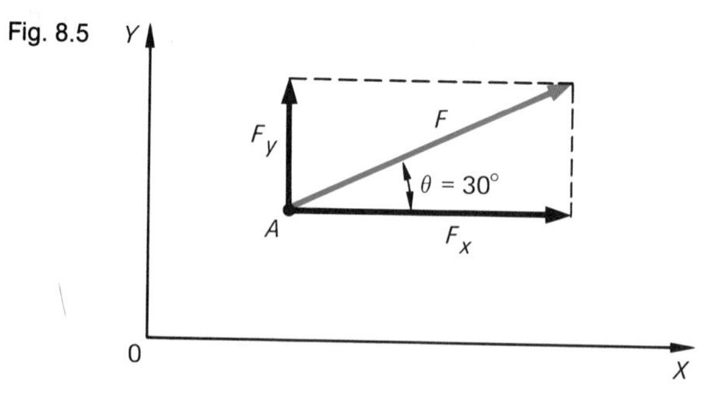

Appendix B to refresh your memory of trigonometric relationships.)

F_x and F_y would be considered positive, following the normal mathematical convention.

In saying that $\overline{F}$ can be replaced by F_x and F_y, we mean that if A is a point on a rigid body, the net effect of a force applied to that body is identical to the combined effects of its components applied at that point. Had F_x and F_y been known initially, it would follow that $\overline{F}$ is the resultant of F_x and F_y. The resultant of a force system acting on a body is the simplest (often a single force) system that can replace the original system without changing the external effect on the body.

Figure 8.6 shows two forces ($\overline{J}$ and $\overline{K}$) acting through point C. The resultant effect of $\overline{J}$ and $\overline{K}$ acting through C is $\overline{R}$. The magnitude and direction of $\overline{R}$ can be determined graphically or by the use of trigonometry. In Fig. 8.6 it can be seen that $\overline{K}'$ has been constructed parallel to and equal in magnitude to $\overline{K}$ from the forward end of $\overline{J}$. Likewise, $\overline{J}'$ has been constructed parallel to and equal in magnitude to $\overline{J}$ from the forward end of $\overline{K}$. $\overline{R}$ then connects C with the intersection of $\overline{J}'$ and $\overline{K}$, giving the magnitude (graphically)

Fig. 8.6

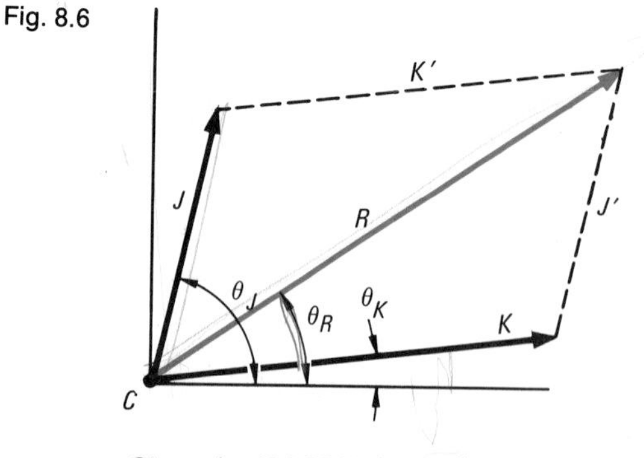

Given: $J = 400.00$ N $\theta_J = 76°10'$
$K = 600.00$ N $\theta_K = 6°30'$

and direction of $\overline{R}$. It can be seen that $\overline{J}'$, $\overline{K}'$, $\overline{J}$, and $\overline{K}$ form a parallelogram, with $\overline{R}$ being one of its diagonals. The method of vector addition used to determine $\overline{R}$ is termed the *parallelogram method.*

If the point of concurrence of force vectors is assumed to be the origin of a cartesian coordinate system, the vectors can be resolved into their x and y components and the components added algebraically. If $\overline{J}$ and $\overline{K}$ are so resolved and added, the result is as follows.

		R_x	R_y
$J_x = 400\ (\cos 76°10')$	=	95.64 N	
$J_y = 400\ (\sin 76°10')$	=		388.40 N
$K_x = 600\ (\cos 6°30')$	=	596.14 N	
$K_y = 600\ (\sin 6°30')$	=		67.92 N
		691.78 N	456.32 N
$R = \sqrt{R_x^2 + R_y^2}$	=	828.73 N	
$\theta_R = \tan^{-1} \dfrac{R_y}{R_x}$	=	33°25′	

A problem of this nature can also be solved by employing the principles of trigonometry. Consider the triangle with sides $KJ'R$, and solve for R by using the law of cosines. See Fig. 8.7.

$$R^2 = K^2 + J'^2 - 2\,KJ' \cos \alpha$$

$$R = 828.73 \text{ N}$$

By employing the law of sines, we can find the value of ϕ or β, with the other found by subtraction from 180°. If R and θ_R are checked with a scale and protractor in Fig. 8.7, it will be noted that the measured values check closely with the calculated values.

If several concurrent-coplanar forces act on a body, the parallelogram method can be extended to determine graphically the resultant of the system.

Consider the forces shown in Fig. 8.8. Forces $\overline{F}_1$, $\overline{F}_2$, $\overline{F}_3$, and $\overline{F}_4$ are coplanar and concurrent through point L, as shown in Fig. 8.8*a*. In Fig. 8.8*b*, all the forces depicted are at the same scale and slope as in *a*. Begin with $\overline{F}_1$ at L and successively add the other forces

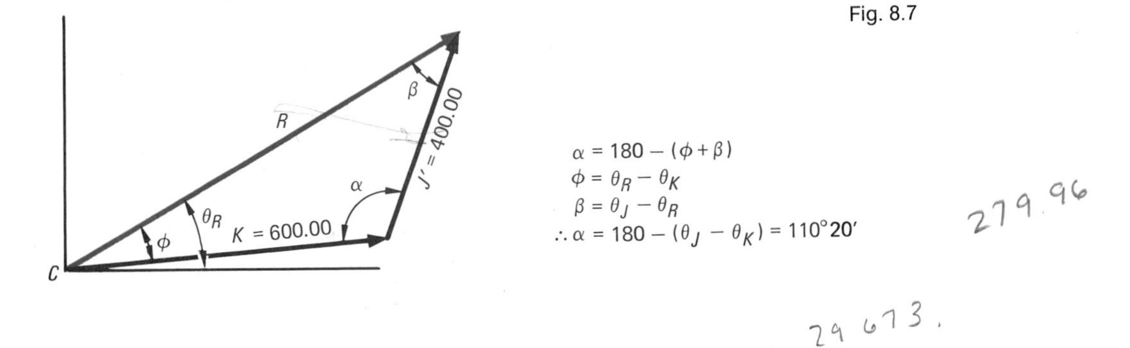

Fig. 8.7

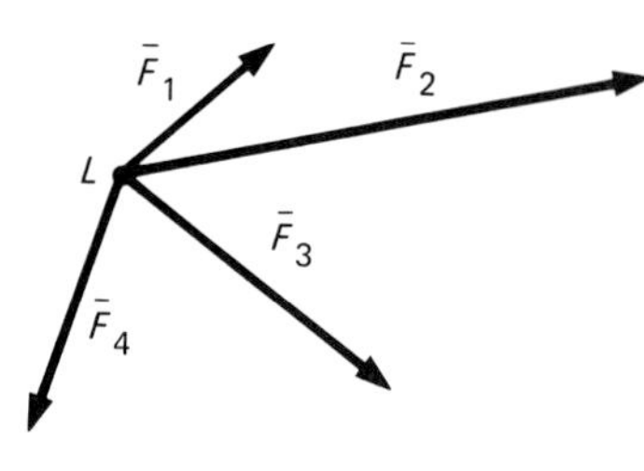

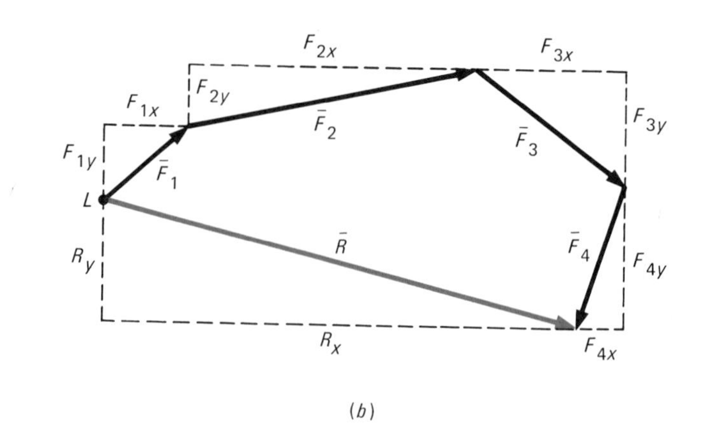

Fig. 8.8

to the end of the one just plotted. When all forces have been plotted, you can then determine the resultant by drawing a line from L to the end of the last force plotted. It makes no difference which force is plotted first, nor in what order the forces are plotted, so long as all the forces are accounted for. The final drawing is called a force polygon.

In Fig. 8.8*b*, the components of the forces and the resultant are shown. By calculating the components and adding them algebraically, you can determine the components of the resultant $\bar{R}$ in the same way as was shown with forces $\bar{J}$ and $\bar{K}$ in Fig. 8.6. You are encouraged to perform this computation and compare your results with those of the graphical method.

8.3 Moments

When was the last time you approached an exit in a public building and pushed on the panic bar of the door only to find that you had chosen the wrong side of the door, the side next to the hinges? No big problem. You simply move your hands to the side opposite the hinges and easily push open the door. You are thereby demonstrating the principle of the turning moment. By definition, the tendency of a force to cause rotation about a point is called the moment of the force relative to that point. The magnitude of the moment is the product of the magnitude of the force and the perpendicular distance from the line of action of the force to the point. With respect to the door just mentioned, the same force may have been exerted in both attempts to open the door. In the second case, however, the moment was greater owing to the fact that you increased the distance from the force to the hinges, the point about which the door turns. Another example is shown in Fig. 8.9. The magnitude of the moment of $\bar{F}$ about point B is Fd_B (clockwise). The magni-

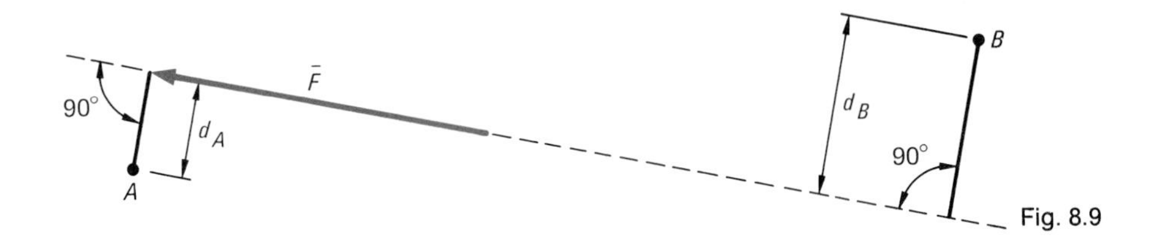

Fig. 8.9

tude of the moment of $\bar{F}$ about point A is Fd_A (counterclockwise). It follows then that the moment of a force *system* about a point is the algebraic sum of the moments of each force about that point. The most commonly used convention is to assign a positive sign to counterclockwise moments and a negative sign to clockwise moments.

8.4 Equilibrium

A body is said to be in equilibrium if the sums of all forces and moments are zero, and it will remain at rest or continue its movement at constant velocity. In this book, the study of statics will consider only bodies at rest. To study a body and the forces acting upon it, one must first determine what the forces are. Some may be unknown in magnitude and/or direction and quite often these unknown magnitudes and directions are information that is being sought. The conditions of equilibrium can be stated in equation from as follows:

$\Sigma F_x = 0$ (the sum of horizontal components of all forces = 0)

$\Sigma F_y = 0$ (the sum of the vertical components of all forces = 0)

$\Sigma M = 0$ (the sum of the moments about any point = 0)

In order to assist in the tabulation and summation of force components and moments, a convention is necessary. In this text, F_x is positive to the right (↦) and F_y is positive upward (↥) and counterclockwise moments are positive (↺).

8.5 Free-body diagrams

For the sake of simplicity in this text, we will work with problems using shapes and configurations that are familiar to you and easy to visualize. In practice, configurations such as a complete bridge or building instead of a single beam, will be more typical. In other cases, we will describe a hinge connection, whereas in actual practice you may have to design a new bearing or a complicated gear. In almost all types of engineering, the necessity of determining what

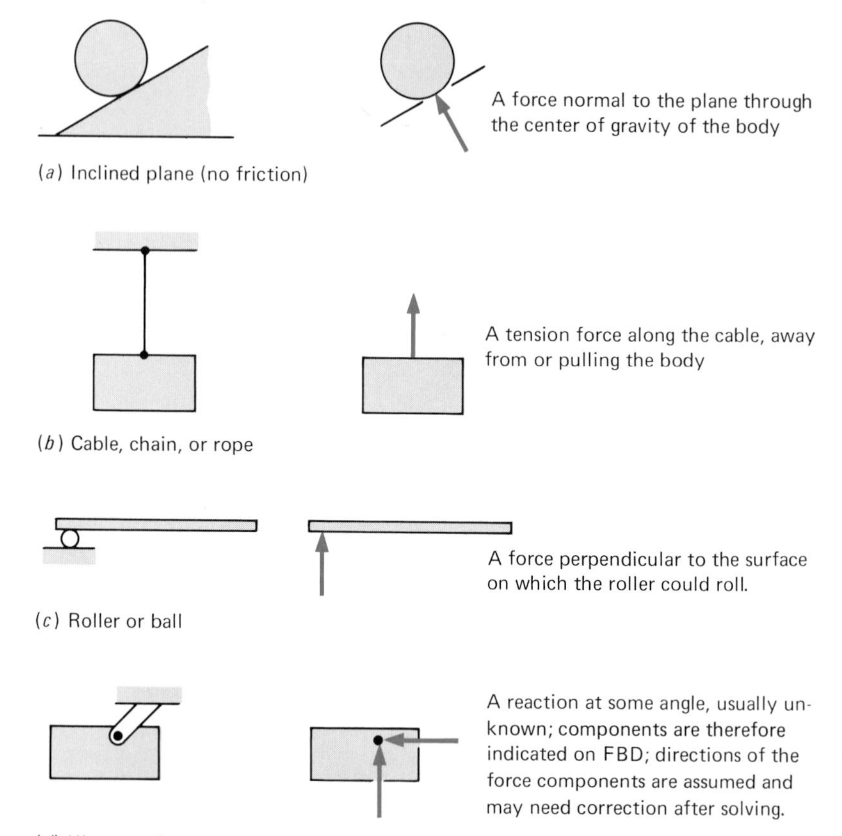

Fig. 8.10 Free-body notations.

forces and moments must be sustained by your design will be part of your analysis.

Normally some form of simple but concise representation of the actual system is needed to help visualize and formulate the problem.

The first step in solving a problem in statics is to draw a sketch of the body, or a portion of the body, and all the forces acting on that body. Such a sketch is called a free-body diagram (FBD). As the name implies, the body is cut free from all others; only forces that act upon it are considered. In drawing the free-body diagram, we remove the body from supports and connectors, so we must have an understanding of the types of reactions that may occur at these supports. See Fig. 8.10.

Example problem 8.1 A uniform 12.00-m beam has a mass of 10.00 kg/m and is supported by two rollers—one at the left end of the beam; and the second, 2.00 m left of the right end. What are the reactions at the two rollers?

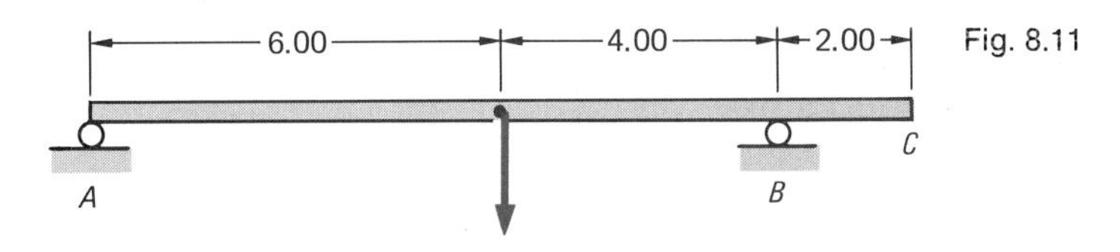

Fig. 8.11

$10.00 \text{ kg/m} \times 12.00 \text{ m} \times 9.807 \text{ m/s}^2 = 1176.8 \text{ N}$
Do not round intermediate figures; save at least one digit that is not significant

Solution

1. Make an accurate sketch of the information given (see Fig. 8.11).

2. Prepare a free-body diagram of the problem, replacing the roller bearings with unknown forces of magnitudes R_A and R_B (see Fig. 8.12).

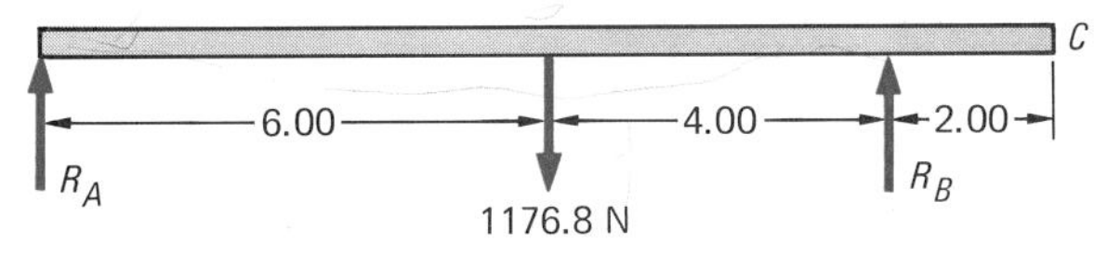

Fig. 8.12

3. Apply the equations of equilibrium.

(*a*) Sum of the horizontal forces must be zero: $\Sigma F_x = 0 \leftrightarrow$.
In this case there are no horizontal forces, since rollers cannot sustain horizontal forces.

(*b*) Sum of the vertical forces must be zero: $\Sigma F_y = 0 \updownarrow$
Therefore,

$$R_A + R_B - 1\,176.8 = 0$$

$$R_A + R_B = 1.1768 \text{ kN}$$

(*c*) The sum of the moments must be zero: $\Sigma M = 0$. Therefore, if moments are calculated about any point, the sum must be zero. Select a point, say the right end of the beam, and write the equation

$$\begin{aligned}
\Sigma Mc &= 0 \circlearrowright \\
&= -12.00R_A + 6.00(1\,176.8) - 2.00R_B = 0 \\
-12.00R_A + 7\,060.8 - (1\,176.8 - R_A)(2.00) &= 0 \\
-12.00R_A + 7\,060.8 - 2\,353.6 + 2.00R_A &= 0 \\
-10.00R_A &= 4\,707.2 \\
R_A &= 470.72 \text{ N} \\
&= 471 \text{ N}
\end{aligned}$$

then

$$
\begin{aligned}
R_B &= 1\,176.8 - R_A \\
&= 1\,176.8 - 470.72 \\
&= 706 \text{ N}
\end{aligned}
$$

It would have simplified the solution if moments had been taken about the left end.

$$
\begin{aligned}
\Sigma M_A &= 0 \circlearrowleft^{+} \\
&= (0)R_A - (6.00)(1\,176.8) + 10.00R_B = 0 \\
&\qquad\qquad\qquad\qquad\qquad 10.00R_B = 7\,060.8 \\
&\qquad\qquad\qquad\qquad\qquad\quad R_B = 706 \text{ N}
\end{aligned}
$$

then

$$
\begin{aligned}
R_A &= 1\,176.8 - 706.08 \\
&= 470.72 \text{ N} \\
&= 471 \text{ N}
\end{aligned}
$$

Fig. 8.13 The mass of metal that is picked up by the electromagnet creates a very large moment about the connection immediately below the operator. (*Square D Company.*)

Example problem 8.2 A 0.320-kg sphere is held by a cable attached to its surface. The sphere rests on an inclined plane that is 25° above the horizontal. The cable in turn is attached to a level plane, and the cable makes an angle of 40° with that plane. What is the tension in the cable, and what is the force on the inclined plane?

Solution

1. Sketch (see Fig. 8.14).

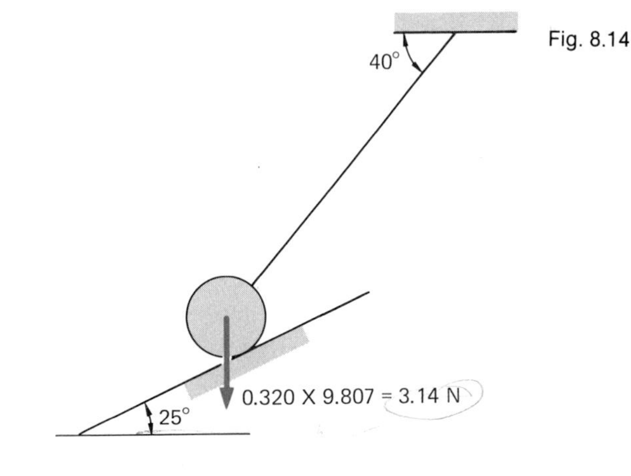

Fig. 8.14

2. Observe.
   (*a*) The weight of the sphere acts vertically through the center of the sphere.
   (*b*) The tension in the cable acts upward along the cable's line and through the center of the sphere.
   (*c*) The force on the inclined plane is resisted by a force normal to the plane and through the center of the sphere.
3. Make a free-body diagram (see Fig. 8.15).
4. Apply equations of equilibrium.

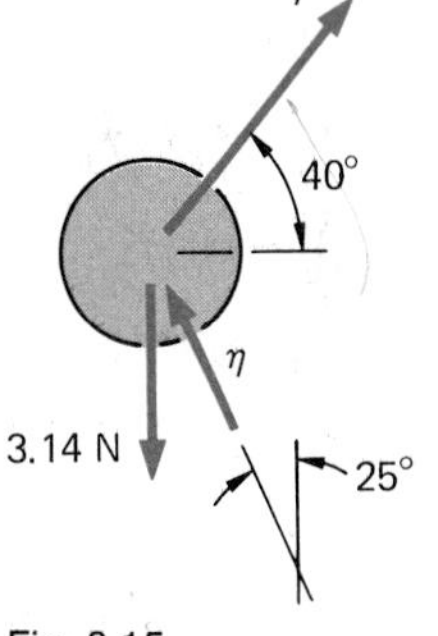

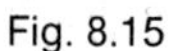

Fig. 8.15

(*a*) $\Sigma F_y = 0 \quad \updownarrow$

$$T_y + \eta_y - 3.14 = 0$$

(*b*) $\Sigma F_x = 0 \quad \leftrightarrow$

$$T_x - \eta_x = 0$$

(*c*) The moments equal zero because the forces are concurrent and if moments are taken about the center of the sphere, the moment arms are all zero. Thus the moment equation does not contribute useful information toward the solution.

(*d*) $T_y = T \sin 40° = 0.642\,8\ T$

$T_x = T \cos 40° = 0.766\,0\ T$

$\eta_y = \eta \cos 25° = 0.906\,3\ \eta$

$\eta_x = \eta \sin 25° = 0.422\,6\ \eta$

5. From these equations, we can write

   (*a*) $\Sigma F_x = 0 \leftrightarrow \quad 0.776\,0\,T - 0.422\,6\,\eta = 0$
   (*b*) $\Sigma F_y = 0 \updownarrow \quad 0.642\,8\,T + 0.906\,3\,\eta = 3.14$

   Multiplying 5*a* by 0.839 2, we get

   (*c*) $0.642\,8\,T - 0.354\,6\,\eta = 0$

   Subtracting 5*c* from 5*b*, we get

   $1.260\,9\,\eta = 3.14$ N

   then

   $\bar{\eta} = 2.49$ N ∠115°

from 5*a*,

$$T = \frac{0.422\,6}{0.766\,0}(2.49)$$

$$= 1.37 \text{ N}$$

Note that force on the inclined plane is $-\bar{\eta}$, so the force of the sphere on the plane = 2.49 N ∠65°

Fig. 8.16 The tension in the cables exceeds 10 MN when they hoist the pressure vessel of a nuclear reactor. (*Chicago Bridge & Iron.*)

Example problem 8.3 A beam is loaded and supported as shown in Fig. 8.17. What are the reactions at *A* and *B*?

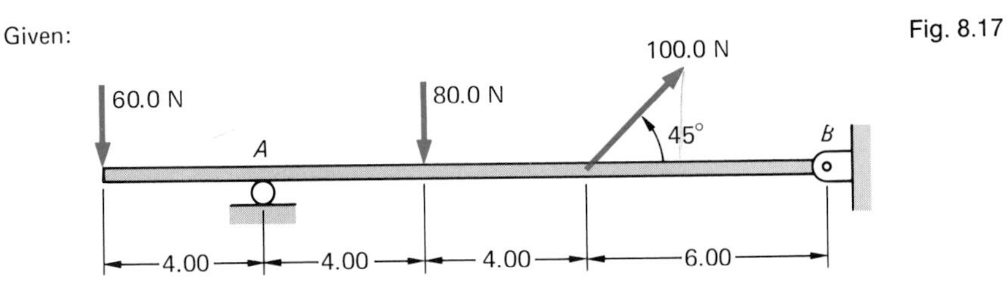

Fig. 8.17

Solution

1. Draw a free-body diagram (see Fig. 8.18).

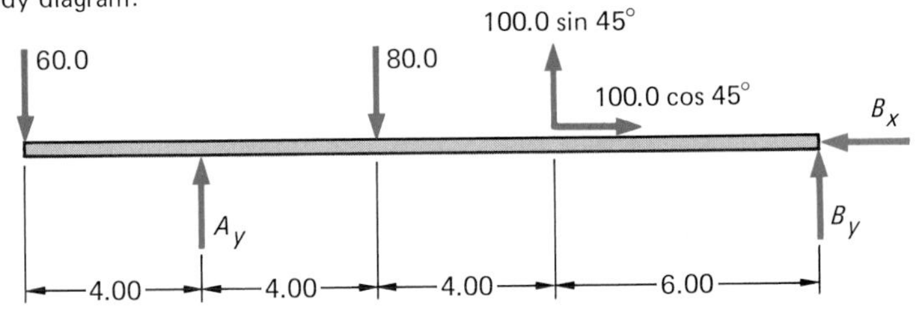

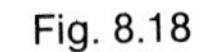
Fig. 8.18

2. Apply the following equations.

$$\Sigma F_x = 0 \quad \overset{+}{\rightarrow}$$

$$100.0 \cos 45° - B_x = 0$$

$$B_x = 70.7 \text{ N}$$

$$\Sigma F_y = 0 \quad \overset{+}{\uparrow}$$

$$A_y + B_y - 60.0 - 80.0 + 70.7 = 0$$

$$A_y + B_y = 69.3 \text{ N}$$

$$\Sigma M_B = 0 \quad \curvearrowleft +$$

$$= 60.0(18.0) - 14.0A_y + 80.0(10.0) - 70.7(6.00)$$

$$14A_y = 1\,080 + 800 - 424.20$$

$$A_y = 104 \text{ N}$$

$$\bar{A} = 104 \text{ N} \ \angle 90°$$

From the vertical force equation,

$$B_y = 69.3 - 104.0 = -34.7 \text{ N}$$

When the computation of an unknown force results in a negative quantity, it means that the assumed direction in the free-body diagram was incorrect.

Combining B_x and B_y gives

$\overline{B} = 78.8$ N 26.1°

To check the results, moments can be taken at another point. The most convenient point would be at A.

$$\Sigma M_A = 0$$

$$= 60.0(4.00) - 80.0(4.00) + 70.7(8.00) + 14.0B_y$$

$$240 - 320 + 565.6 + 14.0B_y = 0$$

$$B_y = -34.7 \text{ N}$$

8.6 Stress

In statics the concepts are limited to rigid bodies. It is obvious that the assumption of perfect rigidity is not likely to be valid. Any force will tend to deform or change the shape of any body, and certainly an extremely large force will cause obvious deformation. In most applications, slight deformations are experienced, but the body returns to its original form after the force is removed. One function of the engineer can thus be to design a structure within limits that allow it to resist permanent change in size and shape so that it can carry or withstand the force (load) and recover.

In statics, forces are represented as having a magnitude or an intensity in a particular direction. Structural members are usually characterized by their mass per unit length or their size in principal dimension, such as width or diameter. Consider the effect on a wire that is 5.00 mm in diameter and 2.50 m in length if it is suspended from a well-constructed table and has a ball whose mass is 30.0 kg attached to its lower end. The force exerted by the mass is (30.0)(9.81) = 294 N. The wire has a cross-sectional area of $\pi(5^2/4) = 19.63$ mm^2. If it is assumed that every square millimeter equally shares the force, then each square millimeter supports 294.3 $\div$ 19.63 = 15.0 N. This can be stated another way: the stress is 15.0 N/mm^2(1.50×10^7 Pa). This is the force in the wire that is literally trying to separate the atoms of the material by overcoming the bonds that hold the material together.

The relationship above is normally expressed as

$$\sigma = F/A,$$

where σ = stress, Pa

F = force, N

A = cross-sectional area, m^2

Since stress is force per unit area, it is obvious that the stress in the wire (ignoring its own weight) is unaffected by the length of the wire. Such a stress is called *tensile stress* (tending to pull the atoms apart). If the wire had been a rod of 5.00 mm diameter resting on a firm surface and the mass applied at its top, the force would have produced the identical stress, but it would be termed *compressive*, for obvious reasons.

A third type of stress is called *shear*. While tension and compression attempt to separate or push atoms together, shear tries to slide layers of atoms in the material across each other. (Imagine removing the top half of a stack of sheets of plywood without lifting.) Consider the bolt in Fig. 8.19 as it resists the force of $1.00(10^5)$ N.

The shear in the bolt is

$$\tau = \frac{F}{A} = \frac{10^5}{\dfrac{(\pi)(2.00 \times 10^{-2})^2(2)}{4}} = 1.59 \times 10^8 \text{ N/m}^2$$

$$= 159 \text{ MPa}$$

Note: This is an example of double shear in that two cross sections of the bolt resist the force; hence, the area is the cross-sectional area of two bolts. The two bolt shear surfaces are indicated in Fig. 8.19(*a*).

To complete the computation, the tensile stress in the bar at the critical section through the bolt hole, as shown in Fig. 8.19(*b*), is

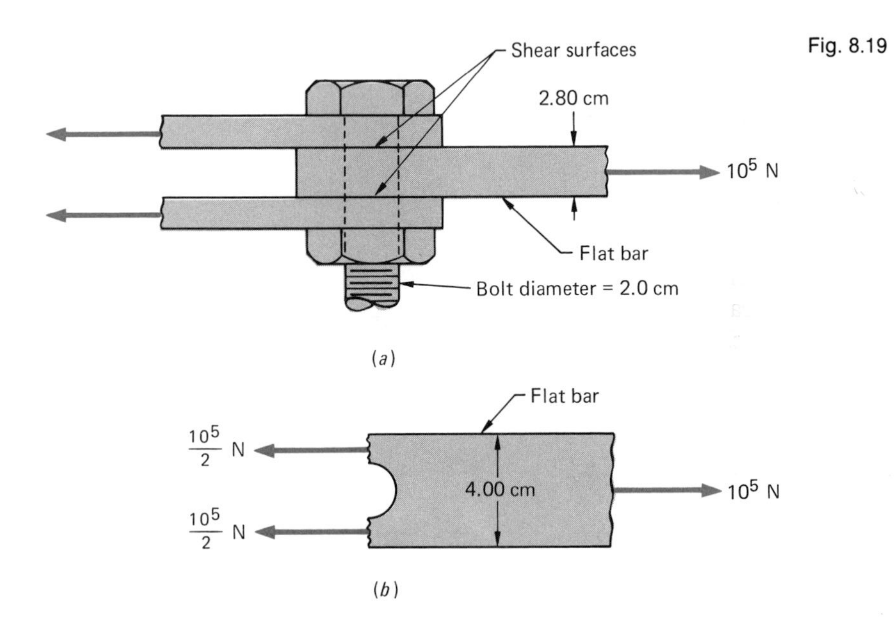

Fig. 8.19

$$\sigma = \frac{F}{A} = \frac{10^5}{(2.00 \times 10^{-2})(2.80 \times 10^{-2})}$$

$$= 179 \text{ MPa}$$

8.7 Strain

The term strain is analogous to deformation; they both refer to the change in dimension due to stress. Most materials used in engineering react to stress first elastically and then plastically. Elasticity is that property of a material that causes it to return to its original size and shape after the stress has been removed. A photograph of a tennis ball striking the strings of a racquet dramatically shows both materials undergoing strain. If you examine them closely a second or so later, you will find no evidence of the deformation. Most materials react similarly, but to varying degrees.

Plastic deformation, on the other hand, occurs when the atoms have been moved beyond their ability to return. When the stress is removed after plastic strain, the atoms attempt to return to their

Fig. 8.20 The analysis of a structure becomes more complex when numerous interconnections are employed. (*American Chain & Cable Co., Inc.*)

original positions but settle into new positions, giving the material new dimensions. Plasticity is particularly important in producing items such as wire, metal members like channels and I beams, and all extruded forms. Knowledge of the material's characteristics is essential to the successful manufacture of a large list of products.

Strain, ϵ, is the dimensionless ratio of the change in length to the original length. Using the example of stress in Sec. 8.6, if the 2.50-m-long wire had stretched to 2.515 m under the 294.3-N force, the strain would be

$$\epsilon = \frac{\Delta \ell}{\ell} = \frac{2.515 - 2.50}{2.50} = 0.006$$

Gross deformation is therefore 1.5 cm. It can be seen that if the wire had been 5.00 m long before loading, the stress and strain would have been unchanged, but the gross deformation would have been twice as much or 3 cm instead of 1.5 cm.

8.8 Modulus of elasticity

About 200 years ago, Robert Hooke discovered that strain is directly proportional to stress for elastic materials in their elastic range. Hooke's law can be written $\epsilon = K\sigma$, where K is a proportionality constant. K is usually replaced by E, the modulus of elasticity, where $E = 1/K$. Values of E for selected materials are given in Table 8.1.

8.9 Yield and tensile strengths

The proportional relationship between stress and strain is valid below a stress value known as the yield point. Figure 8.21 shows a typical stress diagram for a certain grade of steel. The yield point is referred to as a stress that causes the material to strain beyond a true elastic condition. For a limited additional stress, it is difficult to determine whether most materials are undergoing elastic or plastic strain. Then the deformation accelerates rapidly until failure (rupture) occurs.

Table 8.1 Modulus of Elasticity

	psi	MPa
Cold rolled steel	30×10^6	21×10^4
Cast iron	16×10^6	11×10^4
Copper	16×10^6	11×10^4
Aluminum	10×10^6	7×10^4
Stainless steel	27×10^6	19×10^4
Nickel	30×10^6	21×10^4

Fig. 8.21

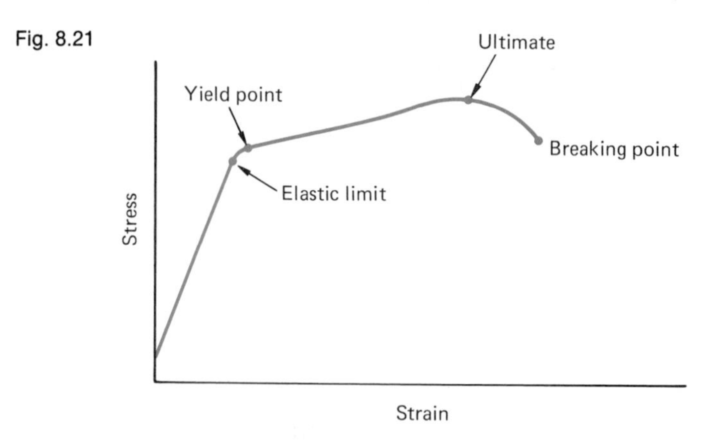

It can be seen in Fig. 8.22 not only that the material is elongated but that it is also reduced in cross-sectional area (necks down) in the region of failure. This is true of materials that are very ductile and not true of brittle materials like concrete.

8.10 Design stress

Obviously, most products or structures that engineers design are not intended to fail or become permanently deformed. The task facing the engineer is to choose the proper type and size of material to perform under the conditions likely to be imposed upon it. Since the safety of the user and the liability of the producer (and of the engineer) are dependent on valid assumptions, the engineer selects a design stress that is less than the yield point. The ratio of the yield point to the design stress is called the *safety factor*. For example, if the yield point is 420 MPa and the design stress is 140 MPa, the safety factor is 3. Care must be exercised in reporting and interpreting safety factors because they are expressed in terms of both yield strength and ultimate (tensile) strength. Table 8.2 lists

Fig. 8.22 Tensile test, showing the "necking-down" effect that occurs prior to actual failure of a steel member.

Table 8.2 Yield and Tensile Strengths

	Tensile strength		Yield strength	
	psi	MPa	psi	MPa
Cast iron	45×10^3	310	30×10^3	210
Wrought iron	50×10^3	345	30×10^3	210
Structural steel	60×10^3	415	35×10^3	240
Stainless steel	90×10^3	620	30×10^3	210
Aluminum	18×10^3	125	12×10^3	85
Copper, hard drawn	66×10^3	455	60×10^3	415

values used in structural design. It should be noted that the United States still lists most of its standards in the English system. Conversions in this and many other areas will be necessary for a decade or so.

Example problem 8.4 A round bar is 40.0 cm long and must withstand a force of 20.0 kN. What diameter must it have if the stress is not to exceed 140.0 MPa?

Solution

$$\sigma = \frac{F}{A}$$

$$A = \frac{F}{\sigma} = \frac{20.0 \times 10^3 \text{ N}}{140.0 \times 10^6 \text{ N/m}^2} \cdot \frac{10^6 \text{ mm}^2}{\text{m}^2} = 143 \text{ mm}^2$$

$$A = \frac{\pi d^2}{4}$$

$$143 = \frac{3.14 d^2}{4}$$

$$d = 13.5 \text{ mm}$$

Example problem 8.5 Assume that in Example Problem 8.4 the allowable elongation is 0.125 mm when the modulus of elasticity (E) is 2.00×10^2 GPa. Determine the diameter (mm) of the bar.
Theory:

$$E = \frac{\sigma}{\epsilon} = \frac{F/A}{\Delta\ell/\ell}$$

so,

$$\Delta\ell = \frac{F\ell}{AE}$$

$\Delta \ell$ is usually written δ.

$$\delta = \frac{F\ell}{AE}$$

so

$$A = \frac{F\ell}{\delta E}$$

Solution

$$A = \frac{(20.0 \times 10^3)\ \text{N}}{125 \times 10^{-6}\ \text{m}} \left| \frac{0.400\ \text{m}}{} \right| \frac{\text{m}^2}{2.00 \times 10^{11}\ \text{N}}$$

$$A = 320\ \text{mm}^2 = \frac{\pi d^2}{4}$$
$$d = 20.2\ \text{mm}$$

Problems

8.1 Find the resultant of two concurrent forces, one is 472 N and acts 20° east of north and the other is 117 N and acts 5°20′ north of east.

8.2 A rope is 10.00 m in length and is attached at points *A* and *B*, which are at the same elevation and 6.000 m apart. If a 50.00-kg mass is attached to the rope 4.000 m from *A*, what is the tension in the two segments of the rope?

8.3 Sam is able to pull a steel cube at constant speed by exerting a force of 50.0 N. His effort is replaced by two people, John pulling at an angle of 32° left of Sam's direction of pull and Joe at 46° right of Sam's direction. If the cube is pulled in the same direction and same speed as Sam pulled it, what are the forces exerted by John and by Joe?

8.4 In Fig. 8.23, if a single guy wire is to be secured to the wall and the hook on the left side of *C*, how long must the wire be (in centimeters)? What will be the guy wire tension?

8.5 Two workers are lifting a 50.0-kg bucket from a pit. The rope held by the first worker makes an angle of 15° with the vertical and that of the second an angle of 25° with the vertical. What is the force exerted by each worker?

8.6 What is the resultant of the force system in Fig. 8.24?

Fig. 8.23

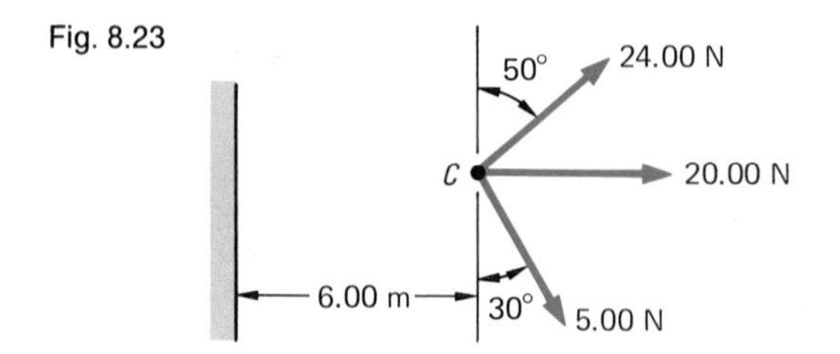

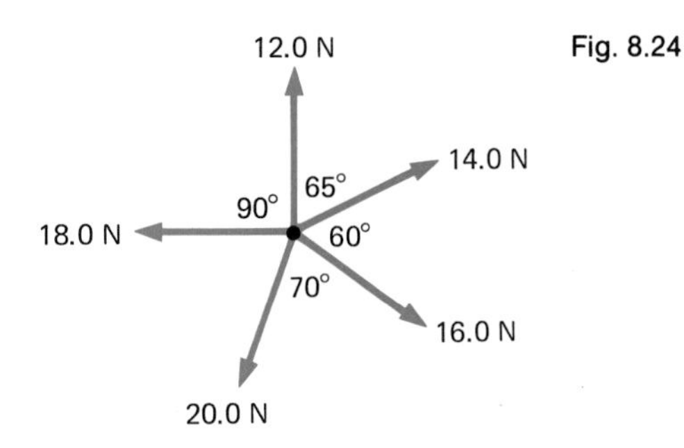

Fig. 8.24

8.7 (*a*) What is the magnitude and direction of the resultant of the force system in Fig. 8.25?
(*b*) What force must be added to the system to put it into equilibrium?

In Probs. 8.8 through 8.12, find the resultant of the concurrent, coplanar force systems. Express each answer as a magnitude and direction.

8.8 $\overline{A}$ = 52.0 N, 32°; $\overline{B}$ = 67.0 N, 105°; $\overline{C}$ = 12.5 N, 182°; $\overline{D}$ = 16.9 N, 235°; and $\overline{E}$ = 41.8 N, 327°.

8.9 $\overline{F}$ = 166.5 N, 77°; $\overline{G}$ = 62.2 N, 90°; $\overline{H}$ = 181.5 N, 312°; and $\overline{I}$ = 52.6 N, 267°.

8.10 $\overline{J}$ = 47.2 N, −37°; $\overline{K}$ = 83.6 N, 215°; $\overline{L}$ = 67.7 N, 85°; and $\overline{M}$ = 90.6 N, 122°.

8.11 $\overline{P}$ = 1.21 kN, 37°; $\overline{Q}$ = 0.194 kN, 141°; $\overline{R}$ = 472 N, 222°; $\overline{S}$ = 92 N, 307°; and $\overline{T}$ = 0.072 kN, 97°.

8.12 $\overline{U}$ = 0.013 kN, 73°; $\overline{V}$ = 77 N, 112°; $\overline{W}$ = 0.112 kN, 167°; $\overline{X}$ = 94 N, 243°; $\overline{Y}$ = 452 N, 298°; and $\overline{Z}$ = 0.498 kN, 350°.

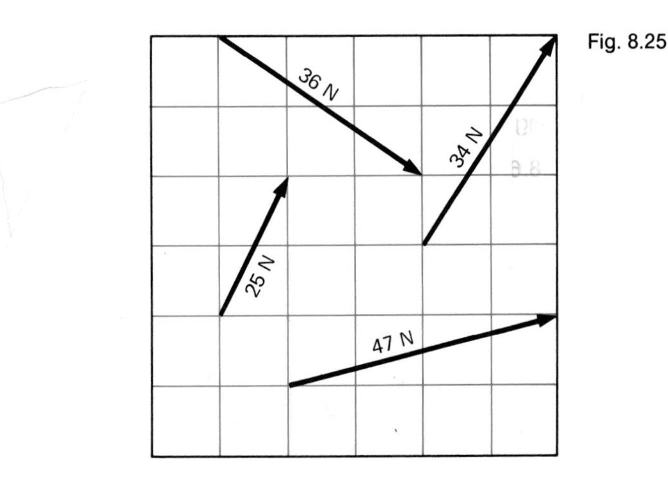

Fig. 8.25

What are the moments of the forces in Probs. 8.13 through 8.19 about point A if all forces go through 0 and make an angle θ with the horizontal? (See Fig. 8.26.)

Fig. 8.26

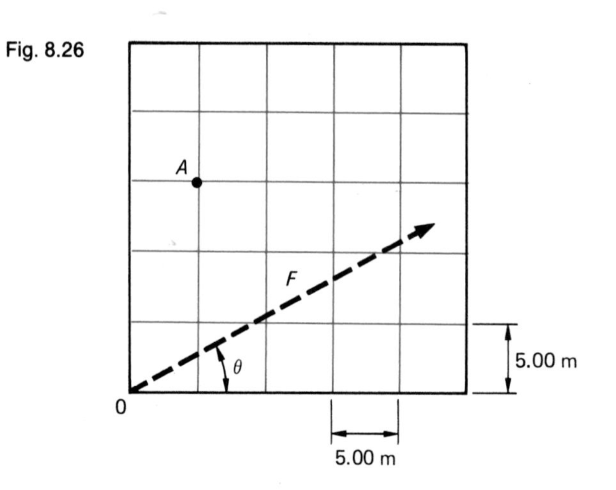

8.13 $\bar{F} = 72.2$ N, $\theta = 30°$.

8.14 $\bar{F} = 6.2$ mN, $\theta = 1.0°$.

8.15 $\bar{F} = 415$ N, $\theta = 82°$.

8.16 $\bar{F} = 40.7$ N, $\theta = 70°$.

8.17 $\bar{F} = 1.74$ N, $\theta = 8°15'$.

8.18 $\bar{F} = 1.627$ mN, $\theta = 27°10'12''$.

8.19 $\bar{F} = 49.2$ N, $\theta = 62°40'$.

Find the moments of the force systems in Probs. 8.20 to 8.23 about hinge B.

8.20 See Fig. 8.27.

Fig. 8.27

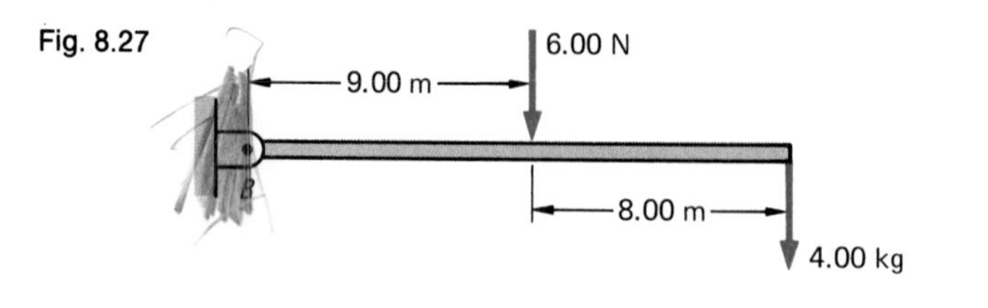

8.21 See Fig. 8.28.

Fig. 8.28

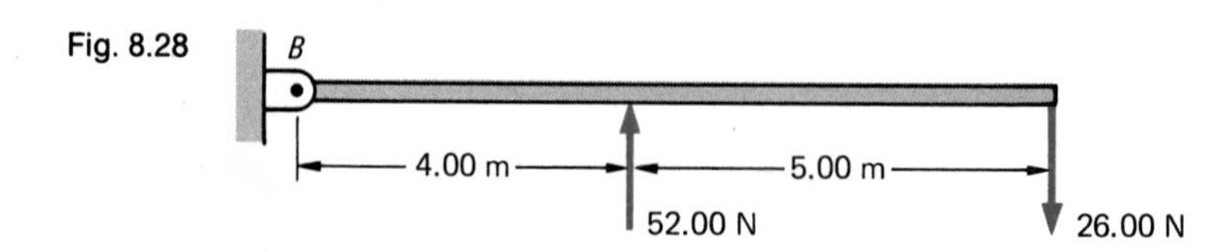

8.22 See Fig. 8.29.

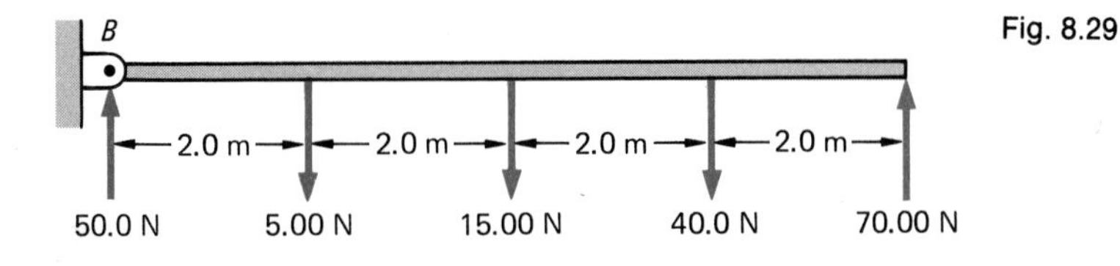

Fig. 8.29

8.23 See Fig. 8.30.

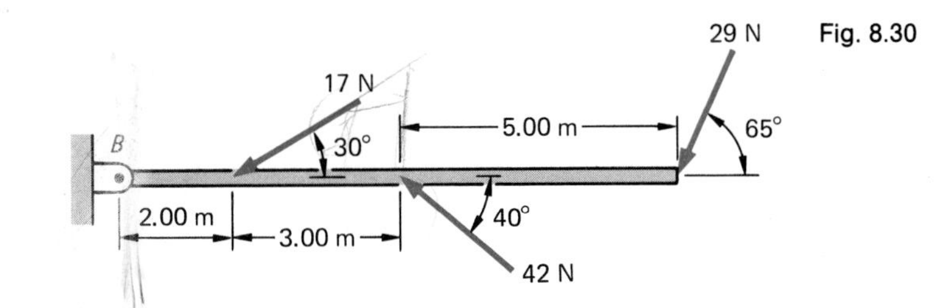

Fig. 8.30

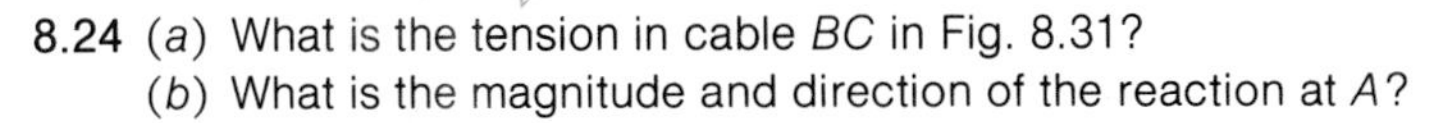

8.24 (*a*) What is the tension in cable *BC* in Fig. 8.31?
(*b*) What is the magnitude and direction of the reaction at *A*?

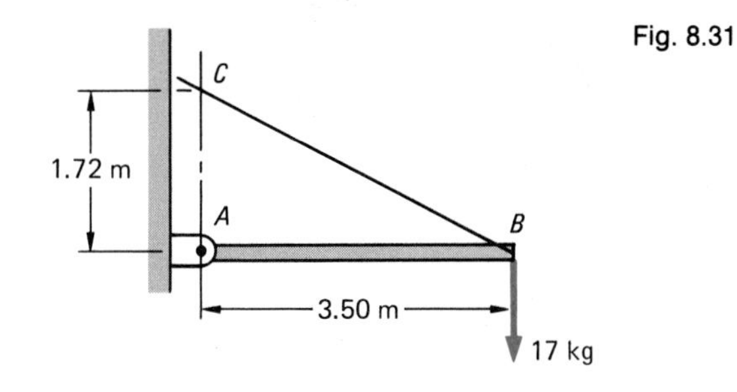

Fig. 8.31

8.25 The beam in Fig. 8.32 is 8.000 m long and has a mass of 2.00 kg/m. What are the reactions at *D* and *E*?

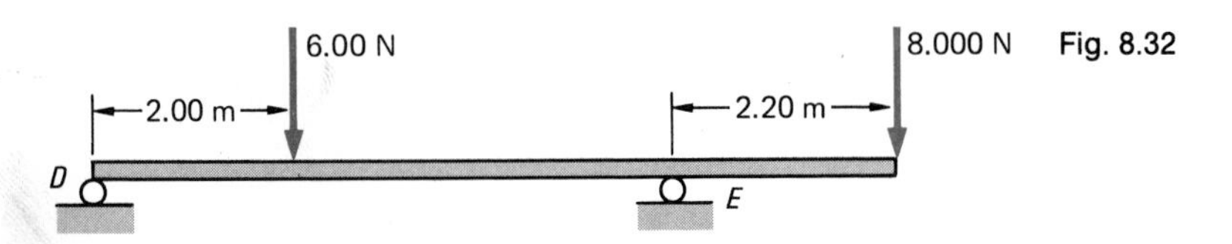

Fig. 8.32

8.26 If the beam in Fig. 8.33 is 20.00 m long and has a mass of 721 kg, what are the reactions at *G* and *H*? Give magnitude and direction.

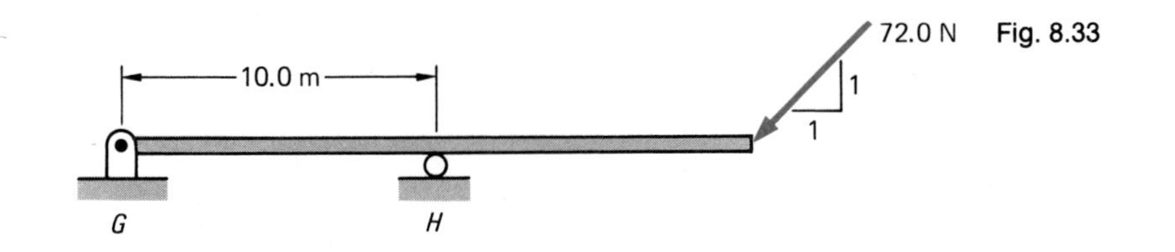

Fig. 8.33

8.27 What are the reactions at P and Q in Fig. 8.34?

Fig. 8.34

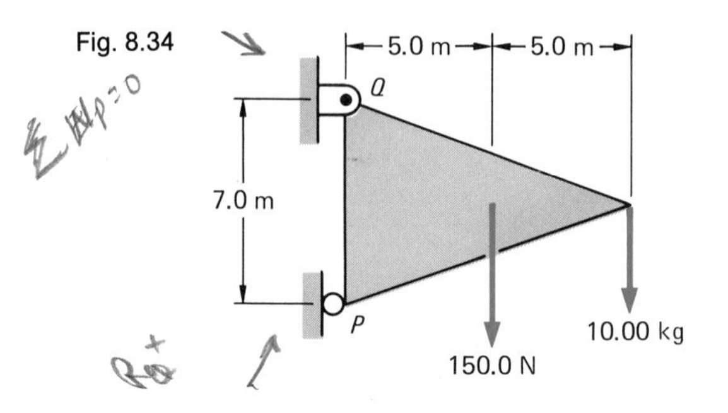

8.28 (*a*) If the beam RT in Fig. 8.35 has a mass of 1.00 kg/m, what is the tension in the cable ST?
(*b*) What is the reaction at R?

Fig. 8.35

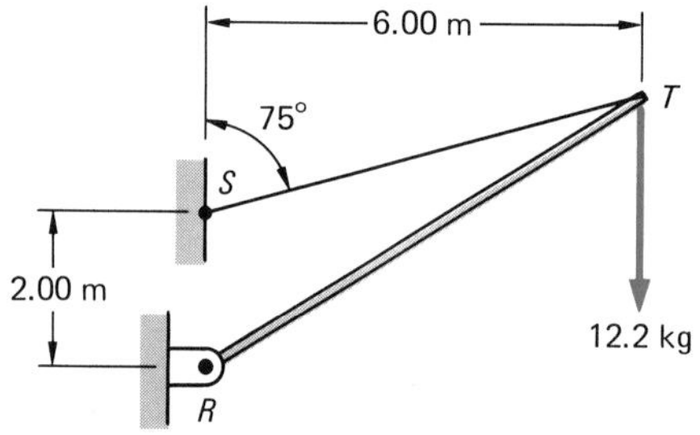

8.29 What are the reactions at A and C in Fig. 8.36?

Fig. 8.36

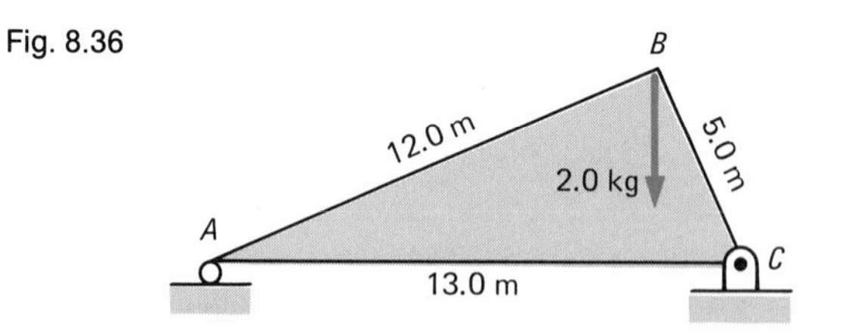

8.30 A string will break when exactly 10.00 kg is suspended from it. A 2.00-m segment of that string is connected to two hooks that are 170.0 cm apart and one hook is 10.0 cm higher than the other. What mass (minimum) attached to the string 50.0 cm from the higher hook will cause the string to break?

8.31 A hot-air balloon has a lifting force of 1.23 kN. If a single rope is used to hold it in a breeze and it makes an angle of 15° with the vertical, what is the tension in the rope?

8.32 In Fig. 8.37, what are the components of the weight perpendicular and parallel to the support surface for the following values of θ: 15°, 27°, 50°, 67°, 80°?

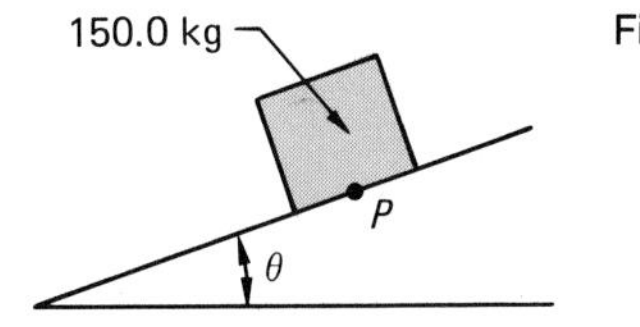

Fig. 8.37

8.33 Find the reaction at A and the cable tension. Assume the beam to be weightless. (See Fig. 8.38.)

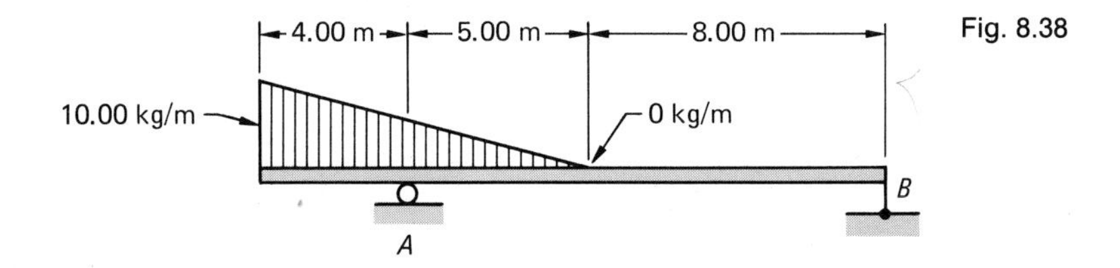

Fig. 8.38

8.34 Find the reactions at D and E in Fig. 8.39.

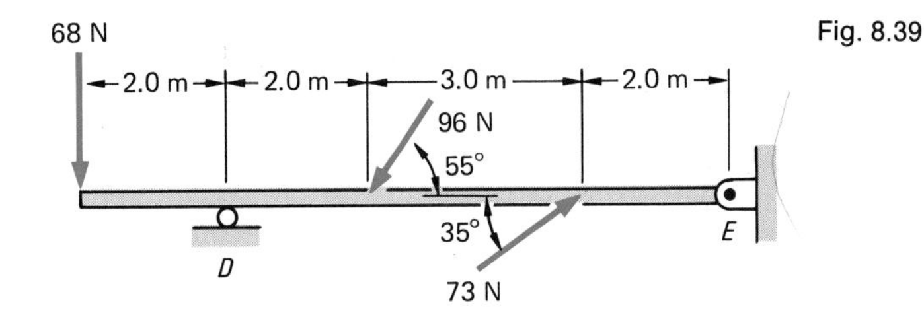

Fig. 8.39

8.35 The container in Fig. 8.40 is 10.0 cm wide and is filled with water. What are the reactions at C and D?

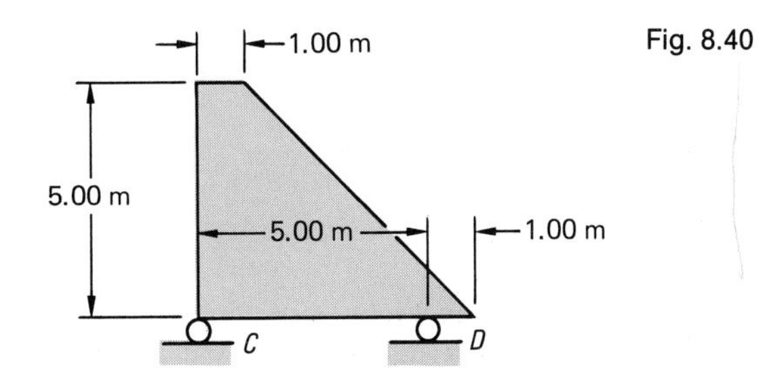

Fig. 8.40

8.36 Find the reaction at G and the cable tension. (See Fig. 8.41.)

Fig. 8.41

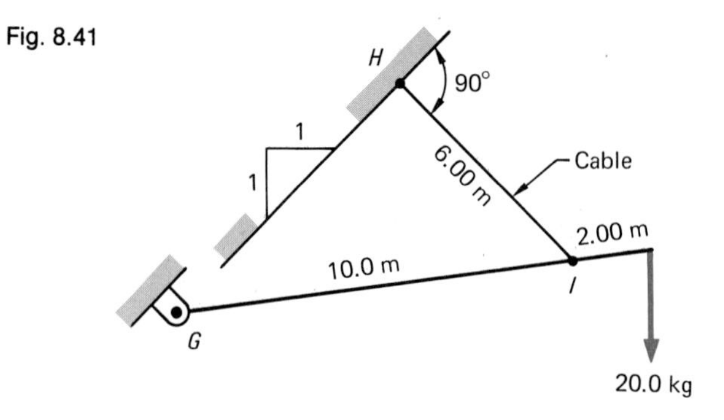

8.37 What mass must D have in order for the system in Fig. 8.42 to be in equilibrium?

Fig. 8.42

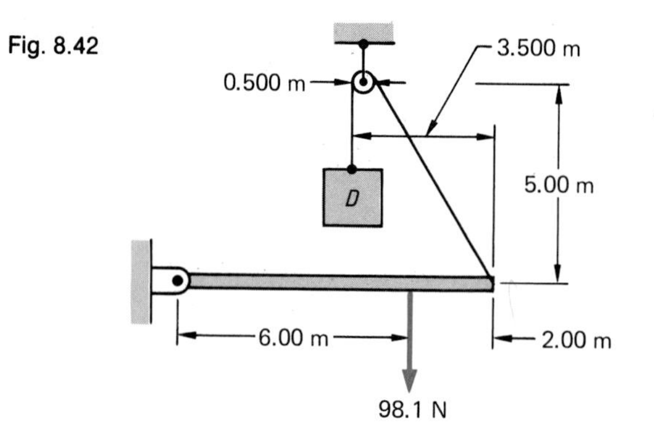

Chemistry–concepts and calculations

CHAPTER 9

9.1 Introduction

Chemistry is one of the foundation sciences upon which engineering is built. Applications of chemistry are obvious in such fields as chemical engineering, ceramic engineering, and petroleum engineering. The use of and need for chemistry in a field such as electrical engineering is somewhat less apparent. For in this field, the transistor and integrated circuit technologies rely heavily on chemistry. Nearly everywhere one looks, some portion of each product or process depends on the application of chemistry.

This chapter discusses some general chemistry concepts and calculation procedures. The coverage of topics is limited to material that is commonly used in engineering courses. The particular concepts selected for discussion are the following:

Fig. 9.1 Student engineers, as well as graduate engineers, work in chemical industries. A process engineer directs the work of an engineering cooperative education student. (*Sun Company, Inc.*)

- Absolute zero
- Temperature scales
- Atomic and molecular weights
- Avogadro's number
- Density
- Boyle's law
- Charles' law
- Equations of state
- Balance of chemical equations
- Composition calculations
- Determination of empirical formulas from composition data

9.2 Chemical concepts

9.2.1 Temperature, temperature scales, and absolute zero

The concept of temperature must be defined by examining the flow of heat from one body to another. Temperature is a physical property that determines which way heat will flow if two bodies are placed in thermal contact with one another. When bodies are in thermal contact, heat will flow from the body with the higher temperature to the one with the lower temperature. It follows then that if the objects have the same temperature, no heat will flow. Temperature can be measured on an empirical scale such as Celsius or Fahrenheit or on an absolute scale such as Kelvin or Rankine. More will be said about these scales later. It is important to point out that temperature is a fundamental, or basic, dimension and therefore cannot be expressed in terms of other fundamental dimensions such as mass, length, or time.

To state a temperature quantitatively, a reproducible scale must be developed. Two such scales, the Celsius scale and the Fahrenheit scale, are in common usage both by technical workers and by the general public. Each of these scales is derived by dividing the difference in temperature between the freezing point and boiling point of water at standard pressure (1 atm or 0.101 MPa) into a specified number of equally spaced divisions.

The Celsius scale is obtained by dividing this temperature span into 100 equal parts. In addition, the freezing point of water is defined to be zero degrees on the scale. The result is that the boiling point of water is 100°C. The scale is linear and can be extended above 100°C for higher temperatures. It can also be extended to temperatures below 0°C by using negative values.

The Fahrenheit scale divides the temperature difference between the freezing point and boiling point of water into 180 parts. Here the freezing point of water is defined to be 32°F. Calculation then gives the boiling point of water as $32 + 180 = 212$°F.

It is apparent then that the number of Fahrenheit units compared with Celsius units is 180 to 100. This makes the Celsius degree larger than the Fahrenheit degree by a factor of 9/5. The conversion from one temperature scale to the other is complicated by the fact that 0°C and 32°F represent the same physical temperature—that of freezing water or melting ice.

To convert from the Celsius scale to the Fahrenheit scale, you must use the following formula.

$$X°\text{F} = \tfrac{9}{5}Y°\text{C} + 32 \qquad 9.1$$

Conversely,

$$Y°\text{C} = \tfrac{5}{9}(X°\text{F} - 32) \qquad 9.2$$

Jacques A. C. Charles, a French physicist, discovered in the late 1780s that the volume of a fixed mass of gas held at constant pressure decreases by about $\frac{1}{273}$ of its volume at 0°C for each degree Celsius that its temperature is decreased. This suggests that at a temperature of −273°C, the volume of the gas will theoretically be reduced to zero. In the actual case, the gas will first liquefy and then solidify with decreasing temperature so that the zero-volume prediction does not actually occur.

This "zero-volume" temperature point is also the temperature where all molecular activity would theoretically cease as the temperature is lowered. (Higher temperature implies greater molecular activity, and vice versa.)

Absolute zero is thus defined in terms of an ideal gas that would follow the same laws near the zero molecular activity point as real gases follow near room temperature. The accepted value of absolute zero is −273.15°C. The corresponding point on the Fahrenheit scale is

$$\tfrac{9}{5}(-273.15°\text{C}) + 32 = -459.67°\text{F}$$

A new set of temperature scales, *absolute temperature scales*, use the absolute zero temperature as the zero point of the scales. Two such scales are commonly used. They are the Kelvin scale based on the Celsius degree and the Rankine scale based on the Fahrenheit degree. Thus the unit size on the Kelvin scale is identical with the Celsius degree; and the unit size on the Rankine scale is identical with the Fahrenheit degree.

1 Kelvin unit = 1 Celsius degree

1 Rankine unit = 1 Fahrenheit degree

The conversion from the Celsius scale to the Kelvin scale is given by

$$Y\,\text{K} = X°\text{C} + 273.15 \qquad 9.3$$

and that from the Fahrenheit scale to the Rankine scale is

$$B°\text{R} = A°\text{F} + 459.67 \quad 9.4$$

Because the Kelvin and Rankine scales use the same zero temperature point, the conversion between them is

$$\text{Kelvin temperature} = \tfrac{5}{9}\,(\text{Rankine temperature}) \quad 9.5$$

A comparison of the four temperature scales discussed here is shown graphically in Fig. 9.2.

Note that in SI units, the kelvin is the base unit for temperature. It is not capitalized nor is it preceded by the symbol for degrees (°). If the unit is abbreviated, it is capital K—again without the degree symbol. For example,

100 kelvins = 100 K

Conventional practice is to use the degree symbol with the other temperature units, such as 20°C, 72°F, or 320°R.

Example problem 9.1 Determine normal body temperature, 98.6°F, on the Celsius, Kelvin, and Rankine scales.

Solution

$$\tfrac{5}{9}(98.6°\text{F} - 32) = 37°\text{C}$$

$$37.0°\text{C} + 273.15 = 3.10 \times 10^2\ \text{K}$$

$$98.6°\text{F} + 459.67 = 5.58 \times 10^2\,°\text{R}$$

9.2.2 Atomic weight and moles

In 1805, John Dalton proposed an atomic theory that all matter is made up of particles called atoms, with atoms of the same element being all alike but differing from atoms of another element. Later

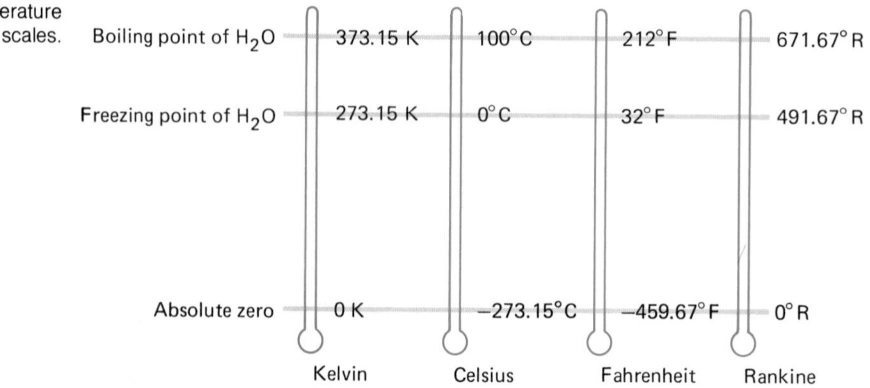

Fig. 9.2 Comparison of temperature scales.

developments have shown that even atoms of the same element may differ. In fact, over three-fourths of all naturally occurring chemical elements are mixtures of between two and ten different kinds of atoms. The atoms of different weight belonging to a given chemical element are called *isotopes* of that element. For example, hydrogen is made up of three isotopes; protium (common hydrogen), deuterium (heavy hydrogen), and tritium.

To better understand isotopes requires an understanding of the basic composition of atoms. *Atoms* are composed of positively charged particles (protons), negatively charged particles (electrons), and electrically neutral particles (neutrons). Electrons, protons, and neutrons are always the same, no matter what element they are a part of; but the number of each of these particles varies from one element to another. The electrons form a revolving cloud around a nucleus of protons and neutrons. The mass of each electron is extremely small compared with the mass of either a proton or a neutron, so that the mass of the nucleus is nearly the same as the mass of the entire atom.

Each isotope of an element contains the same number of electrons and protons, but the number of neutrons differs. This is the factor that causes differences in the weights of the isotopes of the same element.

The number of protons in the nucleus of an atom is called the *atomic number*. The total number of protons and neutrons in the nucleus is the *mass number*. Obviously, the difference between the mass number and the atomic number is the number of neutrons present. From these definitions, isotopes of an element differ in mass number but have the same atomic number. In order to distinguish between the different isotopes of an element, the mass number is added to the elementary symbol of an element as a superscript on the right or left. The symbol ^{12}N or N^{12} represents the nitrogen atom of mass number 12. N^{13}, N^{14}, and N^{15} refer to other isotopes with larger mass numbers (more neutrons in the nucleus).

Tabulated values of the masses of isotopes or elements are not given in absolute mass units such as grams because numerical factors of 10^{-22}, 10^{-23}, and so forth, would have to be used because of the extremely small size of an atom. Masses are instead given in relative units, called *unified atomic mass units*. One unified atomic mass unit is defined to be $\frac{1}{12}$ of the mass of a carbon 12 atom. Therefore, the mass of C^{12} is given as 12. Using this scale, the mass of the nitrogen 14 atom has been measured to be 14.003 07. A tabulation listing the isotope masses of the elements is referred to as a table of nuclidic masses.

Chemical calculations that require knowledge of the masses of the atoms involved normally do not utilize the masses of isotopes.

Usually chemical reactions don't depend on which isotope is present because isotopes have very similar chemical properties. Therefore, since most elements occur in nature (and likewise in chemical reactions) as a mixture of isotopes, the mass used for chemical calculations is a weighted average of the masses of the isotopes involved. The average mass of the atoms of an element is also given in unified mass units. These tabulations of masses are alternately called *relative atomic weights, chemical atomic weights*, or *atomic weights*. It is important to distinguish between nuclidic masses and atomic weights. Atomic weights will be used for calculations in this text.

The extremely small mass of an individual atom makes it inconvenient to make calculations for each atom. Moreover, chemical experiments usually involve many, many atoms. This would suggest that a mass unit be defined which is of the order of magnitude of the mass that might be present in a practical experiment.

The amount of material contained in 12 g of C^{12} is taken as a basis. It is called a *mole*. A mole of any other substance has exactly the same number of atoms that are present in 12 g of C^{12}. That number of atoms, *Avogadro's number*, has been experimentally determined to be 6.023×10^{23}. Therefore, a mole of 0^{16} or a mole of H^1 each contains 6.023×10^{23} atoms, although it is obvious that the mole of 0^{16} has a larger mass than the mole of H^1 because each oxygen atom is more massive than each hydrogen atom.

The definition of 12 g as the mass of one mole of C^{12} makes it possible to calculate the size of the unified mass unit. Since C^{12} has a mass of 12 unified mass units and a mole of C^{12} has a mass of 12 g that contains 6.023×10^{23} atoms, one unified mass unit must be $1/(6.023 \times 10^{23})$ g $= 1.660 \times 10^{-24}$ g.

It is simple to calculate the mass of a mole of any element because the mass of a mole of the element in grams is numerically equal to its atomic weight in unified mass units. Therefore, a mole of H has a mass of 1.007 97 g because 1.007 97 is the atomic weight of hydrogen. A mole of 0 is then 15.9994 g of oxygen. For this reason, atomic weights are sometimes shown in units of grams per gram-atom or grams per mole.

9.2.3 Chemical formulas and equations

Each element is identified by an elementary symbol. When elements react to form new substances, combinations of these symbols (formulas) are used to describe the new substances. An *empirical formula* shows the simplest combination of elements that form the substance. Ordinary water has the empirical formula H_2O; that is, two atoms of the element hydrogen react with one atom of the element oxygen to become water, which has properties quite different from either of the parent elements.

The empirical formula for hydrogen peroxide is HO, meaning that for each atom of hydrogen there is also one atom of oxygen. This, however, is not the formula normally associated with hydrogen peroxide. Rather, the *molecular formula*, H_2O_2, is used, because it gives additional information about the substance. It specifies that the molecule of hydrogen peroxide is actually made up of two atoms each of hydrogen and oxygen. The molecular formula is always a whole number multiple (1, 2, 3, . . .) of the empirical formula.

The weight of the atoms shown in the molecular formula is called the *molecular weight*. It is determined by adding up the atomic weights of each of the atoms involved. In the case of hydrogen peroxide, the molecular weight is $2(1.007\,97) + 2(15.9994) = 34.0147$. A mole of H_2O_2 contains 6.023×10^{23} molecules and therefore has a mass of 34.0147 g (the molecular weight expressed in grams).

In general, a mole of any substance is a collection of 6.023×10^{23} entities, whether they be atoms, molecules, ions, and so on. When the mass of a mole is expressed in grams, it is referred to as a gram-mole.

Chemical reactions take place when many types of substances are brought together under appropriate conditions. The process of combining substances (called reactants) into new substances (called products) can be shown by writing a chemical equation. Convention dictates that the reactants be placed on the left side of the equation and the products on the right side. Arrows are used to separate the two sides and show the direction in which the reaction is occurring. Arrows directed both ways means that some of the products are themselves reacting to form the reactants anew. Chemical formulas represent all substances involved in the reaction; and the coefficients ahead of the formulas specify the relative number of molecules of each substance required.

An example is the exposure of iron to oxygen to form ferric oxide, commonly known as rust.

$$4Fe + 3O_2 \rightarrow 2Fe_2O_3$$

This chemical formula illustrates that both iron (Fe) and oxygen (O_2) must be present as the reactants to form, in this case, the single product ferric oxide (Fe_2O_3). In addition, the arrow to the right shows that the reaction takes place only in that direction (and is not reversible). Also in the equation are coefficients explaining that 4 atoms of Fe and 3 molecules of O_2 combine to form 2 molecules of Fe_2O_3.

The equation does not mean that every iron atom will combine with an appropriate number of oxygen atoms to form ferric oxide,

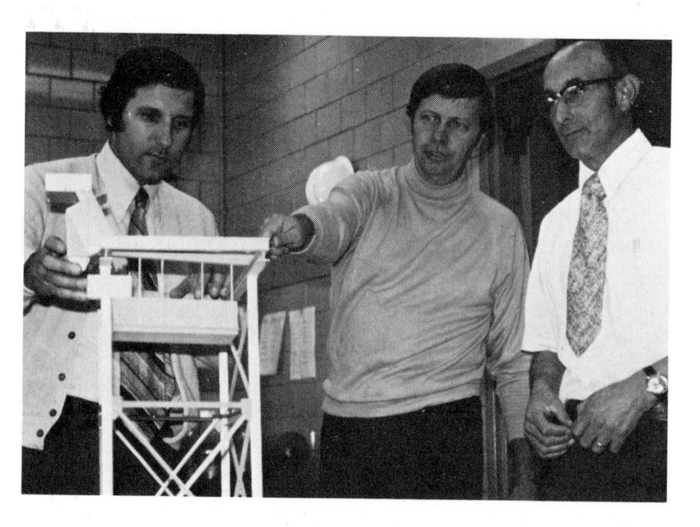

Fig. 9.3 In projects that deal with chemical processes, models are often built to illustrate the physical relationships among the components of the process. These engineers are examining a portion of the model of an experimental machine that reduces the sulfur content of coal. (*Ames Laboratory, U.S. Department of Energy.*)

but rather that if the reaction does take place, the reaction shown will occur. Everyone is aware that iron parts will rust on the surface, but iron atoms in the interior may not react. Moreover, the equation does not give any information about how rapidly the reaction occurs. Some reactions take years to complete, whereas others occur so rapidly that they are explosive. What the equation above does say is that if large quantities of Fe and O_2 are brought together, some of the iron and oxygen will react to form Fe_2O_3.

9.2.4 Density

The density of a substance is a measure of the quantity of material contained in a given volume. The definition is

$$\text{Density} = \frac{\text{mass of substance}}{\text{volume of substance}} = \frac{\text{mass}}{\text{unit volume}} \tag{9.6}$$

Typical units for density would be kilograms per cubic meter (kg/m^3) in SI units or slugs per cubic foot ($slug/ft^3$) in the British system.

Example problem 9.2 What is the density of cast iron if 3.00 m^3 has a mass of 21.6 Mg?

Solution

$$\text{Density} = \frac{\text{mass of cast iron}}{\text{volume of cast iron}} = \frac{21.6\ \text{Mg}}{3.00\ \text{m}^3} = 7.20\ \text{Mg/m}^3$$

Example problem 9.3 If the density of water is 1.000 Mg/m^3, what is the mass of 4 cm^3 of water?

Solution

$$\text{Mass}_{H_2O} = (\text{density}_{H_2O})(\text{volume}_{H_2O})$$

$$= \frac{1.000 \times 10^6\ \text{g}}{\text{m}^3} \cdot \frac{4\ \text{cm}^3}{} \cdot \frac{\text{m}^3}{(1.0 \times 10^2)^3\ \text{cm}^3}$$

$$= 4\ \text{g}$$

Another term used in the measure of density is *specific gravity*, which is the ratio of the mass of a substance to the mass of an equal volume of a standard substance. Water at 4°C is typically taken as the standard for comparison of liquids and solids. The density of water at this temperature is about 1.000 Mg/m^3. Thus, if the specific gravity of a solid or liquid is known, its density can be found by simply multiplying the specific gravity by 1.000 Mg/m^3.

Specific gravity has one advantage in that its value is independent of the system of units being used. Density can be found in any set of units if the density of water is known in those units.

Example problem 9.4 The specific gravity of cast iron is 7.20. What is its density in kilograms per cubic meter?

Solution

$$\text{Density of cast iron} = (\text{specific gravity})(\text{density of water})$$

$$= (7.20)(1.000\ \text{Mg/m}^3)$$

$$= 7.20\ \text{Mg/m}^3$$

$$= 7.20 \times 10^3\ \text{kg/m}^3$$

9.2.5 Perfect-gas relationships

The volume and temperature of a gas is much more sensitive to the pressure exerted on it than is that for a liquid. For many calculations with liquids, the change in pressure can be ignored as long as the pressure is near atmospheric pressure. For gases, however, even small pressure changes can result in significant volume or temperature changes.

The definition of pressure is

$$\text{Pressure} = \frac{\text{force acting on a surface}}{\text{area of surface}} \qquad 9.7$$

Pressure as a force per unit area is typically given in newtons per square meter (N/m^2) or pounds per square inch (lb/in^2). Normal sea-level atmospheric pressure is 0.101 MN/m^2 or 14.7 lb/in^2–the pressure exerted by the column of air above a 1-m^2 or 1-in^2 area of the earth at sea level, repectively. 1 N/m^2 is called a pascal (Pa).

A perfect gas is an idealized behavioral model of a real gas. All gases follow the ideal model at sufficiently low pressures and high temperatures. Although the pressure and temperature range where ideal gas behavior is expected will not be defined numerically here (it depends on the gas, among other things), it can be said that the ideal range is found at conditions removed from the point where the gas becomes a liquid (high pressure, low temperature, or a combination of these). The ideal range is also restricted on the high-temperature side below the temperature where dissociation (breakup of molecules) or ionization (removal of electrons from atoms) begin to occur. Fortunately, the pressures and temperatures where gases are used in many engineering applications are in the ideal behavioral range for the gases.

Three basic laws are used to describe the behavior of a perfect or ideal gas: Boyle's law, Charles' law, and Gay-Lussac's law.

Boyle's law states that pressure and volume are inversely proportional when the temperature of a given mass of gas is held constant. In equation form,

$$P_1 V_1 = P_2 V_2 \qquad 9.8$$

where P_1 and V_1 are the pressure and volume of the gas at state 1 and P_2 and V_2 are the pressure and volume at state 2.

Charles' law relates the temperature and volume of a fixed mass of a perfect gas undergoing a constant-pressure process.

$$\frac{V_1}{T_1} = \frac{V_2}{T_2} \qquad 9.9$$

For a constant-pressure process, an increase in temperature T results in a proportional increase in volume V.

Gay-Lussac's law establishes the relation between pressure and temperature for a constant-volume process.

$$\frac{P_1}{T_1} = \frac{P_2}{T_2} \qquad 9.10$$

Note carefully that the temperature in the perfect-gas laws *must* be expressed as absolute temperature (Kelvin or Rankine scales). Any temperature given in degrees Celsius, or degrees Fahrenheit must first be converted to the appropriate absolute scale before being substituted into the gas-law equations.

The gas laws stated above can be combined and a *general perfect-gas law* can be found which allows for variations in pressure, temperature, and volume, as follows:

$$\frac{P_1 V_1}{T_1} = \frac{P_2 V_2}{T_2} \tag{9.11}$$

or more generally, for a given mass of gas,

$$\frac{PV}{T} = \text{constant} \tag{9.12}$$

9.3 Chemical calculations

Several calculation techniques will be described which use the concepts introduced in Sec. 9.2 to achieve engineering results for practical problems.

9.3.1 Method of balancing chemical equations

Balancing a chemical equation is simply an application of the *law of conservation of mass.* Stated in terms relating to chemical equations, it is as follows: the total number of atoms of each element represented among the reactants must equal the total number of atoms of the same element found in the products.

The balancing process can begin only after you have a skeleton equation. That is, you must know which products are produced from a given set of reactants. Then unknown numbers can be assigned as the coefficients of each reactant and each product. The problem reduces to the determination of this set of unknown numbers such that the number of atoms of each element is conserved and the coefficients are the set of smallest integers that can be found.

The process of producing elemental iron in a blast furnace is given by the following skeleton equation:

$$Fe_2O_3 + CO \rightarrow Fe + CO_2$$

In this form, the proper reactants and products are shown but the relative number of molecules of each needed is not correct, because the law of conservation of mass has been violated. The equation shows two Fe atoms on the left side and only one on the right. Likewise, there are four oxygen atoms on the left and two on the right.

Balancing this or any other equation is a bit of an art, often requiring a trial-and-error process. First, assign an unknown set of numbers as the coefficients of all reactants and products.

$$a\ Fe_2O_3 + b\ CO \rightarrow c\ Fe + d\ CO_2$$

As a general rule, it is best to begin by looking at the most complex molecule. In this case, it is the Fe_2O_3 molecule. Note that there are a minimum of two atoms of Fe among the reactants because a

cannot be smaller than 1. Then as a first trial, choose $a = 1$, which means that $c = 2$, to provide a balance in Fe atoms. It is apparent that carbon atoms appear in CO on the reactant side and CO_2 on the product side only. Therefore, it can be concluded that b and d must be equal.

Selection of $b = 1$ results in four O's on the left side and two on the right. With $b = 2$, there are five O's on the left side and four on the right; but with $b = d = 3$, there are six oxygen atoms on both the left and right sides. The balanced equation is then

$$(1)Fe_2O_3 + (3)CO \rightarrow (2)Fe + (3)CO_2$$

A somewhat more systematic approach to the same problem is to write the set of algebraic equations in the coefficients found by the application of the law of conservation of mass. The equations resulting from a balance of each element are:

$$\text{Fe:} \quad 2a = c$$

$$\text{C:} \quad b = d$$

$$\text{O:} \quad 3a + b = 2d$$

There are three equations in four unknowns, so a choice must be made for one of the unknowns. (The extra unknown is merely a result of the fact that one may multiply through a chemical equation by any number without altering the validity of the equation.) As before, select $a = 1$, then $c = 2a = 2$. Combining the C and O equations results in

$$3a + d = 2d$$

$$d = 3a = 3$$

$$b = d = 3$$

Therefore, the set is $a = 1, b = 3, c = 2$, and $d = 3$, as previously determined.

The number set will not always come out as integers on the first try as it did this time. Just remember, however, that if each coefficient is multiplied by the same number, the relative results are not changed. So, it is always possible to find a multiplier that will give a final set of integers. Finding the multiplier is really the trial-and-error part of the solution.

The simple reaction of hydrogen and oxygen producing water is given by

$$H_2 + O_2 \rightarrow H_2O \qquad \text{(unbalanced)}$$

Assigning coefficients gives

$$r\,H_2 + s\,O_2 \rightarrow t\,H_2O$$

The balanced equations are

$$\text{H:} \quad 2r = 2t$$
$$\text{O:} \quad 2s = t$$

If t is selected as 1, then the equations above give

$$s = \tfrac{1}{2}$$
$$r = t = 1$$

The equation is then

$$(1)\text{H}_2 + (\tfrac{1}{2})\text{O}_2 \rightarrow (1)\text{H}_2\text{O}$$

The coefficients are not all integers, but multiplication of each coefficient by 2 provides the final balanced chemical equation

$$(2)\text{H}_2 + (1)\text{O}_2 \rightarrow (2)\text{H}_2\text{O}$$

9.3.2 Determination of the mass of a constituent from the molecular formula

The molecular formula for a compound specifies the relative numbers of each element contained in the molecule. The atomic weights of the elements are known, so it is possible to calculate the mass of a specified element in a given amount of the compound. Likewise, it is possible to determine the percentage composition by weight (mass) from the formula for the compound and the atomic weights of the elements involved.

Fig. 9.4 A pilot plant may be necessary to demonstrate the feasibility of a new process before full-scale equipment is built as here in the petroleum industry. (*Phillips Petroleum Company.*)

Example problem 9.5 Calculate the mass of potassium in a metric ton (megagram) of potassium permanganate.

Solution

The formula of the compound is $KMnO_4$. The molecular weight is given by

$$\overset{K}{39.102} + \overset{Mn}{54.9380} + \overset{O_4}{4(15.9994)} = 158.02$$

The fraction of $KMnO_4$ that is due to the element potassium is

$$\frac{39.102}{158.02} = 0.247\ 45$$

The amount of potassium in 1.000 Mg (metric ton) of $KMnO_4$ is

$$(1.000 \times 10^3\ \text{kg})(0.247\ 45) = 247.4\ \text{kg}$$

Example problem 9.6 Glycerin is a common substance made of carbon, hydrogen, and oxygen atoms. Its formula is $C_3H_8O_3$. Determine the percentage by weight (mass) of each constituent.

Solution

$$\text{Molecular weight} = \overset{C_3}{3(12.011\ 15)} + \overset{H_8}{8(1.007\ 97)} + \overset{O_3}{3(15.9994)}$$

$$= 92.0954$$

$$\text{Percentage of carbon} = \frac{\text{weight of carbon}}{\text{molecular weight}}(100\%)$$

$$= \frac{3(12.011\ 15)}{92.0954}(100\%)$$

$$= 39.126\%$$

$$\text{Percentage of hydrogen} = \frac{\text{weight of hydrogen}}{\text{molecular weight}}(100\%)$$

$$= \frac{8(1.007\ 97)}{92.0954}(100\%)$$

$$= 8.756\%$$

$$\text{Percentage of oxygen} = \frac{\text{weight of oxygen}}{\text{molecular weight}}(100\%)$$

$$= \frac{3(15.9994)}{92.0954}(100\%)$$

$$= 52.118\%$$

The sum of each constituent percentage = 39.126 + 8.756 + 52.118 = 100.00 percent. This provides a check on the percentage calculations, since the sum must be 100 percent.

The last example could have been asked in terms of the fraction of each constituent element giving 0.391 26, 0.087 56, and 0.521 18 for carbon, hydrogen, and oxygen, respectively.

9.3.3 Determination of the empirical formula of a compound from its percentage composition

Various techniques are available to determine which elements are present in an unknown compound and to find the weights (masses) of each of them in a given amount of the compound. This data can then be used to calculate the empirical formula for the compound. The weight (mass) data may be given in either percent of total or in number of grams of each element.

A procedure for finding the empirical formula is as follows:

1. Choose an amount of compound to use as a basis for the calculations if the data are given in percentage terms. (100 g is an excellent choice.) Use the actual amount if amounts of constituents are given in grams, pounds, etc.
2. List the amount of each element per 100 g or per actual amount if known.
3. Determine and list the atomic weight of each element in grams per gram-atom.
4. Calculate the ratio

$$\frac{\text{mass of element}}{\text{atomic weight of element}}$$

for each element present. (The result is the relative number of gram-atoms of each element.)

5. Choose the smallest ratio found in step 4 and divide all ratios by this number.
6. Multiply all quantities calculated in step 5 by a constant, if each is not already an integer, in order to obtain all integer values. Note that if there are inaccuracies in the quantitative analysis or the calculations, some rounding of figures may be necessary at this point. Be certain to carry out all computations to as many significant figures as possible and reserve the rounding process for the last step of the procedure.
7. Write the empirical formula from the relative gram-atoms of the elements.

Example problem 9.7 Determine the empirical formula for the compound made up of 88.82% oxygen and 11.18% hydrogen.

Solution Select 100 g of compound as a basis.

Element	Amount (g/100 g)	Atomic weight	gram-atoms	Relative gram-atoms (g-atoms/5.5515)
O	88.82 g	15.999 4	5.5515	1.000
H	11.18 g	1.007 97	11.1711	2.012

Round off the number of relative gram-atoms to 1 and 2. The empirical formula is H_2O.

Example problem 9.8 A compound has been found to contain the following percentages of its elements: Sulfur = 24.52%, oxygen = 48.96%, and chromium = 26.52%. What is its empirical formula?

Solution Again select 100 g of compound as a basis with which to perform the calculations.

Element	Amount (g/100 g)	Atomic weight	gram-atoms	Relative gram-atoms (g-atoms/0.5100)
S	24.52 g	32.064	0.7647	1.4994
O	48.96 g	15.9994	3.0601	6.0002
Cr	26.52 g	51.996	0.5100	1.0000

In this case, the relative gram-atoms are not integer values, but if each is multiplied by 2, the S:O:Cr relative values after rounding become 3:12:2. The empirical formula is then written $Cr_2S_3O_{12}$, or $Cr_2(SO_4)_3$.

Fig. 9.5 This desaltation plant in the West Indies, shown during construction (right) and after completion (facing page), demonstrates the kind of impact that engineers who are interested in chemical processes can have on people throughout the world. (*Stanley Consultants.*)

Example problem 9.9 A compound has been experimentally found to contain 68.97 g sodium, 3.024 g hydrogen, 96.19 g sulfur, and 192.0 g oxygen. What is its empirical formula?

Solution In this case, it is unnecessary to assume a basis on which to work because the actual quantities of each element are already available.

Element	Amount, g	Atomic weight	gram-atoms	Relative gram-atoms (g-atoms/3.000)
Na	68.97	22.989 8	3.000	1.000
H	3.024	1.007 97	3.000	1.000
S	96.19	32.064	3.000	1.000
O	192.0	15.999 4	12.000	4.000

The empirical formula for the compound is $NaHSO_4$.

Problems

9.1 What is the equivalent of 72°F on the Celsius and Kelvin scales?

9.2 Convert 600 K to Fahrenheit and Celsius degrees.

9.3 Express the following temperatures in degrees Celsius:
(*a*) 200 K (*b*) 127°F (*c*) 460°R (*d*) 3 000°F
(*e*) 29 K (*f*) 29°F (*g*) 29°R (*h*) 100°F

9.4 Express the following temperatures in kelvins:
(*a*) 200°C (*b*) 460°R (*c*) −4°F (*d*) 4 000°R (*e*) 72°F

9.5 Express the following temperatures in degrees Rankine:
(*a*) 72°F (*b*) −20°C (*c*) 300 K (*d*) 0°F

9.6 Express the following temperatures in degrees Fahrenheit:
(*a*) −20°C (*b*) 200°C (*c*) 440°R (*d*) 300 K

9.7 How many atoms are there in 1 mg of NaCl?

9.8 Determine the formula weight of
(*a*) H_2O (*b*) C_6H_6 (*c*) K_2CO_3 (*d*) $Ba(OH)_2$

9.9 What is the molecular weight of each of the following compounds:
(*a*) CH_2O (*b*) $C_6H_{12}O_6$ (*c*) $(NH_4)_3PO_4$ (*d*) $Al_2(SO_4)_3$

9.10 What is the mass of one mole of
(*a*) H_2 (*b*) H_2SO_4 (*c*) $C_{12}H_{22}O_{11}$ (*d*) $NaHCO_3$

9.11 How many moles of each of the following are contained in 100 g of the pure substance?
(*a*) H_2O (*b*) CH_4 (*c*) C_8H_{18}
(*d*) $CaCO_3$ (*e*) $MgSO_4$ (*f*) $H_3C(CH_2)_{14}CH_2OH$

9.12 In the English system, the density of water is 62.4 lb/ft_3. What is the density of water in grams per cubic centimeter?

9.13 What is the density of aluminum if 60.0 cm^3 has a mass of 162.0 g?

9.14 Determine the weight of 1.0 gal of gasoline if gasoline has a density of 0.71 g/cm^3.

9.15 Calculate the mass of 1.00 m^3 of a substance whose specific gravity is: (*a*) 1.98, (*b*) 2.70, (*c*) 0.178, (*d*) 0.5.

9.16 A part weighs 0.50 N in air and 0.30 N when immersed in water. What is the specific gravity of the part?

9.17 A bottle that weighs 120 N empty weighs 260 N when filled with water and 243 N when filled with ammonia. What is the specific gravity of ammonia? What is the capacity of the bottle?

9.18 What will be the final volume of 2.00 L of a gas at 2.00-atm pressure if the pressure is increased to 6.00 atm? The temperature does not change.

9.19 Calculate the temperature of a fixed volume of gas after its pressure has been increased from 2.57 MPa to 2.96 MPa. Its initial temperature was 20°C.

9.20 Oxygen is heated in a constant-pressure process from 25°C to 170°C. The initial volume was 4.66 m^3. What is the final volume?

9.21 Hydrogen gas can be cooled from 72°F to −20°C by a constant-volume process. The original pressure of 14.7 lb/in^2 has been changed to what pressure in pascals?

9.22 Ammonia occupies 8.9 m^3 at 2.05 MPa. What is its volume at standard pressure? The process is one of constant temperature.

9.23 A constant-pressure process is performed on 6.0 L of neon that results in a volume increase of 2.0 L. What is the final gas temperature if it was initially 25°C?

9.24 Nitrogen is contained in an 11-L tank at 4.2-atm pressure and 100°F. It is allowed to expand into a chamber where the pressure is 1.0 atm and the temperature is 72°F. What volume does the nitrogen occupy now?

9.25 Air, initially at 25°C and 1.0-atm pressure, is compressed to one-half of its original volume. If the new pressure is 1.75 atm, what is the new tempera'ure?

9.26 A gas sample has a volume of 0.54 L measured at 97°C and 0.53 atm. What is its volume at 0°F and 14.7 lb/in^2?

9.27 If a football is inflated to an absolute pressure of 21 lb/in^2 in a room that has a temperature of 25°C, what will the pressure in the ball be during the game at 5°C?

9.28 Balance the following equations:

(*a*) $BCl_3 + P_4 + H_2 \rightarrow BP + HCl$
(*b*) $C_7H_6O_2 + O_2 \rightarrow CO_2 + H_2O$
(*c*) $NH_3 + O_2 \rightarrow N_2 + H_2O$

9.29 Lead carbonate has the formula $PbCO_3$. How much lead is contained in 1 kg of this compound?

9.30 Ferric chloride ($FeCl_3$) is produced where iron (Fe) is heated in an atmosphere of gaseous chlorine (Cl_2). Determine

(*a*) The balanced equation for this reaction
(*b*) The mass of chlorine needed to produce 10 g of ferric chloride

9.31 One component of gasoline, octane, reacts with oxygen according to the following unbalanced equation:

$$C_8H_{18} + O_2 \rightarrow CO_2 + H_2O$$

(*a*) Balance the equation.
(*b*) Calculate the mass of water produced when 4 kg of octane is completely burned.

9.32 Hydrogen gas can be used to reduce iron ore, Fe_2O_3, to its metal as shown in the following unbalanced equation:

$$Fe_2O_3 + H_2 \rightarrow Fe + H_2O$$

(*a*) Balance the equation.
(*b*) Determine the amount of iron ore needed to produce 100 kg of pure iron.
(*c*) Indicate how much hydrogen would be consumed in the process of producing 100 kg of iron.

9.33 A lead storage battery has positive electrodes filled with lead dioxide, PbO_2, and negative electrodes filled with spongy lead, both dipped in a concentrated sulfuric acid solution. Upon discharge, the following reaction takes place:

$PbO_2 + Pb + H_2SO_4 \rightarrow PbSO_4 + H_2O$ (unbalanced)

(*a*) Balance this equation.
(*b*) If 1 g of sulfuric acid is used up, state how much water is produced.

9.34 Liquid hydrogen and liquid oxygen have been a popular fuel combination for large rocket engines. The reaction is simply $2H_2 + O_2 \rightarrow 2H_2O$. If 20 000 kg/s of water must be formed to produce the necessary thrust, how much hydrogen and how much oxygen must be carried aboard to allow a 60-s burn?

9.35 Sodium lauryl sulfate, $C_{12}H_{25}OSO_3Na$, is a detergent produced in a two-step process from lauryl alcohol, $C_{12}H_{25}OH$, sulfuric acid, H_2SO_4, and sodium hydroxide, NaOH.

$$C_{12}H_{25}OH + H_2SO_4 \rightarrow C_{12}H_{25}OSO_3H + H_2O$$

$$C_{12}H_{25}OSO_3H + NaOH \rightarrow C_{12}H_{25}OSO_3Na + H_2O$$

If you as an engineer are operating a process to produce 25 metric tons (t) of detergent per day, what are the minimum amounts of lauryl alcohol, sulfuric acid, and sodium hydroxide that you would need each day?

9.36 Table sugar, $C_{12}H_{22}O_{11}$, can be decomposed into carbon and water.
(*a*) Write the equation for this reaction and balance it.
(*b*) State how many molecules of water are produced by the decomposition of 10 molecules of sugar.
(*c*) Indicate how much carbon and water the decomposition of 1 g of sugar will produce.

9.37 Hydrogen can be produced in the laboratory by reacting CaH_2 with water to produce $Ca(OH)_2$ and H_2.
(*a*) Write a balanced equation for this reaction.
(*b*) State how many grams of H_2 will result from 100 g of CaH_2.

9.38 Acetic acid has the formula CH_3COOH. Determine the percentage by weight of each constituent.

9.39 NH_4NO_3 is the formula for ammonium nitrate. What is the mass percentage of each element it contains?

9.40 Calculate the percentage composition of anhydrous sodium phosphate, Na_2HPO_4.

9.41 What is the percentage of water in hydrated sodium phosphate, $Na_2HPO_4 \cdot 5H_2O$?

9.42 A salt has been found to contain 60.6% chlorine and 39.4% sodium by weight. What is its empirical formula?

9.43 Determine the empirical formula for teflon, which is 70.37% fluorine and 29.63% carbon.

9.44 A compound known as isopropanol has a molecular weight of 60. It is composed of 60% carbon, 13.3% hydrogen, and the remainder oxygen by mass. Write:

(*a*) Its empirical formula
(*b*) Its molecular formula

9.45 Determine the empirical formula for a compound containing 21.6% boron, 22.8% sodium, and 55.9% oxygen by weight.

9.46 Calculate the empirical formula for a compound found to contain 65.3% oxygen, 32.7% sulfur, and 2.04% hydrogen.

9.47 In 30 g of a pure compound, there are 8.73 g of sodium, 9.12 g of oxygen, and 12.15 g of sulfur. What is the simplest formula for this compound?

9.48 10 g of a flammable gas was analyzed and found to be composed of 7.49 g of carbon and 2.51 g of hydrogen. What is its formula?

9.49 A gaseous compound of hydrogen and carbon is burned to provide 1.798 g of water and 6.60 g of carbon dioxide. Find the empirical formula for the compound.

9.50 Derive the empirical formula for a compound containing 56.33% oxygen and 43.67% phosphorus by weight.

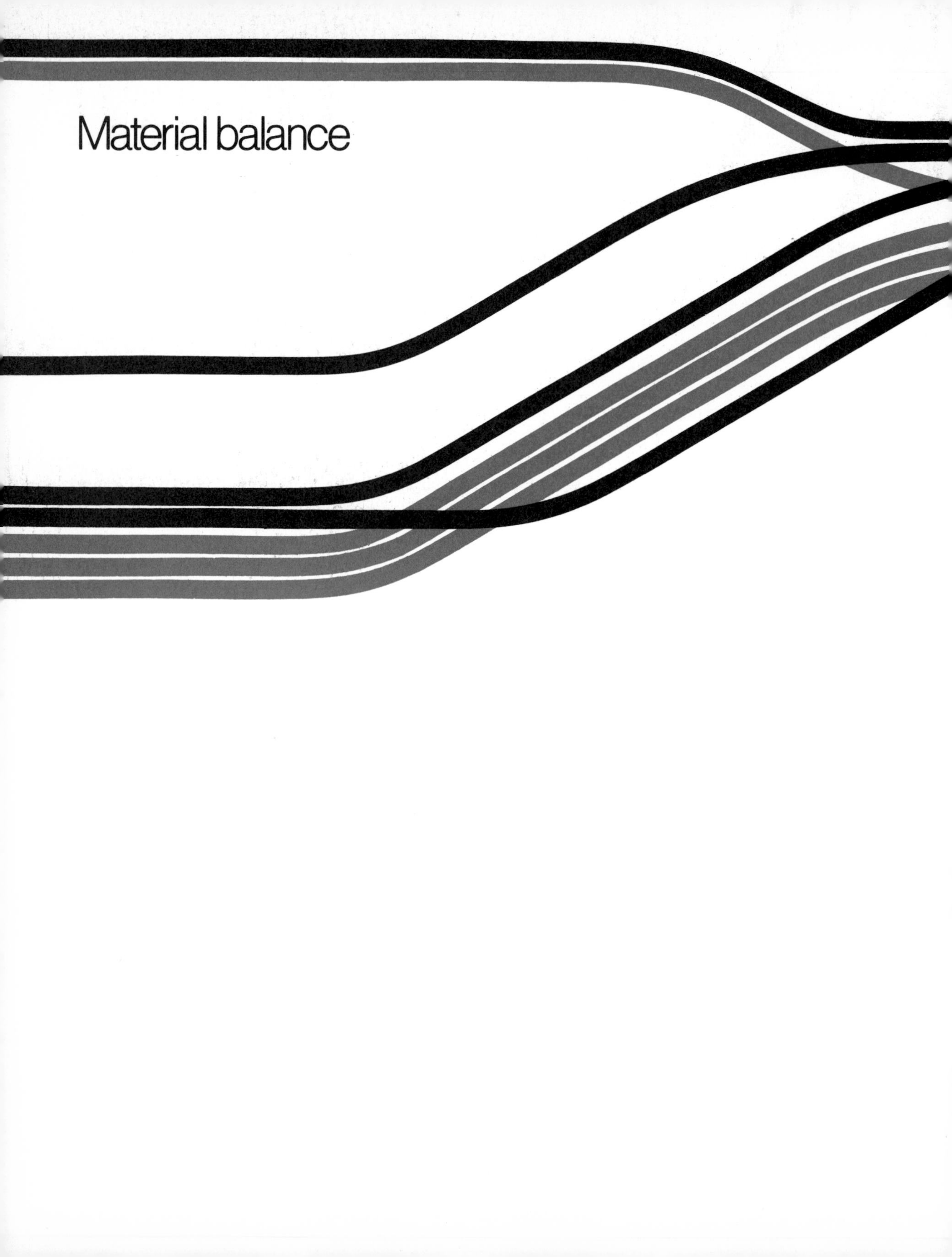

Material balance

CHAPTER 10

10.1 Introduction

We depend a great deal on industries that produce food, household cleaning products, energy for heating and cooling homes and businesses, fertilizers, and many other products and services. These process industries, as they are called, are continually involved with the distribution, routing, blending, mixing, sorting, and separation of materials. (See Fig. 10.1 for one example of a processing system.)

A typical process problem that an engineer might be called on to solve is exemplified by the drying process, shown schematically in Fig. 10.2. A process engineer designing a system to dry grain would most likely know the percent moisture (on a mass basis) of the wet grain, the desired moisture content for the dried grain, and the amount of grain to be dried in a specified amount of time. The engineer would then have to calculate the flow rate of dry heated air required to be forced through the grain. Knowing the air flow rate, the engineer could then design the mechanical system of heaters, motors, blowers, and ducting.

To perform computations involving material flow in a process, you must use an engineering analysis technique called material balance, which is based on the principle of conservation of mass.

Fig. 10.1 An aggregate processing unit that separates quarry material according to exact specifications. It is a portable unit that can perform the required crushing and screening operations as well as function as a stationary plant. (*Iowa Manufacturing Company.*)

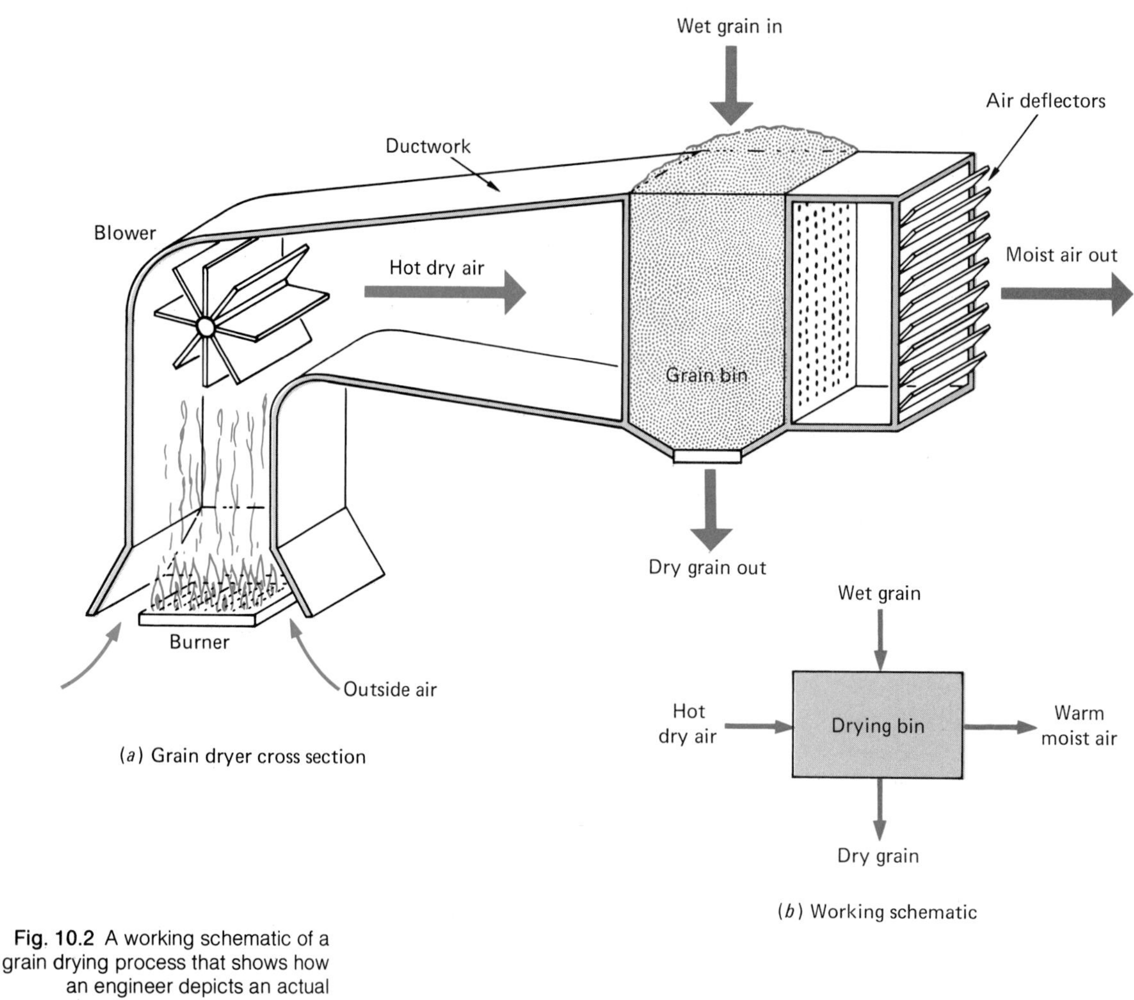

Fig. 10.2 A working schematic of a grain drying process that shows how an engineer depicts an actual process so that necessary calculations can be performed.

10.2 Conservation of mass

The principle of the conservation of mass states simply that mass is neither created nor destroyed, if nuclear reactions are excluded. To apply the principle to a material balance problem, the concept of a "system" is needed. Figure 10.3 illustrates such a system.

A system is any designated portion of the universe with a definable boundary. The remainder of the universe is called the surroundings. Mass may enter and leave the system and also accumulate within the system. The conservation of mass principle applied to a system can be expressed as

$$\text{Input} - \text{output} = \text{accumulation} \qquad 10.1$$

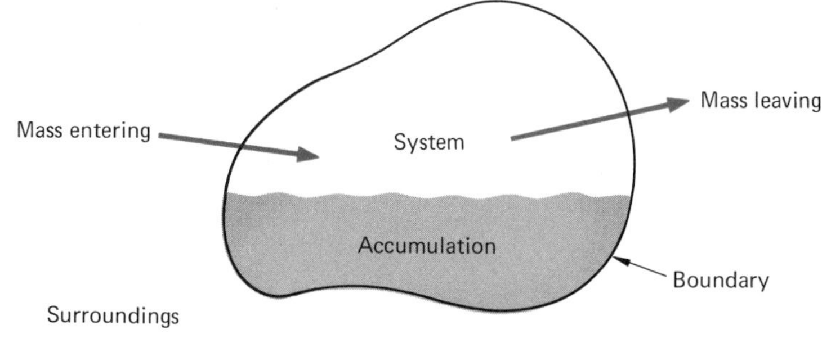

Fig. 10.3 Definition of a system.

That mass may change location but the total amount will remain the same is another way of stating the conservation of mass principle. With respect to the system in Fig. 10.3, material (mass) may be added faster than it is removed, thereby creating an accumulation. Removing material faster than adding it results in a negative accumulation. Equation 10.1 accounts for both possibilities, as well as when input equals output from the system.

Being a general equation, Eq. 10.1 may be applied to any definable system, such as a tank of water, a sewage treatment plant, or even a bank account. When balancing a bank account, the deposits and earned interest constitute the *inputs*, withdrawals and service charges the *outputs*, and the amount by which the balance increases or decreases is the *accumulation*. Needless to say, here the material balance is based on monetary value and not on the mass of coins, bills, checks, etc.

Before using a material balance to analyze a system, you must be able to account for all the constituents that enter and leave the system. This is quite often the most difficult part of a material balance.

Consider, for example, the materials that enter and leave a single-family home during a period of time. Table 10.1 furnishes a partial list from which you can see that accounting for each and

Table 10.1

Material	Examples
Food	Milk, bread, salt
Paper	Newsprint, sacks, food wrappers
Fuel	Coal, oil, gas, firewood
Metal	Cans, tools, toys
Plastic	Furniture, food containers
Textile	Clothing, drapes
Glass	Cups, bottles
Soil	Dust, dirt
Water	City water supply, distilled
Gas	Air, Freon

Table 10.2

Material	Examples
Water	Sewage carrier, lawn
Garbage	Empty containers, used paper
Gas	Carbon dioxide, air
Others	Fireplace ashes, dirt, food waste

every item of material entering the average home is almost impossible—and certainly prohibitive in cost. Appropriate assumptions must be made that probably neglect many of the constituents or, at best, merely estimate their amounts.

The situation becomes even more complicated when items leaving the home are considered. Table 10.2 gives a partial listing of such items.

This abbreviated list points out another difficulty with material balance: much of the material under consideration changes form as a result of mixing, absorption, or chemical reactions. Direct measurement of the mass of each constituent is nearly impossible. The ability of an engineer to perform an analytical material balance on a system is thus quite important.

10.3 Processes

Two types of processes typically analyzed with a material balance are the *batch process* and the *rate process*. A batch process is similar to a recipe in which a fixed amount of end product results. Rate (continuous-flow) processes involve a mass flow per unit of time.

Fig. 10.4 This complicated arrangement of ducts directs the air flow in a building to provide comfortable working conditions for employees. (*Stanley Consultants.*)

An example would be a pipe delivering water to a tank at a rate of 2 kg/s.

For a batch process, Eq. 10.1 would properly be written with total quantities.

$$\text{Total input} - \text{total output} = \text{total accumulation} \qquad 10.2$$

Figure 10.5 is an example of a batch process for a concrete mixer. Note that there is no accumulation in the mixer.

For a rate process, the terms in Eq. 10.1 must be on a rate basis.

$$\text{Rate of input} - \text{rate of output} = \text{rate of accumulation} \qquad 10.3$$

Figure 10.6 illustrates a rate process for a coal dewatering process.

Rate processes may be classified as either uniform or nonuniform, steady or unsteady. A process is uniform if the input rate equals the output rate. It is steady if the rates do not vary with time. Solution of material balance problems involving nonuniform and/or unsteady flows may require the use of differential equations (that is, mathematical expressions that contain derivatives of functions as variables). However, many important processes can be classified as uniform and steady and thus be analyzed with more elementary mathematical skills. For a uniform, steady rate process,

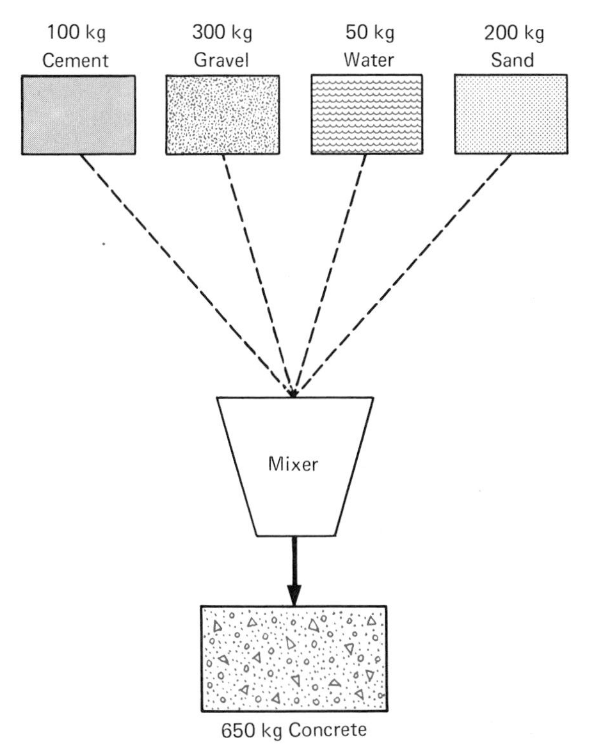

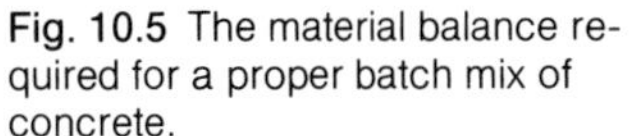

Fig. 10.5 The material balance required for a proper batch mix of concrete.

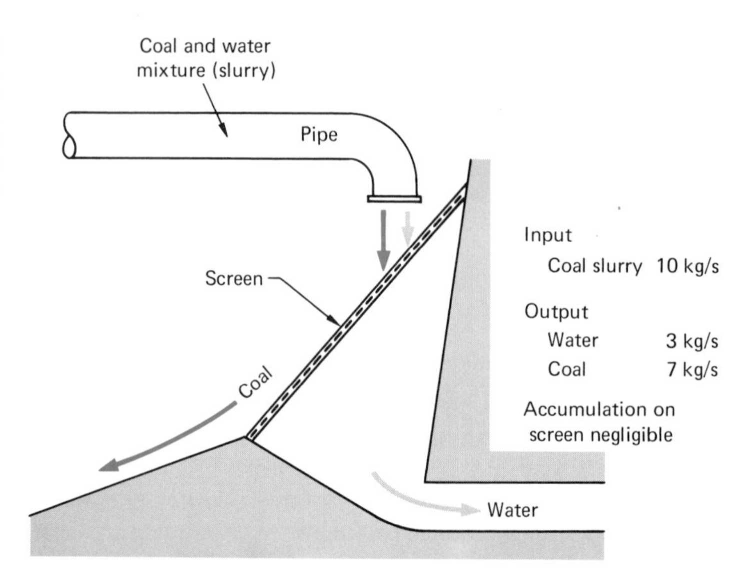

Fig. 10.6 A coal dewatering process shown as a rate process. Sampling the downstream portion would provide a more detailed material balance because the coal would carry some water with it and the water side would contain some fine coal particles.

the input equals the output; and Eq. 10.2 and 10.3 reduce to

$$\text{Total input} = \text{total output} \tag{10.4}$$

$$\text{Rate of input} = \text{rate of output} \tag{10.5}$$

Fig. 10.7 A large-scale rate process that turns trona ore into refined soda ash at the rate of 2.5 million tons per year. (*FMC Corporation*)

Although Eqs. 10.4 and 10.5 seem so overly simple as to be of little practical use, application to a given problem may be complicated by the need to account for several inputs and outputs as well as many constituents in each input or output. The simplicity of the equations is in fact the advantage of a material balance approach, because order is brought to seemingly disordered data.

10.4 Performing a material balance

Material balance computations require the manipulation of a substantial amount of information. Therefore, it is essential that a systematic problem layout be developed and strictly followed. If a systematic approach is used, material balance equations can be written and solved correctly in a straightforward manner. The following list of steps is recommended as a procedure for solving material balance problems.

1. Identify the system(s) involved.
2. Determine whether the process is batch or rate and whether a chemical reaction is involved. If no reactions occur, the material balance will involve compounds. Elements must be balanced if a reaction occurs. In a process involving chemical reactions additional equations involving chemical composition may also be required.
3. Construct a diagram showing the feeds (inputs) and products (outputs) schematically.
4. Label known material quantities or rates of flow.
5. Identify with an appropriate symbol each unknown input and output.
6. Apply Eq. 10.4 or 10.5 for each constituent as well as for the overall process. Care must be taken to include only independent equations.
7. Solve the equations for the desired unknowns and express the results in suitable form.

The following examples will serve to illustrate this procedure.

Example problem 10.1 Freshwater can be obtained from seawater by partially freezing the seawater to create salt-free ice and a saltwater solution. If seawater is 3.50 percent salt by mass and the saltwater solution (brine) is to be an 8.00 percent concentrate by mass, determine how many kilograms of seawater must be processed to form 2.00 kg of ice.

Solution The freezing operation is a batch process because a fixed amount of product (ice) is required. (A diagram of the process is shown in Fig. 10.8.) There are two constituents (salt and water) involved in the process and thus three material balance

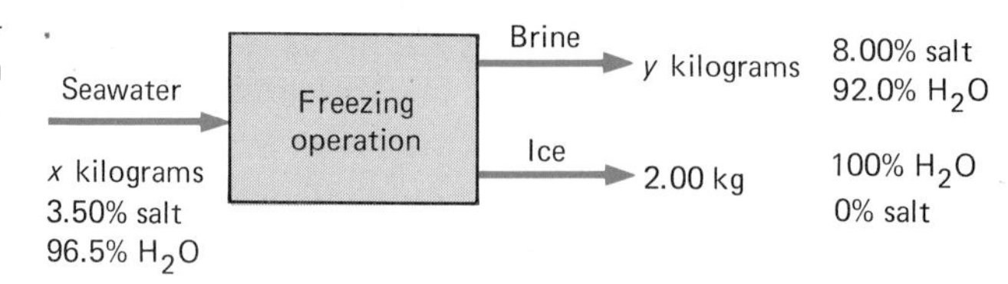

Fig. 10.8 Material balance for obtaining freshwater from seawater.

equations can be written using Eq. 10.4, with accumulation being zero.

Conservation of mass: Input = output

Overall balance:	$x = y + 2$
Salt balance:	$0.035x = 0.08y$
H_2O balance:	$0.965x = 0.920y + 2$

Note that only two of the equations are independent, since the salt and H_2O balance equations can be added to obtain the overall balance equation.

Solving the first two equations by substitution, we get

$$0.035(y + 2.00) = 0.080y$$

$$0.070 = 0.045y$$

$$y = 1.56 \text{ kg}$$

$$x = 3.56 \text{ kg}$$

As a check, the values of x and y can be substituted into the H_2O balance:

Fig. 10.9 Engineers measure flow rates of waste material from an industrial plant. (*Black and Veatch Consulting Engineers.*)

$$0.965(3.56) \stackrel{?}{=} 0.920(1.56) + 2.00$$

$$3.44 = 3.44$$

Example problem 10.2 Leftover acid from a nitrating process (i.e., combining with nitric acid) contains 24% HNO_3, 55% H_2SO_4, and 21% H_2O (mass percents). The acid is to be concentrated (strengthened in acid content) by adding sulfuric acid with 92% H_2SO_4 and nitric acid containing 89% HNO_3. The final product is to contain 28% HNO_3 and 61% H_2SO_4. Compute the mass of the initial acid solution and the mass of concentrated acids that must be combined to obtain $1.0(10^3)$ kg of the desired mixture.

Solution The system in this problem would be the tank in which the proper quantities of the acids are mixed. Since $1.0(10^3)$ kg of a mixture is desired, the operation is a batch process. No chemical reactions will take place, so the material balance can be performed with the compounds. If we let

x = mass of initial solution

y = mass of concentrated H_2SO_4

z = mass of concentrated HNO_3

then the diagram shown in Fig. 10.10 indicates all known and unknown quantities involved in the problem.

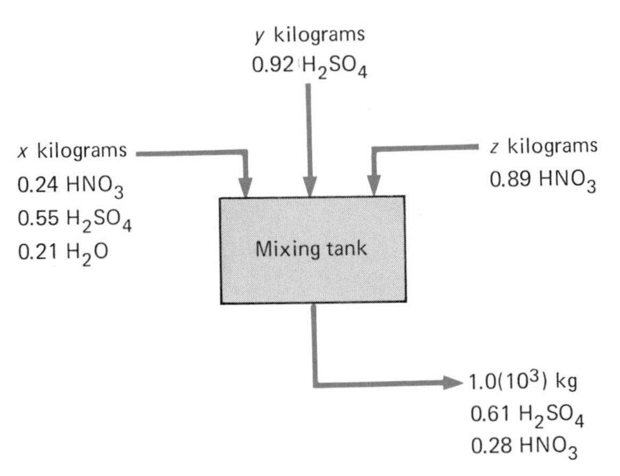

Fig. 10.10 Material balance for an acid concentration process.

Applying Eq. 10.4 to the system and each of the constituents yields

Overall balance: $x + y + z = 1\,000$

H_2SO_4 balance: $0.55x + 0.92y = 610$

HNO_3 balance: $0.24x + 0.89z = 280$

H_2O balance: $0.21x + 0.08y + 0.11z = 110$

There are four equations in three unknowns, but the equations are not independent, because the latter three add up to the first. We will solve the first three for x, y, and z and use the H_2O balance as a check. Solving the H_2SO_4 and HNO_3 balances for y and z, respectively, and substituting into the overall balance yields

$$x + \left(\frac{610 - 0.55x}{0.92}\right) + \left(\frac{280 - 0.24x}{0.89}\right) = 1\,000$$

$$x = 170 \text{ kg}$$

Solving for y and z gives

$$y = 560 \text{ kg}$$

$$z = 270 \text{ kg}$$

Checking the H_2O balance, we get

$$0.21(170) + 0.08(560) + 0.11(270) \stackrel{?}{=} 110$$

$$110 = 110$$

Example problem 10.3 A dryer processes 4 500 kg/h of a wood pulp. If the pulp contains 36% water before it enters the dryer and 12 000 kg of water are removed in an 8-h day, what is the final moisture content of the pulp?

Solution We can consider this either a batch process with 36 000 kg of wood pulp having 12 Mg of water removed or a rate process with 1 500 kg/h of water removed from 4 500 kg/h of wood pulp. Figure 10.11 shows a diagram of the drying process with all quantities labeled appropriately. A rate process is depicted, with x representing the final moisture content of the pulp.

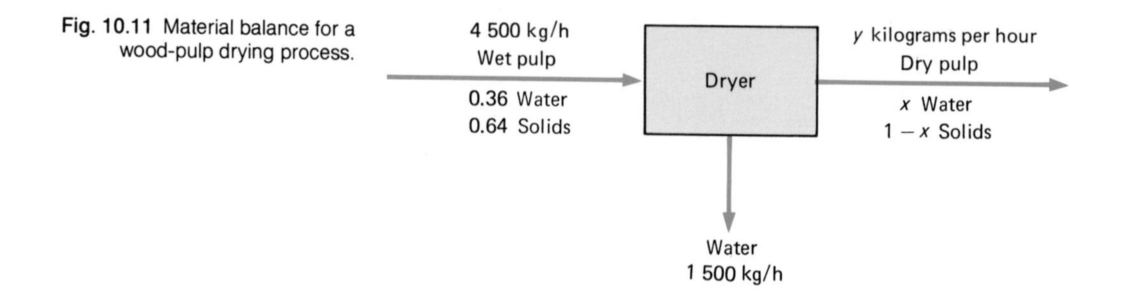

Fig. 10.11 Material balance for a wood-pulp drying process.

Applying Equation 10.5, we get

Overall balance: $4\,500 = y + 1\,500$

$y = 3\,000$ kg/h (dried pulp)

Water balance: $0.36(4\,500) = x(3\,000) + 1\,500$

$$x = \frac{120}{3\,000}$$

$$= 4.0\%$$

Example problem 10.4 Streams A, B, and C of waste materials are fed together into a mixing basin and leave as a single stream D, with no accumulation. Stream A has a flow rate of 5.000 L/s and is 0.12% solids, with the solids consisting of 62.00% carbon. Stream B has not been measured but is known to contain 1.05% solids. Stream C flows at 2.000 L/s and contains 0.82% solids, of which 90.00% is carbon. After mixing, stream D is measured and flows at a rate of 9.67 L/s and contains 0.52% solids that are 71.00% carbon. Determine the amount of carbon in stream B, expressed as a percent of the flow rate of the solids.

Solution Since flow rates are on a volume basis, assume that water is the carrying stream and each stream has a density of 1 kg/L. The mass flow rates, percent solids, and carbon percents are shown in Fig. 10.12.

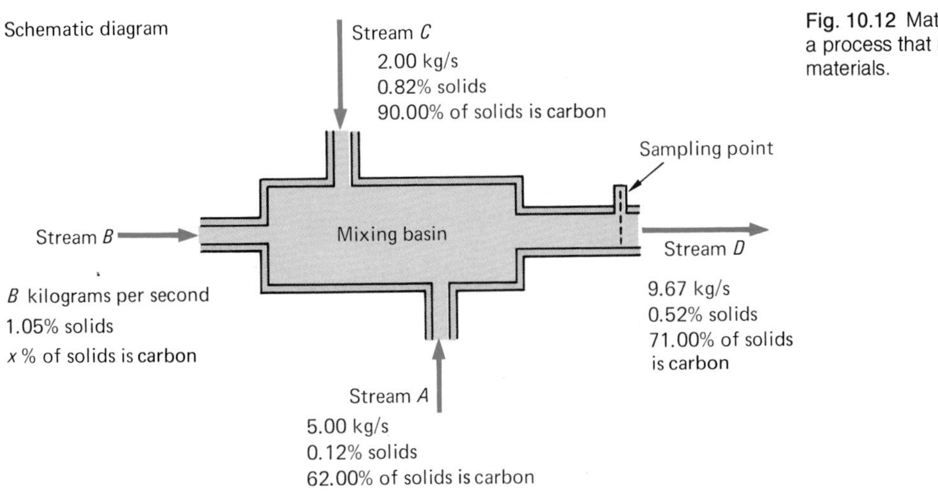

Fig. 10.12 Material balance for a process that mixes waste materials.

Since this is a rate process, Eq. 10.5 applies. This example again illustrates the value of constructing a diagram of the process and carefully labeling. The problem has a great amount of information given but the required calculations are direct.

Overall balance: $A + B + C = D$

$$B = D - (A + C)$$

$$B = 9.67 - (5.00 + 2.00)$$

$$B = 2.67 \text{ kg/s}$$

Carbon balance: Carbon A + carbon B + carbon C = carbon D

$$\text{Carbon } B = \text{carbon } D - \text{carbon } A - \text{carbon } C$$

$$= 9.67(0.005\,2)(0.71) - (5.00)(0.001\,2)(0.62) - 2.00(0.008\,2)(0.90)$$

$$= 0.035\,7 - 0.003\,7 - 0.014\,8$$

$$= 0.017\,2 \text{ kg/s}$$

Note: Carbon flow rate = (stream flow rate)(solids in stream)(carbon in solids). Therefore,

$$x = \frac{\text{Carbon } B}{(\text{Flow rate, } B)(\text{Solids in } B)} 100$$

$$= \frac{0.017\,2}{(2.67)(0.010\,5)} 100$$

$$= 61.35\%$$

Fig. 10.13 A scale model of a waste treatment facility illustrating the material handling systems required for proper operation. (*Stanley Consultants.*)

Problems

Note: Unless otherwise specified, all percentages are given on a mass basis.

10.1 How much pure sugar must be added to 60.0 kg of 10% syrup to produce a syrup that is 18% sugar?

10.2 A beaker holds 962 g of a brine solution that is 6.20% salt. If 123 g of water are evaporated from the beaker, how much salt must be added to have an 8.60% salt solution?

10.3 A continuous batch processor starts with 60.0 kg of brine and evaporates 20.0 kg of water, leaving a mixture that is 90.0% salt. Compute the percent of salt in the original brine.

10.4 A brine solution is 20.0% salt and has 70.0 kg of water evaporated from it per hour. To produce 195 kg of pure salt (0% moisture) per day, how long should the process operate each day and how much brine must be fed per hour?

10.5 Liquids *A* and *B* are mixed together. *A* is 3.00% solids, the remainder water. *B* is 8.00% solids, the remainder water. To the mixture is added 1.52 g of bone dry solids producing 168 g of final product containing 94.5% water. What were the initial amounts of liquids *A* and *B*?

10.6 If 10^5 kg/h of a 20 percent sugar solution is being concentrated by passing though an evaporator that removes $7.2\,(10^5)$ kg of water every 24 h, determine the percent of sugar in the discharge solution.

10.7 A polluting material is discharged into a stream at the rate of 100 kg/h. At a point downstream where mixing is complete, the concentration of pollutant is measured as 100 ppm (parts per million) by mass. Determine the flow rate of the stream in liters per second.

10.8 Ether is used in the process of extracting cod-liver oil from livers. The following information is known about the oil extraction process. Livers enter the extractor at 1 050 kg/h and consist of 33.0% oil and 67.0% inert material. The solvent enters the extractor at 2 150 kg/h and consists of 1.5% oil and 98.5% ether. The extract leaves the process at 1 850 kg/h and consists of 19% oil and 81% ether. Determine the flow rate and composition of the leaving processed livers.

10.9 In a kerosene vat, 20.0 t/d (metric tons per day) of machine parts are to be degreased. The parts average 3.00 kg of grease per $1.00(10^2)$ kg of parts. The used kerosene, containing 11.0% grease, is sent to a solvent recovery process. How many liters of kerosene must be added to the vat per day?

10.10 A quarry mixes different sized stones together for particular applications. Initially a mixture consisting of 40% stones (by mass) greater than $\frac{1}{2}$ in in diameter, 25% equal to $\frac{1}{2}$ in in diameter, and 35% less than $\frac{1}{2}$ in in diameter are fed to a screening complex. All stones less than $\frac{1}{2}$ in are screened out; some stones greater than $\frac{1}{2}$ in are added; and the resulting mixture includes 78% stones greater than $\frac{1}{2}$ in. If the final mixture needs to be 35 tons, what must the initial mixture weigh?

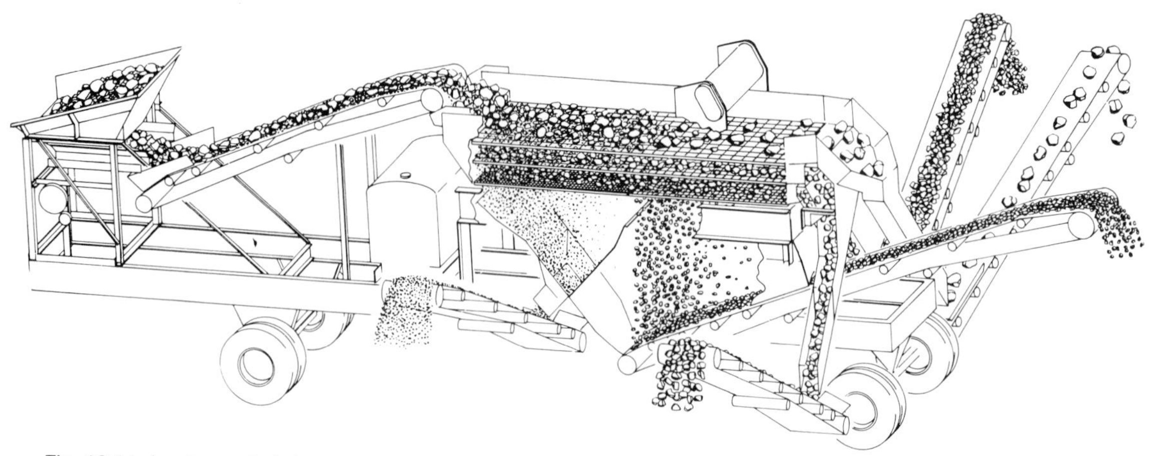

Fig. 10.14 A cutaway that shows the operation of a screening conveyor used to separate various-sized aggregate. (*Iowa Manufacturing Company.*)

10.11 A friend makes homemade cherry wine that is 12% alcohol and costs 30 cents/L to produce. To raise the alcohol content to 20%, the friend learns that cherry brandy, which is 80 proof (40% alcohol) and sells for $3 per fifth of a gallon, and that 190-proof grain alcohol (95% alcohol), which sells for $50 per gallon, are available. If 500 L of the homemade wine are on hand, which of the higher concentrations would be most economical to increase the alcohol content if

(*a*) The alcohol contents are considered as mass percent

(*b*) The alcohol contents are considered volume percent and the specific gravity of the alcohol is 0.79

10.12 The leftover acid from a nitrating process contains 21% HNO_3, 55% H_2SO_4, and 24% H_2O (mass percents). The acid is to be concentrated by the addition of concentrated sulfuric acid containing 93% H_2SO_4 and concentrated nitric acid containing 90% HNO_3. The desired final product is to contain 28% HNO_3 and 62% H_2SO_4. What amounts of leftover and concentrated acids must be combined to obtain 2 100 kg of the desired mixture?

10.13 The feed to a distillation column consists of 20% toluene, the remainder being benzene and water. The overhead (product) contains 95% benzene and 5% water and the bottoms (waste) contain 35% water and 10% benzene. If the overhead flow rate is $2.00(10^2)$ kg/h, what is the required feed rate to the column and what is the percent benzene in the feed?

10.14 The feed to a distillation column contains 35% benzene, the remainder being toluene. The overhead (product) contains 54% benzene and the bottoms contain 4% benzene. Calculate the percentage of the total feed that leaves as product.

10.15 A laundry can buy liquid soap containing 31% water for a price of $15.50 per 45 kg container f.o.b. the factory. The same manufacturer offers another soap containing 6% water. The shipping cost is $2.40 per 100 kg.

What is the maximum price the laundry should pay the manufacturer if the laundry decides to purchase the soap containing 6% water?

10.16 In a mechanical coal-washing process a portion of the inorganic sulfur and ash material in the mined coal can be removed. A feed stream to the process consists of 75% coal, 10% sulfur, 10% ash, and 5% water (to suppress dust). The "clean" coal side of the process consists of 90% coal, 5% sulfur, 3% water, and 2% ash. The refuse side consists of 30% sulfur, with the remainder being ash, coal, and water. If 65 t/h of clean coal are required, determine the feed rate required for the mined coal and the composition of the refuse material.

10.17 A dewatering process consists of a centrifuge and dryer. If 25 t/h of a mixture containing 35% solids is centrifuged to form a sludge containing 65% solids and then the sludge is dried to 3% moisture, how much water must be removed from the centrifuge and dryer during each 8-h workshift?

10.18 A 20.0-m^3 tank is filled with a 25.0% solution of ethylene glycol and water. A 37.0% concentrated mixture is desired. How much must be drained from the tank and be replaced by 90.0% ethylene glycol solution to obtain the desired mixture?

10.19 A tank contains an alcohol-water mixture containing 15% alcohol and 85% water. When 15 kg of a solution containing 75% alcohol and 25% water is added to the mixture, a 23% alcohol mixture results. Determine

(*a*) The mass of the original solution
(*b*) The mass of the alcohol in the final solution

10.20 A slurry of crushed mine coal and water is pumped to a cyclone separator. The mine coal is 75% pure coal and 25% pyrite. Recovery of pure coal from the top of the separator is desired at a rate of 6.0 t/h. The waste from the bottom of the separator contains 4% pure coal, 36% pyrite, and 60% water. Compute the required feeds of coal and water.

Electrical theory

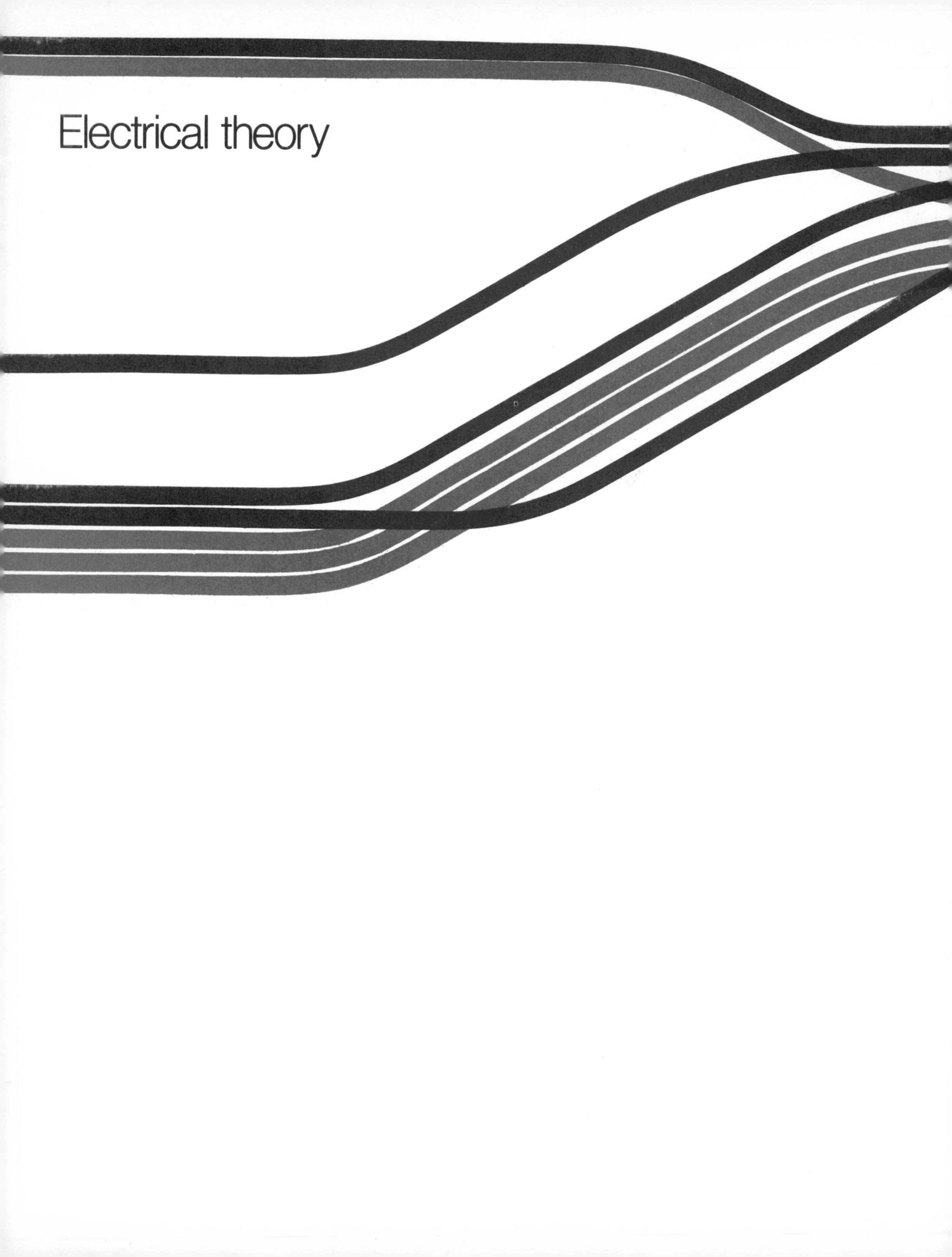

CHAPTER 11

11.1 Introduction

Electricity is one of our most powerful and useful forms of energy. It is extremely convenient; and because of its relative ease of transport, it has had an enormous impact on society.

Electric energy is difficult to store, however, so that demand for it must be met by constant production. Recent history indicates that the consumption of electricity in the United States has been doubling every 10 years; and worldwide use of electricity is also increasing rapidly. For society to meet such an increasing demand, a multitude of energy sources must be considered. Electric energy sources of the future may include fission, fusion, solar, and geothermal, as well as other techniques that are currently in developmental stages.

At the present time, the largest supply of electric energy comes from power plants fired by fossil fuels such as coal, oil, and natural gas. Hydroelectric power is an excellent source; however, less than 5 percent of today's energy in the United States is provided by

Fig. 11.1 Fossil-fuel-fixed electric generating plant. (*Black and Veatch Consulting Engineers.*)

hydropower. The problem with hydropower, unfortunately, has to do with the inaccessibility to locations like the Andes and the Himalayas. Large industry is normally not located near these sources and transmission lines from mountain ranges to industrial concentrations are currently not feasible owing to line losses and construction costs.

Coal, oil, and natural gas have been and continue to be very acceptable commodities in the production of energy. However, as natural resources, these commodities are finite and will eventually be entirely consumed.

Engineers and scientists around the world must continue to research and work toward the development of new ways to produce electric energy. Engineers are obviously concerned not only with preparing for future technologies but also with meeting the demand for energy today. This involves a variety of additional problems, such as environmental pollution associated with thermal energy, smoke, and radioactivity. In fact, reduction of the effect of large power plants on the environment is indeed a major challenge if human beings are to control and balance the ecological system.

There is no doubt that serious problems lie ahead as a result of the continuing demand for more and more electric energy, but the

Fig. 11.2 Transformers necessary to distribute electricity. (*McGraw-Edison.*)

significant role that cheap power has played in our technological and industrialized society is hard to minimize. Certainly a keystone in any such society is abundant low-cost electric power.

This chapter introduces a large and complex subject. Since electric energy has played such an important part in the development of the world, and since it will continue to be a most significant form of energy in the future, some of the more fundamental concepts associated with electrical theory should be understood by engineering students.

11.2 Structure of electricity

First we must ask what electricity is. Most people today accept the idea that all matter consists of minute particles called molecules. *Molecules* are the smallest particles into which a substance can be divided and still retain all the characteristics of the original substance. Each of these particles will differ according to the type of matter to which they belong. Thus, a molecule of iron will be different from a molecule of paper.

Looking more closely at the molecule, we find that it can be divided into still smaller parts called *atoms*. Each atom has a central core, or nucleus, that contains both *protons* and *neutrons*. In a somewhat spherical motion around the nucleus are particles of extremely small mass called *electrons*. In fact the entire mass of the atom is practically the same as that of its nucleus, since the proton is approximately 2.0×10^3 times more massive than the electron.

To understand how electricity works bear in mind that electrons possess a negative electric charge, and protons a positive electric charge. Their values are opposite in sign but numerically the same.

The neutron is considered neutral, being neither positive nor negative.

The atom in its entirety has no electric charge because the positive charge of the nucleus is exactly balanced by the negative charge of the surrounding electron cloud. That is, each atom contains as many electrons orbiting the nucleus as there are protons inside the nucleus.

The actual number of protons depends on the element of which the atom is a part. Thus, hydrogen (H) has the simplest structure, with one proton in its nucleus and one orbital electron. Helium (He) has two protons and two neutrons in the nucleus; and since the neutrons exhibit a neutral charge, there are two orbital electrons. More complex elements have many more protons, electrons, and neutrons. For example, gold (Au) has 79 protons and 118 neutrons, with 79 orbital electrons.

As the elements become more complex, the orbiting electrons arrange themselves into regions, or "shells," around the nucleus.

The maximum number of electrons in any one region is uniquely defined. The shell closest to the nucleus contains two electrons, the next eight, etc. There are a maximum of six shells, but the last two shells are never completely filled. Atoms can therefore combine by sharing their outer orbital electrons and thereby fill certain voids and establish unique patterns or molecules.

Atoms are extremely minute. In fact, it may be very difficult for one to imagine the size of an atom, since a grain of table salt is estimated to contain 10^{18} atoms. However, it is possible to understand the relation between the nucleus and the orbital electrons. Assume that the hydrogen nucleus is a 1-mm-diameter sphere; its orbiting electron appears to be the same size, but at an average distance of 25 m from the nucleus. Although the relative distance is significant, the electron is prevented from leaving the atom by an electric force of attraction that exists because the proton has a positive charge and the electron a negative charge.

How closely the millions upon millions of atoms and molecules are packed together will determine the state (solid, liquid, or gas) of a given substance. In solids, the atoms are packed closely together, generally in a very orderly manner. The atoms are held in a specific, lattice structure but vibrate around this fixed position. Depending on the substance, some electrons may be free to move from one atom to another.

It is the nature of the molecular structure with the availability of free electrons that results in electricity.

11.3 Static electricity

History indicates that the word "electricity" was first used by the Greeks. They discovered that after rubbing certain items together, the materials would exert a force on one another. It was concluded that during the rubbing process, the bodies were "charged" with some unknown element, which they called electricity. For example, it was believed that by rubbing silk over glass, electricity was added to each substance. We realize today that during the process of rubbing, electrons are removed from some of the surface atoms of the glass and added to the surface atoms of the silk.

The branch of science concerned with static or stationary charged bodies is called *electrostatics*.

The charge on an electron can be measured, but the amount is extremely small. In fact, it is inconvenient to use such a small quantity as a unit of electric charge. A larger and more practical unit, the coulomb, has thus been selected to denote electric charge. A *coulomb* is defined in terms of the force exerted between unit charges. A charge of one coulomb (C) will exert on an equal charge, placed 1 m away in air, a force of 8.988×10^9 N.

11.4 Electric current

Earlier we noted that electrons are prevented from leaving the atom by the positive proton force of the nucleus. It is entirely possible, however, for an electron to become temporarily separated from an atom. These free electrons drift around randomly in the space between atoms. During their random travel, many of them will collide with other atoms; and when they do so with sufficient force, they dislodge electrons from new atoms. Since atoms are frequently colliding with other atoms, there is a continuous movement of free electrons in a solid. If the electrons drift in a particular direction instead of moving randomly, there is movement of electricity through the solid. This continuous movement of electrons in a direction is called an *electric current.*

The ease with which electrons can be dislodged by collision as well as the number of free electrons available varies with the substance. Materials in which the drift of electrons can be easily produced are good conductors; those in which it is difficult to produce an electron drift are good insulators. For example, copper is a good conductor, whereas glass is a good insulator. Practically all metals are good conductors. Silver is perhaps the best, but it is expensive. Copper and aluminum are also good conductors and are commonly used in electric wire.

11.5 Potential

From research and experimentation it is known that like charges repel and unlike charges attract one another. Consequently, to bring like charges together, an external force is necessary. The amount of work required to bring a positive charge near another positive charge from a large distance is used as a measure of the electrical potential at that point. This amount of potential is measured in units of work per unit charge or joules per coulomb; and one joule per coulomb is one volt, a unit of electromotive force.

Certain devices, such as electric batteries or generators, are capable of producing a difference in electrical potential between two points. Such equipment is said to produce an electromotive force (emf), which is rated in terms of the potential difference produced in volts. Thus, *emf* is the driving force actually causing the flow of electricity; whereas *current* is the rate of flow. The rate of

Fig. 11.3 High-voltage dc transformers. (*General Electric.*)

flow, of course, depends on the difference in electrical potential between the negative and positive poles.

11.6 Simple electric circuit

When electric charge and current were initially being explored, scientists thought that current flow was from positive to negative. They had no knowledge of electron drift. By the time it was discovered that the electron flow was from negative to positive, the previous theory had become so well established that it was decided not to change the convention. (See Fig. 11.4.)

When the two terminals are connected to a conducting material, the battery or generator emf will create a potential difference across the load. It follows that the random movement of the negatively charged electrons will have a drift direction induced by this potential difference. The resulting effect will be the movement of electrons away from the negative terminal. Electrons will travel in a continuous cycle around the circuit, leaving the conductor to enter

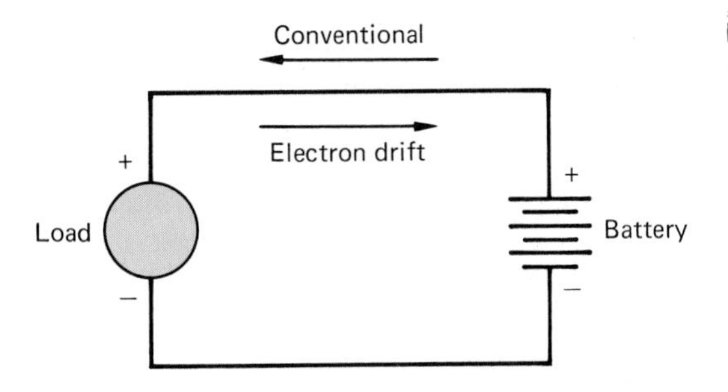

Fig. 11.4 Electric circuit.

the positive terminal and reentering the conductor from the negative terminal.

The speed by which any individual electron moves is relatively slow, less than 1 mm/s. Once a potential difference is connected into a circuit, the "flow" of electrons starts almost instantaneously at all points. Individual electrons at all locations begin their erratic movement around the circuit, colliding frequently with other atoms in the conductor. Because electron activity starts at all points practically simultaneously, electric current appears to travel about 3×10^8 m/s.

Electric current is really nothing more than the rate at which electrons pass through a conductor. The number of electrons involved is gigantic in magnitude. It is estimated that 6 280 000 000 000 000 000 electrons pass a point per second per ampere. Owing to the difficulty of working with such large numbers, it is not practical to use the rate of electron flow as a unit of current measurement. Instead, current is measured in terms of the total electric charge (coulomb) that passes a certain point in a unit of time (second). This unit of current is called the ampere. That is, one ampere equals one coulomb per second.

11.7 Resistance

Ohm, a German scientist, investigated the relation between electric current and potential difference. He found that for a metal, the potential difference across the conductor was directly proportional to the current. This important relationship is concisely stated as Ohm's law:

At constant temperature, the current I in a conductor is directly proportional to the potential difference between its ends, E.

The ratio E/I is called *resistance* and is denoted by the symbol R.

$$R = \frac{E}{I} \tag{11.1}$$

or

$$\text{Resistance, or } R(\text{ohms, } \Omega) = \frac{\text{emf, or } E \text{ (volts, V)}}{\text{current, or } I \text{ (amperes, A)}}$$

This is one of the simplest but most important relations used in electric circuit theory. A conductor has a resistance R of one ohm when the current I through the conductor is one ampere and the potential difference E across it is one volt.

The reciprocal of resistance is called conductance G, or

$$G = \frac{1}{R} \tag{11.2}$$

Conductance is measured in siemens (S).

Example problem 11.1 The current in an electric aircraft instrument heater is 2.0 A, with a power supply of 60.0 V. Calculate the resistance of the heater if the current is from a dc source.

Solution

$$E = RI$$

$$R = E/I$$

$$= \frac{60.0}{2.0}$$

$$= 3.0 \times 10^1 \ \Omega$$

11.8 Circuit concepts

A considerable amount of information about electric circuits can be presented in a condensed form by means of circuit diagrams. Figure 11.5 illustrates some typical symbols that are frequently used in such diagrams.

Often, circuits have resistors that are connected either end to end or in parallel to each other. When resistors are attached end to end, they are said to be connected in series.

A constant electric current will pass through each resistor, so the rate of flow will be the same at all points.

For the series circuit illustrated in Fig. 11.6, there will be a potential across each resistor, since a potential must exist between the ends of a conductor if current is to flow. This potential difference E

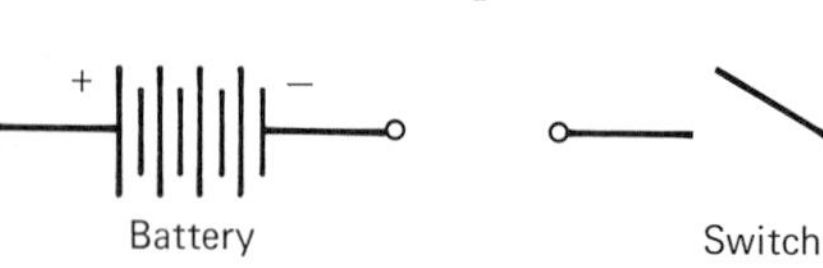

Fig. 11.5 Symbols.

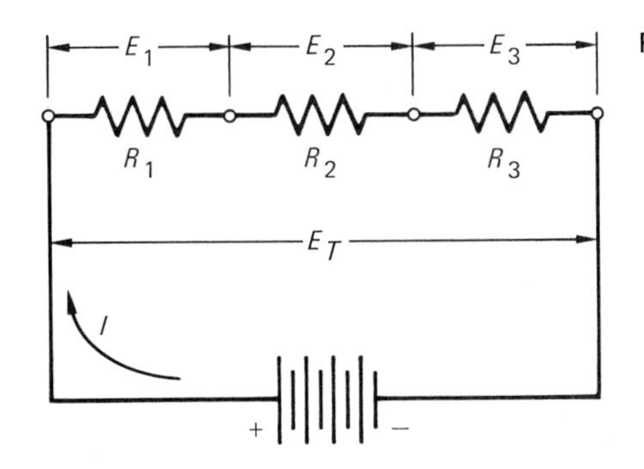

Fig. 11.6 Series circuit.

is related to current and resistance by Ohm's law. For each unknown voltage, we can write

$$E_1 = R_1 I \qquad E_2 = R_2 I \qquad E_3 = R_3 I$$

Since the total voltage drop across the three resistors is the sum of the individual drops, and since the current is constant, then

$$\begin{aligned} E_T &= E_1 + E_2 + E_3 \\ &= IR_1 + IR_2 + IR_3 \\ &= I(R_1 + R_2 + R_3) \end{aligned}$$

or

$$R_1 + R_2 + R_3 = \frac{E_T}{I}$$

Since E_T is the total potential difference across the circuit, and I is the circuit current, then the total resistance must be

$$R_T = \frac{E_T}{I}$$

Therefore,

$$R_T = R_1 + R_2 + R_3 \qquad 11.3$$

These steps demonstrate that when any number of resistors are connected in series, their combined resistance is the sum of their individual values.

Example problem 11.2 The circuit in Fig. 11.7 has three resistors connected in series with a 12.0-V source. Determine the line current and the voltage drop across each resistor.

Solution

$$R_T = R_1 + R_2 + R_3$$

Fig. 11.7 Series circuit.

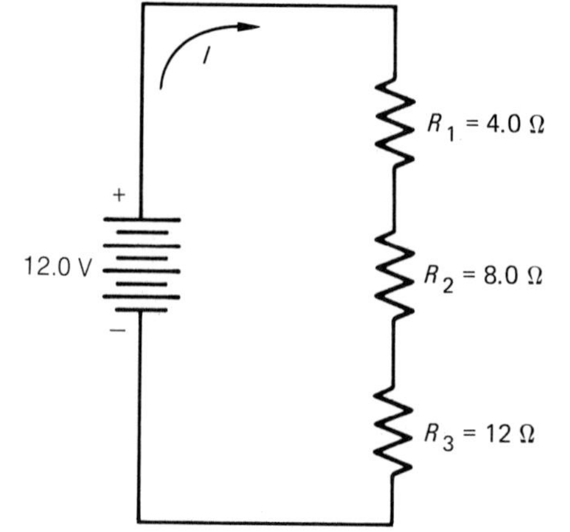

$$R_T = 4.0 + 8.0 + 12$$
$$= 24\ \Omega$$
$$E = RI$$
$$I = \frac{E_T}{R_T} = \frac{12.0}{24} = 0.50\ \text{A}$$
$$E_1 = 0.50(4.0)$$
$$= 2.0\ \text{V}$$
$$E_2 = 0.50(8.0)$$
$$= 4.0\ \text{V}$$
$$E_3 = 0.50(12)$$
$$= 6.0\ \text{V}$$

Check: $E_T = 12.0 = 2.0 + 4.0 + 6.0$

When several resistors are connected between the same two points, they are in parallel. Figure 11.8 illustrates three resistors in parallel.

The current between points 1 and 2 divides among the various branches formed by the resistors. Since each resistor is connected between the same two points, the potential difference across each of the resistors is the same.

In analyzing the problem, let I_T be the total circuit passing through the points 1 and 2 with I_1, I_2, and I_3 representing the branch circuits through R_1, R_2, and R_3, respectively.

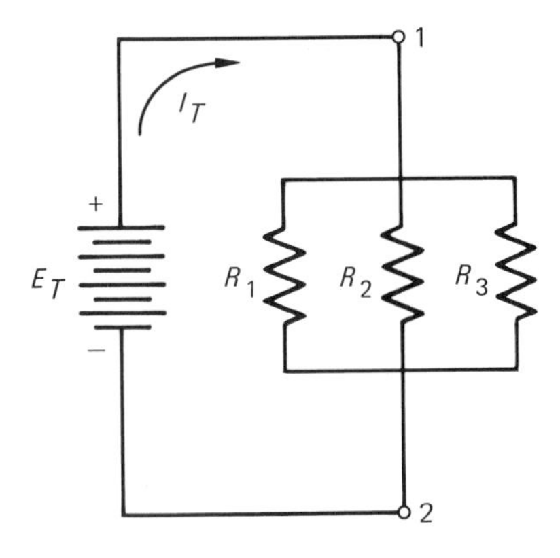

Fig.11.8 Parallel circuit.

Using Ohm's law we can write

$$I_1 = \frac{E_T}{R_1} \qquad I_2 = \frac{E_T}{R_2} \qquad I_3 = \frac{E_T}{R_3}$$

or

$$I_1 + I_2 + I_3 = E_T\left(\frac{1}{R_1} + \frac{1}{R_2} + \frac{1}{R_3}\right)$$

But we know that

$$I_T = I_1 + I_2 + I_3$$

Applying Ohm's law to total circuit values reveals that

$$I_T = \frac{E_T}{R_T}$$

Therefore,

$$\frac{E_T}{R_T} = E_T\left(\frac{1}{R_1} + \frac{1}{R_2} + \frac{1}{R_3}\right)$$

or

$$\frac{1}{R_T} = \frac{1}{R_1} + \frac{1}{R_2} + \frac{1}{R_3} \qquad 11.4$$

This equation indicates that when a group of resistors are connected in parallel, the reciprocal of their combined resistance is equal to the sum of the reciprocals of their separate resistances.

Example problem 11.3 In Fig. 11.9 three resistors are connected in parallel across a 6.0×10^1 V battery. What is the equivalent resistance and the line current?

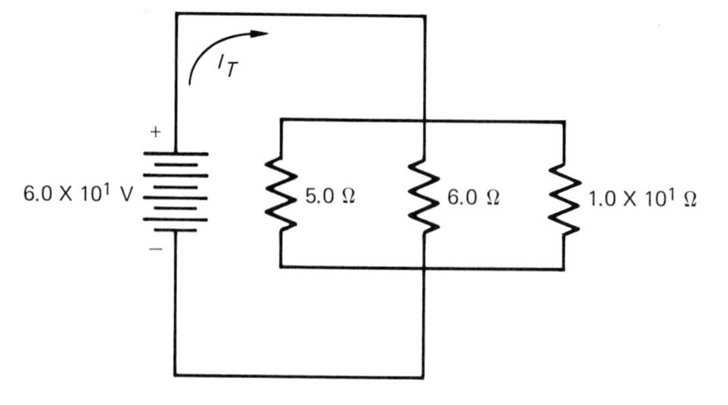

Fig. 11.9 Parallel circuit.

Solution

$$\frac{1}{R_T} = \frac{1}{R_1} + \frac{1}{R_2} + \frac{1}{R_3}$$

$$\frac{1}{R_T} = \frac{1}{5.0 \times 10^0} + \frac{1}{6.0 \times 10^0} + \frac{1}{1.0 \times 10^1}$$

$$\frac{1}{R_T} = \frac{28}{6.0 \times 10^1}$$

$$R_T = \frac{6.0 \times 10^1}{2.8 \times 10^1}$$

$$= 2.1\ \Omega$$

$$E = RI$$

$$E_T = R_T I_T$$

so,

$$I_T = \frac{E_T}{R_T}$$

$$= 6.0 \times 10^1 \frac{2.8 \times 10^1}{6.0 \times 10^1}$$

$$= 28\ \text{A}$$

Many electric circuits involve resistors in both parallel and series. The next example problem demonstrates a solution of that nature.

Example problem 11.4 Determine the line current, the circuit equivalent resistance, and the voltage drop across R_4 for the circuit in Fig. 11.10.

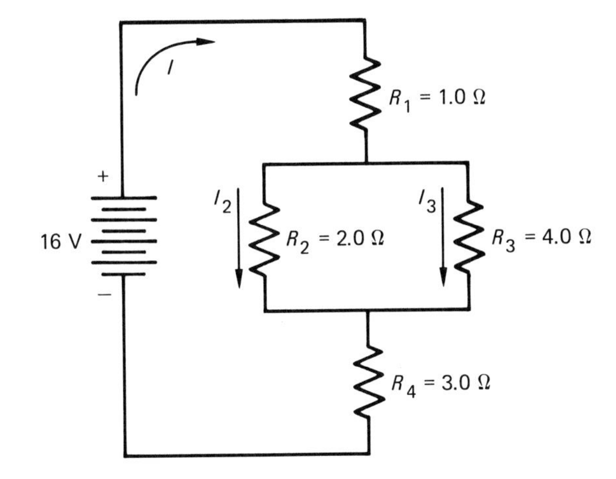

Fig. 11.10 Series-parallel circuit.

Solution

Theory

$$E = RI$$

$$R_T = R_1 + R_2 + \cdots + R_N \qquad \text{(series)}$$

$$\frac{1}{R_T} = \frac{1}{R_1} + \frac{1}{R_2} + \cdots + \frac{1}{R_N} \qquad \text{(parallel)}$$

Parallel resistors:

$$\frac{1}{R_T} = \frac{1}{R_2} + \frac{1}{R_3}$$

$$= \frac{1}{2.0} + \frac{1}{4.0}$$

$$R_T = \frac{4.0}{3.0}\Omega$$

Series resistors:

$$R_E = R_1 + R_T + R_4$$

$$= 1.0 + \frac{4.0}{3.0} + 3.0$$

$$= \frac{16}{3.0}$$

$$= 5.3\Omega$$

Line current:

$$E = RI$$

$$I = E\left(\frac{1}{R_E}\right)$$

$$= 16\left(\frac{3.0}{16}\right)$$

$$= 3.0 \text{ A}$$

Voltage drop:

$$E_4 = R_4 I$$

$$= 3.0(3.0)$$

$$= 9.0 \text{ V}$$

11.9 Electric power

If a current flows for some time period as a result of a potential difference, then an amount of energy is produced or released as follows:

$$\text{Energy} = EIt \qquad 11.5$$

Energy, measured in joules, can be converted into heat energy by resistors in a circuit, or it can be lost through heat dissipation to the surroundings.

Power is the time rate at which energy is supplied or consumed. It is expressed in joules per second, but this unit has been given a special name after the scientist James Watt. Thus, one watt equals one joule per second.

Power can be expressed as

$$\text{Power} = \frac{\text{Energy}}{\text{Time}} = \frac{EIt}{t} = EI \qquad 11.6$$

By applying Ohm's law, we can express power in two other convenient forms:

$$P = \frac{E^2}{R} \qquad 11.7$$

and

$$P = I^2 R \qquad 11.8$$

In all the discussion so far, direct and alternating current have not been mentioned. Most applications that we are familiar with involve alternating current (ac), not direct current (dc). In most homes and offices, the wall outlets are either 110 V or 220 V ac.

Fig. 11.11 A management simulator for a programmable controller. (*Stanley Consultants.*)

It is not our intention to discuss alternating current at length in this chapter, but a brief reference is necessary to ensure a clear understanding of the section.

Ohm's law can be used as it was presented in Sec. 11.7 for circuits involving alternating currents if they contain only resistors. When coils and capacitors are added to ac circuits, the problems become more complex. (They will be covered in other engineering courses.) Ohm's law does apply to dc circuit applications.

Example problem 11.5 An iron in your home consumes 6.0 A. What is the power required? Assume that $E = 110$ V.

Solution

$$P = EI$$
$$= 110(6.0)$$
$$= 660 \text{ W}$$

Example problem 11.6 A current of 4.0 A flows through a 50.0-Ω resistor. What is the power in watts?

Solution

$$P = I^2R$$
$$= (4.0)^2(50.0)$$
$$= 8.0 \times 10^2 \text{ W}$$

Table 11.1

Power, W	Voltage, V	Price, $
4 000	220	110
3 000	220	90
2 000	110	70
1 000	110	60

These examples demonstrate the use of the basic equation, but let's consider a more practical situation.

Suppose that you recently purchased a small cabin in Canada. It has been wired for electricity but does not have any heating device. Heat-loss calculations show that to maintain an indoor temperature of 70°F (21°C) when the outdoor temperature is −20°F (−29°C), will require 13 000 Btu/h (3 800 W).

In a nearby town there is a sale on recessed wall heaters with fans and thermostatic temperature controls. Table 11.1 lists the wall heaters available according to size and price.

The heating requirement would indicate that the 4.0-kW unit would be the most appropriate purchase.

Let's further assume that the unit would need to run half of the time to maintain the cabin at 70°F during the month of December. If the cost of electric energy is $0.05/kWh, we can calculate the power cost for the month.

$$\text{Cost} = \frac{\$0.05 \mid 4.0\text{ kW} \mid 24\text{ h} \mid 1 \mid 31\text{ d}}{\text{kWh} \mid \qquad \mid \text{d} \mid 2 \mid} = \$74.40$$

11.10 Terminal voltage

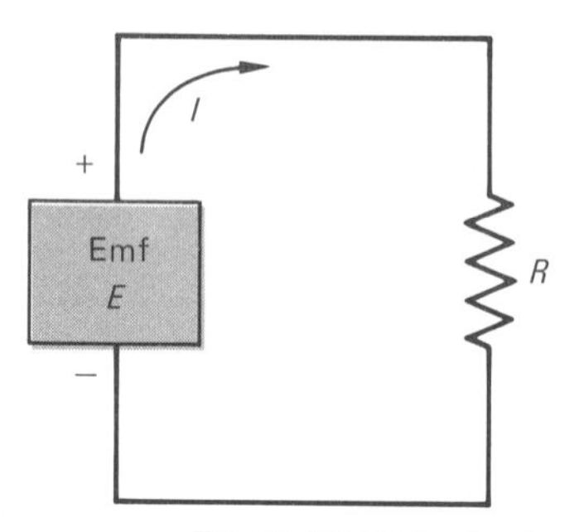

Fig. 11.12 Basic circuit.

Figure 11.12 is a basic circuit, but there are different sources of electrical potential (emf). The storage battery and the electric generator are two familiar examples.

The emf does work of amount E in joules per coulomb on charges passing through the source from the negative to the positive terminal. This results in a difference of potential (E) across the resistor (R) which causes current to flow in the circuit. The energy furnished by the source reappears as heat in the resistor.

Current can travel in either direction through an emf source. When the current moves from the negative to the positive terminal, some other form of energy (e.g., chemical energy) is converted into electric energy. If we were to impose a higher emf in the external circuit, for example, forcing current backward through the emf source, the electric energy would be changed or converted to some other form. When current is sent backward through a battery, electric energy is converted into chemical energy (which can,

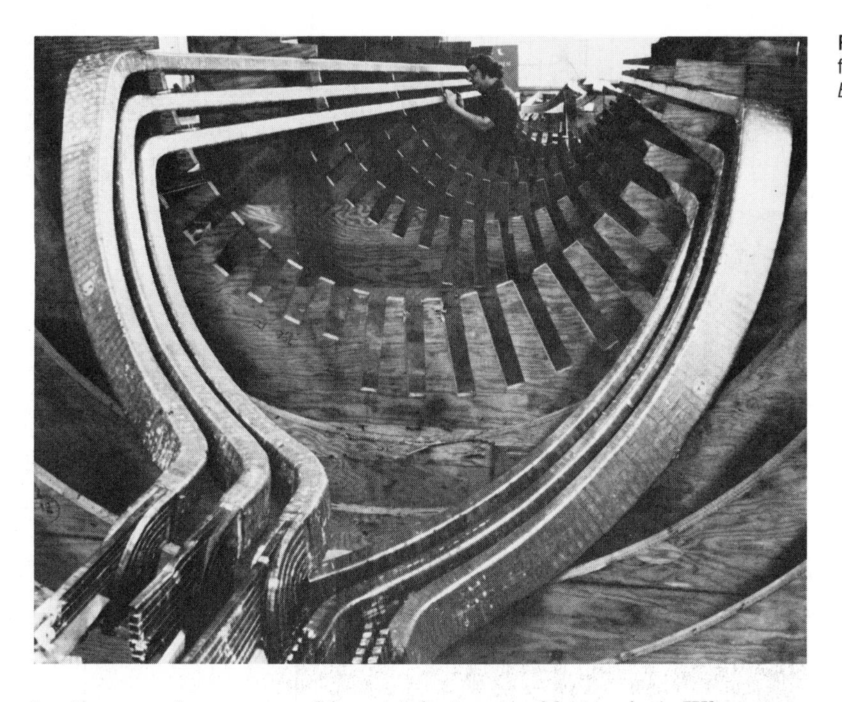

Fig. 11.13 Inspecting electric coils for a large generator. (*McGraw-Edison.*)

by the way, be recovered in certain types of batteries). When current is sent backward through a generator, the device becomes a motor.

A resistor, on the other hand, converts electric energy into heat no matter what the direction of current. However, it is impossible to reverse the process and regain electric energy from the heat.

Electrical potential always drops by the amount IR as you travel through a resistor in the direction of the current.

In a generator or battery, the positive terminal will be E_x above the negative terminal minus the voltage drop due to internal resistance between terminals. There will always be some energy converted to heat inside a battery or generator no matter which direction the current.

When a battery or a generator is discharging, the internal current passes from the negative to the positive terminal. Each coulomb of charge gains energy E_x from chemical or mechanical energy but loses IR in heat dissipation. The net gain in joules per coulomb can be determined by $E_x - IR$.

For a motor or a battery being charged, the opposite is true, because the internal current passes from the positive to the negative terminal. Each coulomb loses energy E_x and IR. The combined loss can be determined by $E_x + IR$.

In the case of a motor, the quantity E_x is commonly called back-emf, since it represents a voltage that is in a direction opposite the current flow.

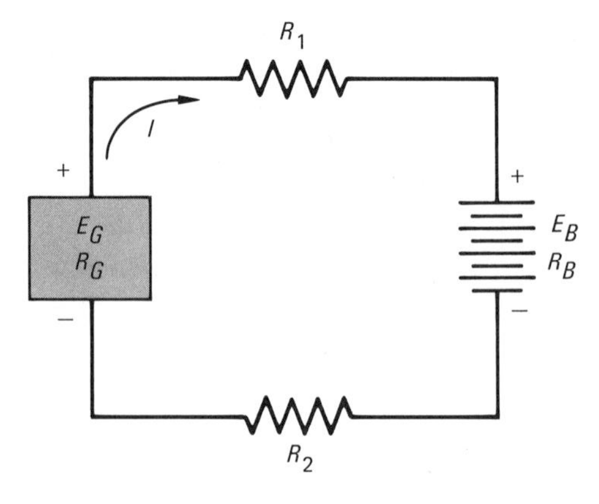

Fig. 11.14 Battery-generator circuit.

Figure 11.14 shows a circuit wherein the battery is being charged by a generator. E_G, E_B, R_G, and R_B indicate the emf's and internal resistances of the generator and battery. Each coulomb that flows around the circuit in the direction of the current I gains energy E_G from the generator and loses E_B in the form of chemical energy to the battery. Heat dissipation is realized as IR_G, IR_1, IR_B, and IR_2.

11.11 Kirchhoff's laws

Two fundamental principles that are frequently used in dc circuit analysis are known as Kirchhoff's laws:

1. The summation of the potential differences around a closed loop must be zero.
2. The summation of the currents at a junction must be zero.

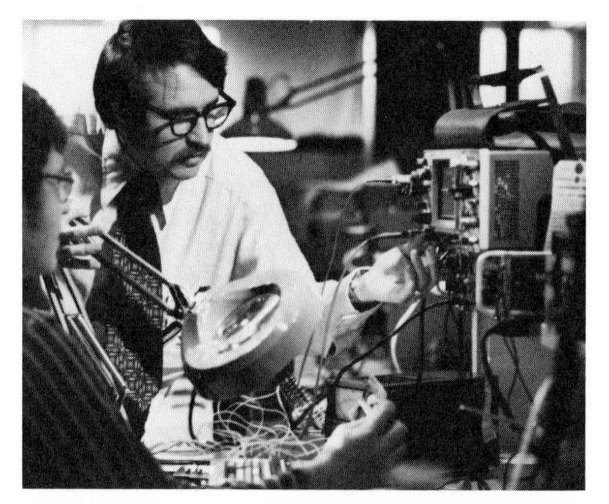

Fig. 11.15 An engineer-technician team isolate a noise-rating problem in a digital board. (*General Electric.*)

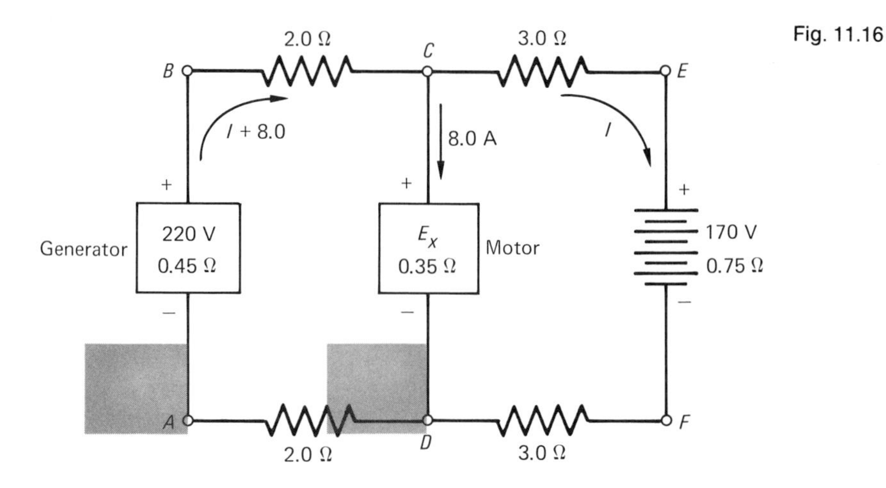

Fig. 11.16

To demonstrate these principles consider the following problem, as illustrated in Fig. 11.16.

Example problem 11.7 A 220-V generator is driving a motor drawing 8.0 A and charging a 170-V battery. The unknowns in this problem are the back-emf of the motor (E_x) and the charging current of the battery (I). Since the current through the motor is given as 8.0 A, we can see by applying Kirchhoff's second law to junction C that the current into the junction is $I + 8.0$ and the current through the battery is I.

Kirchhoff's first law dictates that we select a beginning point and travel completely around a closed loop back to the starting point, thereby arriving at the same electrical potential. As a path is selected and followed, note carefully all changes in potential. Once a loop has been completely traveled and all potential changes noted, the sum is set equal to zero.

The following voltage summation for the circuit $ABCDA$ in Fig. 11.16 results from Kirchhoff's first law.

$$(E_{A-B})_{\text{emf}} + (E_{A-B})_{\text{loss}} + (E_{B-C}) + (E_{C-D})_{\text{emf}} + (E_{C-D})_{\text{loss}} + (E_{D-A}) = 0$$

Substituting correct algebraic signs according to the established convention we get the results given in Table 11.2.

From the table, values can be correctly interpreted and the equation becomes

$$220 - 0.45(I + 8.0) - 2.0(I + 8.0) - E_x - 0.35(8.0) - 2.0(I + 8.0) = 0$$

Table 11.2

Symbols	Quantities	Notes
$(E_{A-B})_{emf}$	$+220$ V	Emf of generator
$(E_{A-B})_{loss}$	$-0.45(I + 8.0)$	Loss in generator
(E_{B-C})	$-2.0(I + 8.0)$	Loss in line
$(E_{C-D})_{emf}$	$-E_x$	Back-emf of motor
$(E_{C-D})_{loss}$	$-0.35(8.0)$	Loss in motor
(E_{D-A})	$-2.0(I + 8.0)$	Loss in line

Simplifying, we get

$$-4.45\,I - E_x + 181.6 = 0$$

This equation cannot be solved, because there are two unknowns. However, a second equation can be written around a different loop of the circuit.

$$(E_{D-C})_{emf} + (E_{D-C})_{loss} + (E_{C-E}) + (E_{E-F})_{emf} + (E_{E-F})_{loss} + (E_{F-D}) = 0$$

From this we can develop Table 11.3, and therefore a second equation.

Table 11.3

Symbols	Quantities	Notes
$(E_{D-C})_{emf}$	$+E_x$	Back-emf of motor
$(E_{D-C})_{loss}$	$+0.35(8.0)$	*IR* rise in motor
(E_{C-E})	$-3.0\,I$	Loss in line
$(E_{E-F})_{emf}$	-170 V	Drop across battery
$(E_{E-F})_{loss}$	$-0.75\,I$	Loss in battery
(E_{F-D})	$-3.0\,I$	Loss in line

$$E_x + 0.35(8.0) - 3.0\,I - 170 - 0.75\,I - 3.0\,I = 0$$

$$E_x - 6.75\,I - 167.2 = 0$$

Solving these two equations by substitution, we obtain the following results

$$I = 1.3 \text{ A}$$

$$E_x = 1.8 \times 10^2 \text{ V}$$

These values can be checked by writing a third equation around the outside loop.

By the procedure above, a set of simultaneous equations may be found that will solve any similar problem, provided the number of unknowns is not greater than the number of circuit paths or loops.

Fig. 11.17 A development engineer uses a computer to assist a test run on a linear induction motor. (*General Electric.*)

The following general procedure is outlined as a guide to systematically applying Kirchhoff's laws.

1. Sketch a circuit diagram and label all known voltages, resistances, etc. Show + and − signs on emf's.
2. Assign letters to all unknown quantities.
3. Assume a current direction in all branches of the circuit. If the direction is not known, choose a direction. A negative current solution will indicate an incorrect assumption.
4. Assign symbols to all unknown currents and apply Kirchhoff's second law at junctions.
5. Apply Kirchhoff's first law to as many circuit loops as necessary.
6. Solve the resulting set of equations.

Problems

11.1 What is the current in a wire if there are 6 000 C of charge flowing every 5 min?

11.2 How many coulombs are supplied by a battery in 36 h if it is supplying current at the rate of 1.5 A?

11.3 If electric energy costs 5 cents/kWh, what is the cost of operating five 150-W lamps for 24 h?

11.4 If a storage battery is supplying 45 A and has an emf of 12 V, what power is it delivering?

11.5 A 120-V generator delivers 12 kW to an electric furnace. What current is the generator supplying?

11.6 The difference in potential across a wire is 35 V and the current is 5.0 A. What is the resistance of the wire?

11.7 What is the resistance of a 550-W bulb in a 110-V dc power line?

11.8 A dc power supply of 95 V is connected across three resistors in series. $R_1 = 12\ \Omega$, $R_2 = 15\ \Omega$, $R_3 = 25\ \Omega$.

(*a*) Draw a circuit diagram.
(*b*) Determine the equivalent resistance.
(*c*) What is the line current?
(*d*) Calculate the voltage drop across each resistor.

11.9 Two resistors are connected in parallel across a 75-V voltage supply. $R_1 = 15\ \Omega$, $R_2 = 25\ \Omega$.

(*a*) Draw a circuit diagram.
(*b*) Determine the equivalent resistance.
(*c*) What is the line current?
(*d*) Calculate the current through each resistor.

11.10 Given the circuit diagram and values in Fig. 11.18, determine the current through the 12-Ω resistor.

Fig. 11.18

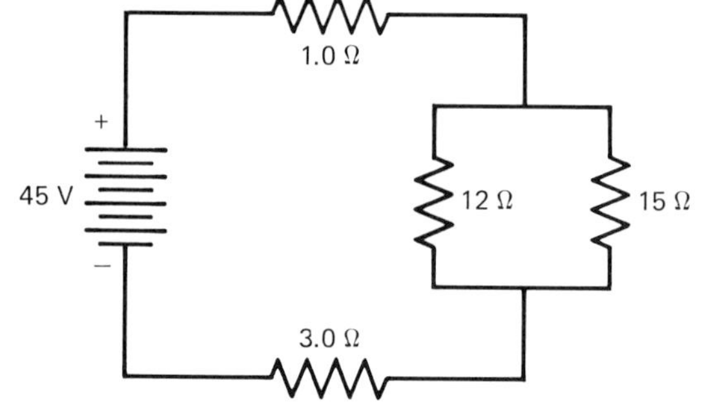

11.11 A 24-V storage battery has an internal resistance of 0.10 Ω but delivers a current of 45 A.

(*a*) What is the terminal voltage?
(*b*) What is the power delivered?

11.12 A dc motor has an internal resistance of 0.20 Ω. It draws 55 A at full load from a 110-V line.

(*a*) What is the back-emf of the motor?
(*b*) What is the power drawn by the motor?
(*c*) How much heat is generated in internal resistance?

11.13 In Fig. 11.19, determine the current and the potential at points *S*, *T*, *U*, and *V* when $R_1 = 5.0\ \Omega$, $R_2 = 2.6\ \Omega$, $E_G = 120$ V, $R_G = 0.20\ \Omega$, $E_B = 24$ V, and $R_B = 0.20\ \Omega$.

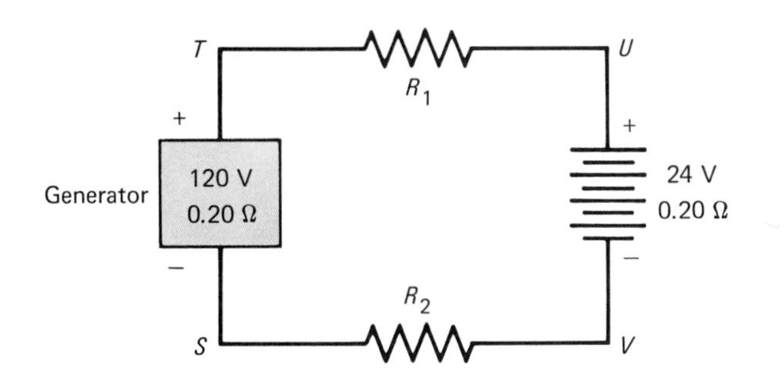

Fig. 11.19

11.14 A 155-V generator in Fig. 11.20 is charging a 110-V battery and driving a motor. Determine the charging current and back-emf of the motor. Assume no internal resistances in the generator, motor, and battery.

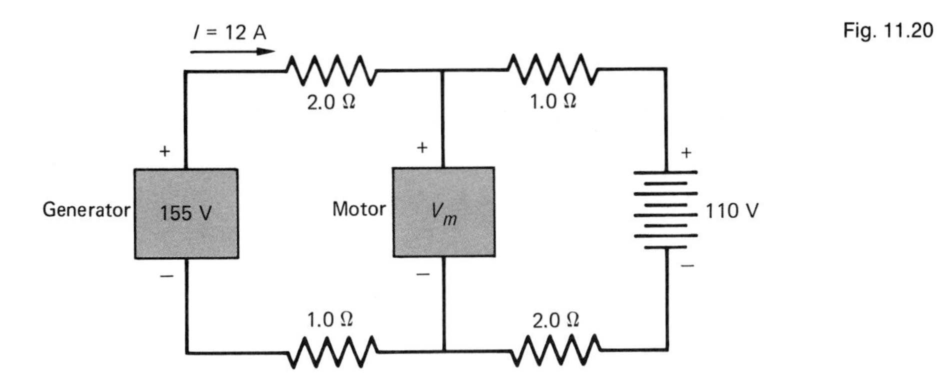

Fig. 11.20

11.15 A 120-V generator and a 24-V battery are driving a motor. The current through the motor is 12 A. Determine the back-emf of the motor and the current through the battery (see Fig. 11.21).

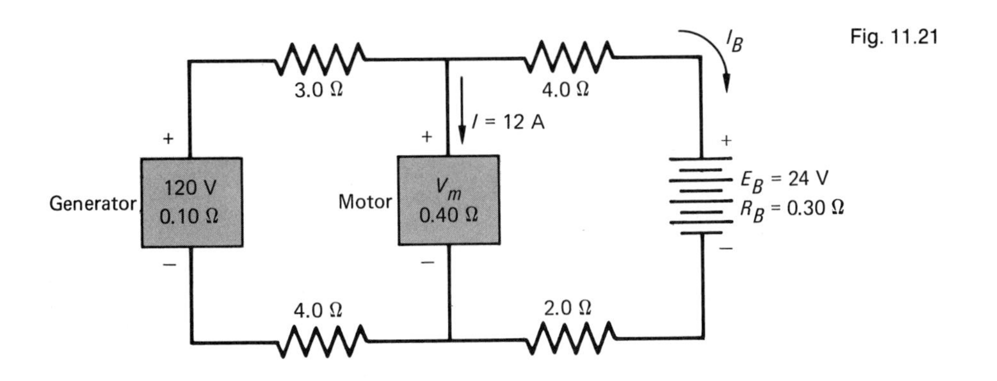

Fig. 11.21

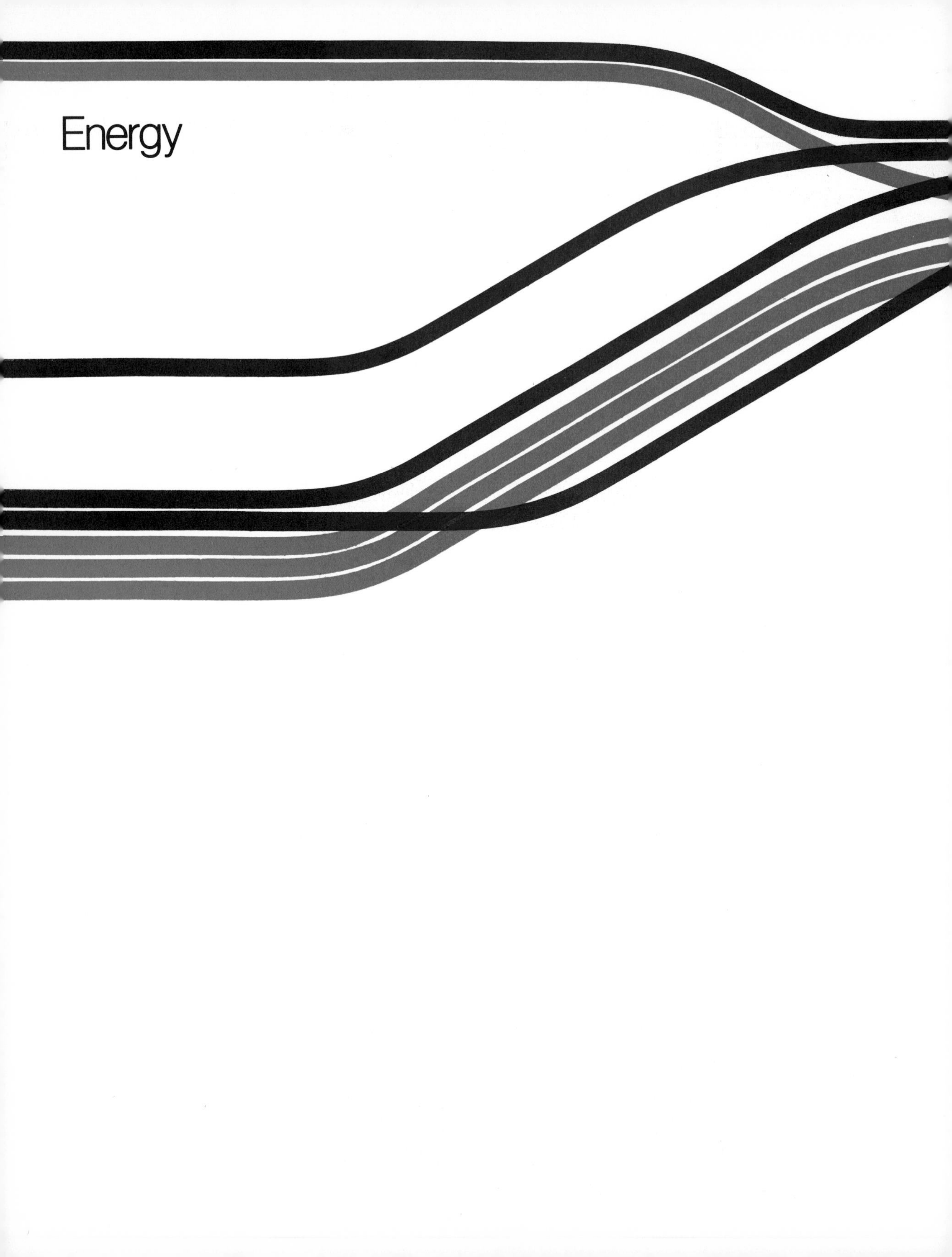

Energy

CHAPTER 12

12.1 Introduction

Today energy is the world's most important commodity. One way of characterizing the development of society during the past 200 years would be in the substitution of machine power for muscle power. This transformation has been helped by the rapid development of the natural sources of energy, namely, fossil fuels, water power, and the atom (see Fig. 12.1). Energy from these natural sources must be converted into forms that can be transported, stored, and applied at the appropriate time and place. The degree of industrial development of society can be determined by the extent of energy usage. There is an excellent correlation between productivity of a nation and its capability to generate energy.

What is energy? Energy cannot be seen; it has no mass or defining characteristics; it is distinguished only by what it can produce. In a broad sense, *energy* may be defined as an ability to produce an effect (change) or the capacity for producing an effect on matter. Energy may be within an object or move from one to another. Thus, energy is usually spoken of as being either stored or in transit.

When supplied, energy can transform natural resources into products and services beneficial to society. We observe the effects of energy when physical changes occur in objects.

Fig. 12.1 Coal remains the largest source of fossil energy. Here a giant dragline is used to strip mine coal in the western United States. *(Sun Company, Inc.)*

12.2 Stored energy

Stored energy exists in five distinct forms: potential, kinetic, internal, chemical, and nuclear. The ultimate usefulness of stored energy depends on how efficiently the energy is converted into a form that produces a desirable result.

When an object of mass m is elevated to a height h in a gravitational field, a certain amount of work must be done to overcome the gravitational pull. Work is energy associated with a force producing motion of an object. (Work will be considered further in Sec. 12.3.) The object gains an amount of energy equal to the work required to elevate it to the new position. Its potential energy has been increased. The quantity of work done is the amount of force multiplied by the distance moved in the direction in which the force acts. It is a result of a given mass going from one condition to another. *Potential energy* is thus stored-up energy. It is derived from force and height from a reference plane, not from the means by which the height was attained. In letting mass m store up energy when being elevated to height h, it will do work when it comes back to its starting point. When an object is raised, the force needed is that of gravity (weight) and the distance is the change in elevation. Thus,

$$\begin{aligned} \text{PE} &= (\text{weight})(\text{height}) \\ &= Wh \\ &= mgh \end{aligned} \qquad 12.1$$

If the units for mass (m) are kilograms; those for acceleration of gravity (g), meters per second squared; and those for height (h), meters; the units for potential energy, or PE, are newton-meters, or joules.

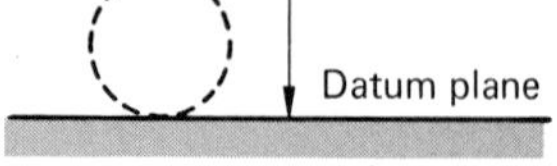

It is unnecessary in most cases to evaluate the total energy of an object, but it is essential to evaluate the energy change when there is a change in the physical state of the object. In the case of potential energy, this is accomplished by establishing a datum plane (see Fig. 12.2) and evaluating the energy possessed by objects in excess of that possessed at the datum plane. Any convenient location, such as sea level, may be chosen as the datum plane. The potential energy at any other elevation is then equal to the work required to elevate the object from the datum plane.

Fig. 12.2 Potential energy.

The energy possessed by an object by virtue of its velocity is called *kinetic energy*. It is equal to the energy required to accelerate the object from rest to its given velocity (considering the earth's velocity to be the datum plane). In equation form,

$$\text{KE} = \frac{1}{2}mv^2 \qquad 12.2$$

where m is mass, in kilograms; v is velocity, in meters per second; and KE is kinetic energy, in joules. See Fig. 12.3.

In many situations in nature, an exchange of the forms of energy is common. Consider a ball thrown into the air. It is given kinetic energy when thrown upward. When the ball reaches its maximum altitude, its vertical velocity is zero, but it now possesses a higher potential energy because of the increase in altitude (height). Then as it begins to fall, the potential energy decreases and the kinetic energy increases until it is caught. If other effects such as air friction are neglected and the ball is caught at the same altitude from which it was thrown, there is no change in energy. This phenomena is called conservation of energy, which is discussed in Sec. 12.4.

All matter is composed of molecules that, at finite temperatures, are in continuous motion. In addition, there are intermolecular attractions which vary as the distance between molecules changes. The energy possessed by the molecules as a composite whole is called *internal energy*, designated by the symbol U, which is largely dependent on temperature.

When a fuel is burned, energy is released. When food is consumed, it is converted into energy that sustains human efforts. The energy that is stored in a lump of coal or a loaf of bread is called *chemical energy*. This energy is transformed by a natural process called photosynthesis, that is, a process that forms chemical compounds with the aid of light. The stored energy in combustible fuels is generally measured in terms of heat of combustion, or heating value. For example, gasoline has a heating value of $47.7(10^6)$ J/kg, or $20.5(10^3)$ Btu/lb.

Certain processes change the atomic structure of matter. During the processes of *nuclear fission* (breaking the nucleus into two

Fig. 12.3 Kinetic energy.

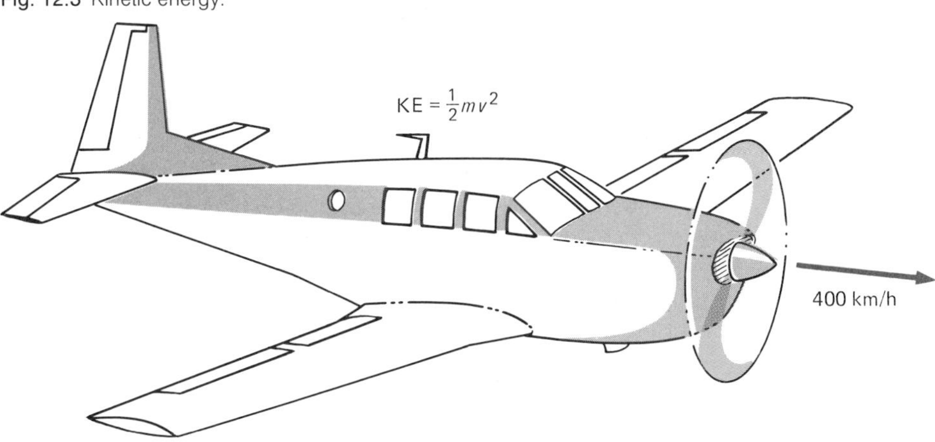

parts, which releases high amounts of energy) and *fusion* (combining lightweight nuclei into heavier ones, which also releases energy), mass is transformed into energy. The stored energy in atoms is called *nuclear energy,* which will play an ever-increasing role in the technology of the future.

12.3 Energy in transit

Energy is transferred from one place or form to another during many processes, such as the burning of fuels to run engines or the converting of electric energy to heat by passing a current through a resistance. Like all other transfer processes, there must be a driving force or potential difference in order to effect the transfer. The character of the driving force distinguishes the different forms of energy in transit: work, heat, electric energy, and radiant energy.

If there is an imbalance of forces on an object and movement against a resistance, energy is required for the operation. The transferred energy associated with mechanical forces is defined as *work,* expressed in equation form as

$$W = (\text{force})(\text{distance})$$

$$W = Fd \qquad 12.3$$

where force is in the direction of motion. If force is in newtons and distance in meters, then work is in joules.

Heat is defined as energy that is transferred from one region to another by virtue of a temperature difference. The unit of heat is the joule. Because of the large numbers that oftentimes must be used in analyzing energy-transfer processes, the megajoule (MJ) is frequently used. 1 MJ $= 10^6$ joules. The symbol used for heat energy is Q.

Another form of energy in transit is *electric energy,* which is transferred through a conducting medium when a difference in electromotive force (emf) exists in the medium.

The greatest source of natural energy is *radiant energy,* which reaches us from the sun. This energy is transferred in the form of electromagnetic waves within which energy appears as light, radio waves, ultraviolet radiation, infrared radiation, x-rays, and gamma rays. The conversion of radiant energy into other forms of energy represents a challenge for the future. The available solar energy totals many times the energy demands that will conceivably ever exist.

Several examples of computations involving energy quantities are given below. Particular attention should be given to the units and unit conversions in the examples.

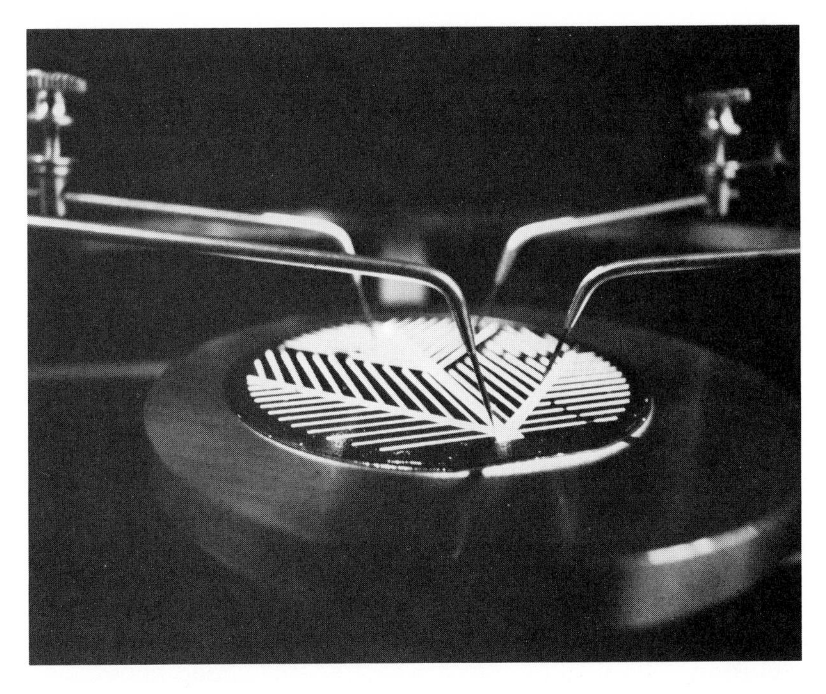

Fig. 12.4 A solar wafer is shown under test conditions. This wafer can serve as an energy source for recharging storage batteries. (*McGraw-Edison.*)

Example problem 12.1 An object with mass of 1.000 Mg rests on a ledge 200.0 m above sea level. What type of energy does the object possess and what is its magnitude?

Solution The object possesses potential energy, so from Eq. 12.1,

$$PE = mgh$$

$$= (1.000\ \text{Mg})(9.807\ \text{m/s}^2)(200.0\ \text{m})$$

Note that $1\ \text{N} = 1\ \text{kg} \cdot \text{m/s}^2$, so PE becomes

$$PE = 1.961(10^3)\text{Mg} \cdot \text{m}^2/\text{s}^2$$

$$= 1.961(10^6)\ \text{Nm}$$

$$= 1.961(10^6)\ \text{J}$$

$$= 1.961\ \text{MJ}$$

Example problem 12.2 A $2.00(10^4)$ lb truck is traveling at sea level with a speed of 50.0 mi/h. Determine the energy form and magnitude possessed by the truck.

Solution Note that the units are in the English system and will be converted to SI units so that energy can be expressed in joules.

Converting to SI units, we get

$$2.00(10^4)\ \text{lb} = 2.00(10^4)\ \text{lb}(0.4536\ \text{kg/lb}) = 9\,072\ \text{kg}$$

$$50.0\ \text{mi/h} = \frac{50.0\ \text{mi}}{\text{h}} \left| \frac{1\,609\ \text{m}}{\text{mi}} \right| \frac{1\ \text{h}}{3\,600\ \text{s}} = 22.3\ \text{m/s}$$

The truck possesses kinetic energy; therefore, from Eq. 12.2,

$$\begin{aligned} KE &= \tfrac{1}{2}mv^2 \\ &= \tfrac{1}{2}(9\,072\ \text{kg})(22.3\ \text{m/s})^2 \\ &= 2.26(10^6)\ \text{kg}\cdot\text{m}^2/\text{s}^2 \\ &= 2.26(10^6)\ \text{Nm} \\ &= 2.26\ \text{MJ} \end{aligned}$$

Example problem 12.3 A mass of water is heated from 10.0°C to 20.0°C by the addition of 5.00(10³) Btu of energy. What is the final form of the energy? Express the final form of the energy in megajoules.

Solution The final form of the heat energy added appears as increased internal energy of the water. If state 1 of the water is prior to heating and state 2 is after heat has been added, then $U_2 - U_1 = 5\,000$ Btu.

Converting to SI, we get

$$\begin{aligned} U_2 - U_1 &= 5.00(10^3)\ \text{Btu} = (5.00(10^3)\ \text{Btu})(1\,005\ \text{J/Btu}) \\ &= 5.28(10^6)\ \text{J} \\ &= 5.28\ \text{MJ} \end{aligned}$$

Example problem 12.4 A force of 2.00(10²) N acting at an angle of 20° with the horizontal is required to move block *A* along the horizontal surface (see Fig. 12.5). How much work is done if the block is moved 1.00(10²) m?

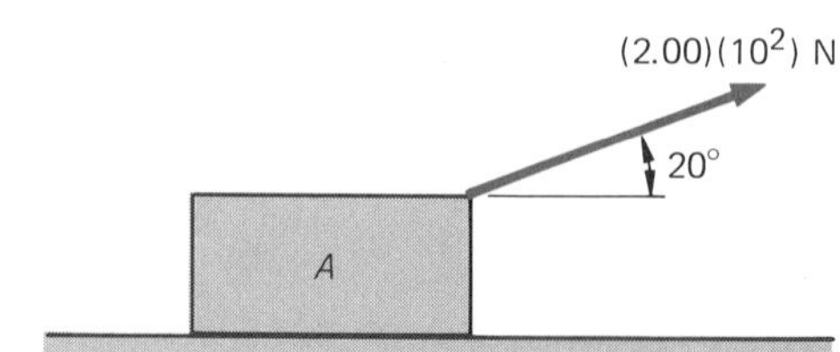

Fig. 12.5 An applied force doing work.

Solution Work is computed as the product of the force in the direction of motion and the distance moved, as in Eq. 12.3.

$$W = Fd$$

$$= (2.00 \times 10^2 \text{ N})(\cos 20°)(1.00 \times 10^2 \text{ m})$$

$$= 18\,800 \text{ Nm}$$

$$= 18\,800 \text{ J}$$

$$= 18.8 \text{ kJ}$$

Where did the energy of the work go? There is no increase in potential energy, since height was not changed. There is no velocity change, so kinetic-energy change is zero. The energy of the work in this example is dissipated as heat of friction between the block and surface.

12.4 Conservation of energy

The law of the conservation of energy states simply that energy can never be created or destroyed but only transformed (in nonnuclear processes). In effect, energy is converted from one form to another without loss. Careful measurements have shown that during energy transformations there is a definite relationship in the quantitative amounts of energy transformed. This relationship —the first law of thermodynamics—is a restatement of the law of the conservation of energy.

12.5 Concepts from thermodynamics

Thermodynamics is one of the major areas of engineering science. It is usually introduced to engineering students in a one-quarter or one-semester course, with students in energy-related disciplines continuing with one or more advanced courses. But it is necessary in this introduction to energy to introduce briefly some concepts that will help you solve some basic problems involving the transformation and consumption of energy. Our discussion is limited to the first law for closed systems; and the second law, efficiency and power.

12.5.1 First law of thermodynamics

When applying the first law to substances undergoing energy changes, it is necessary to define a system and write a mathematical expression for the law (see Sec. 10.2). In Fig. 12.6, a generalized closed system is illustrated. In a closed system, no material may

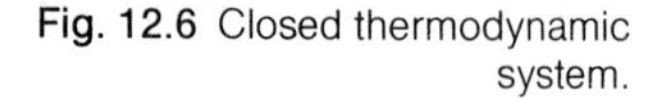

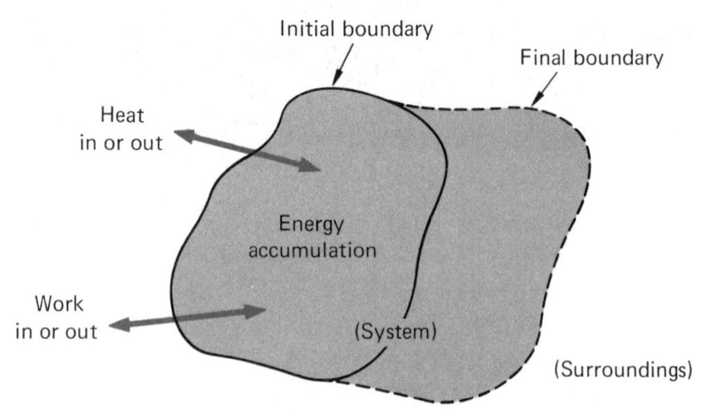

Fig. 12.6 Closed thermodynamic system.

cross the boundaries, but the boundaries may change shape. Energy, however, may cross the boundaries and/or the system may be moved intact to another position.

Some applications involve analysis of a system in which mass crosses the defined boundaries and a portion of the energy transformation is carried in or out of the system with the mass. An example is an air compressor, which takes atmospheric air, compresses it, and delivers it to a storage tank. (Because of the complexity of analysis, problems involving open systems will not be considered here.)

For the generalized closed system, the first law can be written as

Energy in + energy stored at condition 1
= Energy out + energy stored at condition 2

where conditions 1 and 2 refer to initial and final states of the system.

Energy in any of its forms may cross the boundaries of the system. If the assumption is made that there is no nuclear, chemical, or electrical energy involved and that no changes will occur in potential and kinetic energy, then the first law becomes

Heat in + work in + internal energy at condition 1
= Heat out + work out + internal energy at condition 2

or combining heat, work, and internal energy quantities at conditions 1 and 2.

$$ {}_1Q_2 = U_2 - U_1 + {}_1W_2 \qquad 12.4 $$

${}_1Q_2$ = Heat *added* to system

${}_1W_2$ = Mechanical work *removed* from system

$U_2 - U_1$ = Change in internal energy

Example problem 12.5 The internal energy of a system decreases by 108 J while 175 J of work are transferred to the surroundings. Determine if heat is added to or taken from the system.

Solution

$${}_1Q_2 = U_2 - U_1 + {}_1W_2$$

$$= -108 + 175$$

$$= +67 \text{ J} \qquad \text{(heat is added to system)}$$

Example problem 12.6 Analyze the energy transformations that can take place when the piston in Fig. 12.7 moves in either direction and/or heat is transferred.

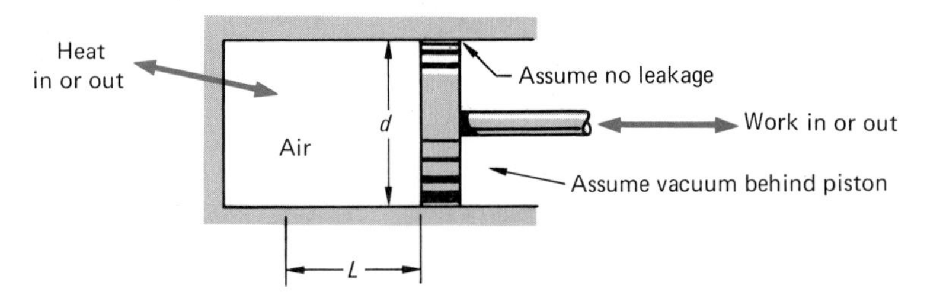

Fig. 12.7 Piston-cylinder combination.

Solution The air is considered to be the system. If the position shown is condition 1, there must be a force applied to the piston toward the left to hold its position. The magnitude of the force is

$$F = \text{(pressure of air)(area of piston)}$$

$$= PA \qquad \text{newtons}$$

At this position, with no change in volume of the air, heat could be added or removed, which would increase or decrease, respectively, the internal energy of the system. In terms of the first law,

$${}_1Q_2 = U_2 - U_1$$

Note the absence of the work term. Work cannot take place without action of a force through a distance. The force (PA) is present, but no movement takes place; thus no work is transferred.

If the force to the left is increased slightly, then the piston will move left; and by Newton's third principle (action equals reaction), the pressure of the air will increase. The air temperature will change depending on how much heat is removed from the air. The first law for this situation is written

$${}_1Q_2 = U_2 - U_1 + {}_1W_2$$

The ${}_1W_2$ will be negative, since work is being added to the system.

For the case of the piston moving to the right, the first law is still

$${}_1Q_2 = U_2 - U_1 + {}_1W_2$$

but ${}_1W_2$ will be positive.

If no heat is allowed to transfer across the boundaries, then the first law becomes

$${}_1W_2 + U_2 - U_1 = 0$$

This process, where no heat crosses the boundary, is called an *adiabatic process.*

A constant-temperature (*isothermal*) process implies that there is no change in internal energy; and therefore the first law for an isothermal process becomes

$${}_1Q_2 = {}_1W_2$$

Example problem 12.7 Using the previous example and Fig. 12.7, let the piston be moved so that the volume of air is reduced to one-half its original value. During this process, 116 MJ of heat leaves the air. If the process is isothermal, how much work is done on the system?

Solution

$${}_1Q_2 = {}_1W_2$$

$$-116 \text{ MJ} = {}_1W_2$$

The negative sign indicates that work has been done on the system.

Example problem 12.8 Determine the pressure of the air in Fig. 12.7 if the force on the piston is 1.00(10²) N and the diameter d is 0.100 m.

Solution

$$F = PA$$

$$P = \frac{F}{A}$$

$$= \frac{1.00(10^2)}{\pi d^2/4}$$

$$= \frac{1.00(10^2)(4)}{0.01\pi}$$

$$= 1.27(10^4) \text{ N/m}^2$$

$= 1.27(10^4)$ Pa

$= 12.7$ kPa

12.5.2 Second law of thermodynamics

The first law states that heat is a form of energy; it may be turned into work, and conversely, work can be returned to heat. However, there is a limitation in the conversion of heat into work. Stated simply, heat cannot perform work except when it passes from a higher to a lower temperature. In other words, heat will not flow spontaneously from a colder to a hotter substance. This limitation on heat conversion forms the basis of the second law of thermodynamics. Developed from the second law are the concepts of available energy and entropy, which state that since all energy transformations result in the production of heat, eventually all energy will be dissipated as heat and become unavailable where it is wanted. However, as long as the sun continues to shine, the energy "lost" as heat will be replenished. No energy actually disappears (first law); it becomes unavailable (second law) for transformation into a useful form.

Another use of the second law is for determination of the maximum efficiency of any device that converts heat into work. Such devices, called heat engines, cannot attain 100 percent efficiency because of the second law. Sadi Carnot proposed in 1824 an ideal engine cycle that had the highest attainable efficiency within thermodynamic laws. In reality, an engine following the Carnot cycle cannot be constructed, but the theory provides a basis of comparison for practical engines. The Carnot efficiency can be shown to be

$$\text{Carnot efficiency} = 1 - \frac{T_2}{T_1} \qquad 12.5$$

where T_1 is the absolute temperature at which the engine receives heat (high temperature) and T_2 is the absolute temperature at which the engine rejects heat (low temperature) after performing work.

Example problem 12.9 A steam engine is designed to accept steam at 300°C and exhaust at 100°C. What is its maximum possible efficiency?

Solution

$$\text{Carnot efficiency} = 1 - \frac{T_2}{T_1}$$

$$= 1 - \left(\frac{100 + 273}{300 + 273}\right)$$

$$= 1 - \left(\frac{373}{573}\right)$$

$$= 35 \text{ percent}$$

12.5.3 Efficiency

All the energy put into a system does not end up producing useful results. According to the second law, a certain amount of energy is unavailable for productive work. In addition, the available energy does not perform an equivalent amount of work because of losses incurred during the transfer of energy from one form to another. An automobile engine converts chemical energy in gasoline to mechanical energy at the axle; however, some of the energy is lost through bearing friction, incomplete combustion, cooling water, and other thermodynamic and mechanical losses. The engineer designing units to convert heat into work must be concerned with the overall efficiency of the proposed unit. This is an estimate of actual performance, not ideal performance. The overall efficiency for a heat engine may be written as

$$\text{Overall efficiency} = \frac{\text{useful output}}{\text{total input}} \qquad 12.6$$

From Eq. 12.5, it is clear that the maximum (Carnot) efficiency can be increased by lowering the exhaust temperature T_2 and/or increasing the input temperature T_1. In theory this is true, but practical design considerations must include available materials for construction of the heat engine. The engineer thus attempts to obtain the highest possible efficiency using existing technology. Examples of overall efficiency are 17 to 23 percent for automobile engines (gasoline), 26 to 38 percent for diesel engines, and 20 to 33 percent for turbojet aircraft engines. Thus, in the case of the gasoline automobile engine, for every 80-L (21-gal) tank of gasoline, only 20 L (5.3 gal) end up moving the automobile.

Care must be exercised in the calculation and use of efficiencies. To illustrate this point, consider Fig. 12.8, which depicts a steam power plant from the burning of fuel for steam generation to driving an electric generator. Efficiencies of each stage or combinations of stages in the power plant may be calculated by comparing energies available before and after the particular operations. For example, combustion efficiency can be calculated as $0.80/1.00 =$ 80 percent. The turbine efficiency is $0.33/0.50 = 66$ percent. The overall power plant efficiency up to the electric generator is $0.33/1.00 = 33$ percent. By no means is the 33 percent a measure of the efficiency of generation of electricity for use in a residential home. There will be losses in the generator and line losses in the

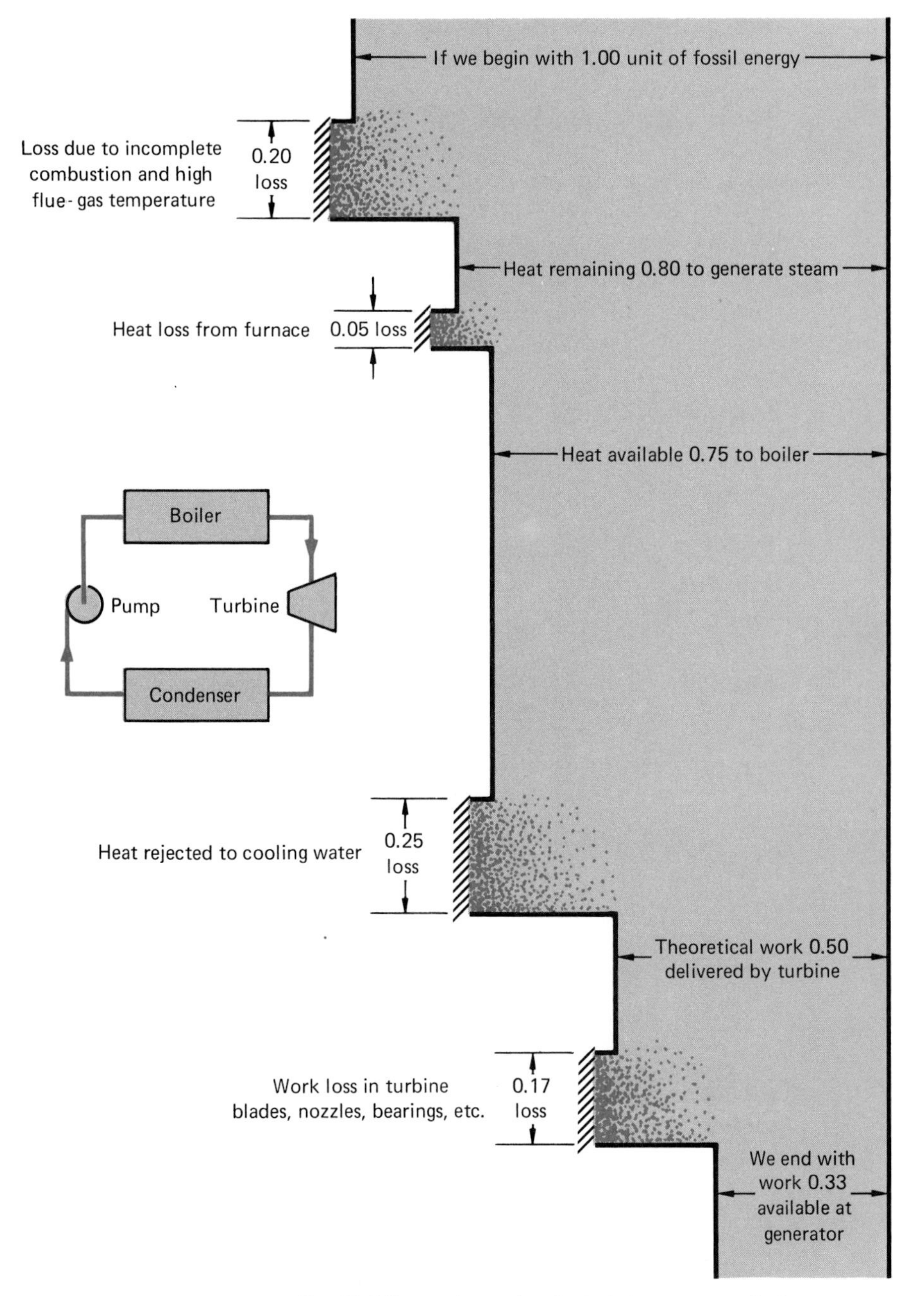

Fig. 12.8 Energy losses in a typical steam power plant.

transmission of the electricity from the power plant to the home. The overall efficiency from fuel at the power plant to the electric oven in the kitchen may run as low as 25 to 30 percent. It is perhaps ironic to note that the cycle is complete when the oven converts electricity back into heat, which is where the entire process began.

12.5.4 Power

Power is the rate at which energy is transferred, generated, or used. In many applications it may be more convenient to work with power quantities rather than energy quantities. The unit of power in the SI is the watt, but many problems will have units of horsepower (hp) and/or foot-pound per second (ft · lb/s) as units, so conversions will be necessary.

Example problem 12.10 A steam turbine at a power plant produces 3 500 hp at the shaft. The plant uses coal as fuel (12 000 Btu/lb). Using Fig. 12.8 to obtain an overall efficiency, determine how many metric tons of coal must be burned in a 24-h period to run the turbine.

Solution From Fig. 12.8, efficiency is 0.33/1.00 = 33 percent From Eq. 12.6,

$$\text{Total input} = \frac{\text{useful output}}{\text{overall efficiency}}$$

$$= \frac{3\ 500\ \text{hp}}{0.33} \left| \frac{2\ 545\ \text{Btu}}{\text{hp} \cdot \text{h}} \right.$$

$$= 2.7(10^7)\ \text{Btu/h}$$

The amount of coal needed for one day of operation is therefore

$$\text{Coal required} = \frac{2.7(10^7)\ \text{Btu}}{\text{h}} \left| \frac{\text{lb}}{12\ 000\ \text{Btu}} \right| \frac{\text{kg}}{2.205\ \text{lb}} \left| \frac{\text{metric ton}}{10^3\ \text{kg}} \right| \frac{24\ \text{h}}{\text{d}}$$

$$= 24\ \text{metric ton/day}$$

A measure of efficiency commonly used in the refrigeration and air-conditioning fields is called the energy efficiency ratio, abbreviated EER. In essence, a refrigerating machine is a reversed heat engine, that is, heat is moved from a low-temperature region to a high-temperature region, requiring a work input to the reversed heat engine. The expression for efficiency of a reversible heat engine is

$$\text{Refrigeration efficiency} = \frac{\text{refrigerating effect}}{\text{work input}} \qquad 12.7$$

Refrigeration efficiency is also called coefficient of performance (CP). The numerical value for CP will be greater than 1.

Refrigeration units are measured in tons of cooling capacity. The ton unit originated with early refrigerating machines and was defined as the amount of refrigeration produced by melting one ton of ice in a twenty-four-hour period. If the latent heat of ice is taken into account, then a ton of refrigeration will be equivalent to 12 000 Btu/h.

The EER takes into account the normal designations for refrigerating effect and work input and expresses these as power rather than energy. Thus, EER is the ratio of a refrigerating unit's capacity to its power requirements.

$$\text{EER} = \frac{\text{Refrigerating effect(Btu/h)}}{\text{Power input(W)}}$$

Example problem 12.11 Compute the cost of running a 3-ton air-conditioning unit an average of 8 h/d for 1 month if electricity sells for 4.5¢/kWh. The unit has an EER = 8.

Solution

$$\text{EER} = \frac{\text{Refrigerating effect}}{\text{Power input}}$$

$$8 = \frac{3(12\ 000)}{\text{Power input}}$$

$$\text{Power input} = 4\ 500\ \text{W}$$

$$\frac{\$}{\text{month}} = \frac{4\ 500\text{W}}{} \left| \frac{8\ \text{h}}{\text{d}} \right| \frac{30\ \text{d}}{\text{month}} \left| \frac{\$0.045}{\text{kWh}} \right| \frac{\text{kW}}{10^3\ \text{W}}$$

$$= \$48.60/\text{month}$$

Fig. 12.9 New designs can save energy. This automated laundry and dry-cleaning system reduces operating costs and brings savings to users. (*McGraw-Edison.*)

Fig. 12.10 An offshore oil-storage platform assists human beings in making efficient use of remaining fossil energy sources. (*Phillips Petroleum Company.*)

12.6 Energy sources in the future

The engineer will continue to design devices that will perform work by conversion of energy. Only now the source of energy and/or the conversion efficiency will play a much greater role in the design procedure. The energy demand today is greater than the discovery or practical use of additional sources. The following are brief discussions of possible sources that will play an important role in satisfying energy demands in years to come.

12.6.1 Wind power

Wind power is not a new concept (see Fig. 12.11). There have been many attempts to harness its energy. Research is being conducted

Fig. 12.11 Wind may again become an important source of energy for many of us in the future. (*Sun Company, Inc.*)

Fig. 12.12 Electricity generated from water power continues to play an important role in our search for energy. The Folsom Dam and Powerplant provides electricity for parts of California. (*U.S. Bureau of Reclamation.*)

that is aimed at more efficient windmill and storage systems. It appears that the major application of wind power could be for small generating units, such as for farms, rather than for large cities or industrial complexes. Moreover, to be efficient, wind machines must be located where there is a high minimum average wind velocity.

12.6.2 Water, or hydroelectric, power

Generation of electricity by using the energy of falling water constitutes about five percent of the electricity requirements in the United States (see Fig. 12.12). The limit on this highly efficient method is the number of dams that can be built on the world's rivers. Thus, hydroelectric power is not going to be the total answer to growing energy needs.

Example problem 12.12 How many megawatts of electric power can be obtained from the water of the Zambezi River over Victoria Falls if $1.0(10^5)$ m^3/s of water passes over the falls and drops $1.0(10^2)$ m. Assume a conversion efficiency of 80 percent and that the water turbines can be placed at the bottom of the falls.

Solution Equation 12.6 may be used with an estimated efficiency of 0.80: Output = (0.80)(input)

The input may be calculated as the potential energy of the water that would be converted to kinetic energy during the 100-m drop. Equation 12.1 is used with the mass equal to the volume multiplied by the density of the water. Conversion of input energy to power is accomplished by dividing by 1 s.

$$\begin{aligned} PE &= mgh \\ &= (V\rho)gh \\ &= (10^5\ m^3)(10^3\ kg/m^3)(9.81\ m/s^2)(100\ m) \\ &= (9.81)(10^{10}) kg \cdot m^2/s^2 \\ &= 9.81(10^{10})\ J \end{aligned}$$

The input power is then

$$\begin{aligned} \text{Input} &= \frac{9.81(10^{10})\ J}{1\ s} \\ &= 9.81(10^{10})\ W \end{aligned}$$

and the output is

$$\begin{aligned} \text{Output} &= (0.80)(9.81)(10^{10})\ W \\ &= 7.8(10^4)\ MW \end{aligned}$$

Recently, interest has grown in harnessing the tides for power. To make tidal energy conversion economically feasible, the range of the high to low tide must be at least 10 m and there must be sufficient inlets that can be dammed. Sites fulfilling both conditions exist at 5 percent of the world's coastlines, with most of these in remote areas.

12.6.3 Geothermal power

In areas of the world where natural geysers, or hot springs, exist, a tremendous heat energy source can be tapped. (See Fig. 12.13.) In the western states, the output of electricity generated by using high-pressure ground steam reached 1 000 MW in the late 1970s and is increasing by about 100 MW/year. At the San Andreas fault in southern California, there is enough natural steam to produce 4 000 – 5 000 MW of low-cost electric power.

Another method of obtaining energy from within the earth is receiving a great deal of attention. A well is drilled into the hot subterranean rock beds, and water is injected into the well, is heated, and returns as steam. The steam then drives turbines on the surface.

12.6.4 Solar power

Heat is transferred at a rate of about 1 kW/m² to the earth's surface whenever the sun is shining (see Fig. 12.14). The problem of

Fig. 12.13 Geothermal energy is a potential source of power for areas where this energy exists. (*Phillips Petroleum Company.*)

harnessing this power lies with collecting, converting, and storing the energy. Several research projects are being carried out in areas of the world with a high percentage of sunshine in an attempt to develop devices that will convert solar energy to one of the other forms in an economically feasible manner. A great number of solar

Fig. 12.14 Solar energy is the greatest source of energy available. This artist's conception of a solar energy conversion station depicts the research being conducted in the attempt to harness the sun's energy. (*Black and Veatch Consulting Engineers.*)

collectors are on the market today for home use. Some research is being conducted in extracting the solar heat from the oceans.

Example problem 12.13 Estimate the area over which solar energy must be collected to provide the electric energy needed by a small community of 4 500 persons in a 24-h period. Assume a conversion efficiency of 8 percent.

Solution This example requires some assumptions in order to obtain a meaningful result.

1. Each home consumes about 13 000 kWh of electricity on the average each year.
2. An average of three persons lives in each home.
3. The sun shines an average of 8 h/d.

Equation 12.6 may be used.

$$\text{Input} = \frac{\text{output}}{\text{efficiency}}$$

$$= (\text{area})(\text{solar heat-transfer rate})$$

$$= \text{area } (1 \text{ kW/m}^2)$$

Therefore, the area can be computed as

$$\text{Area} = \frac{\text{m}^2 \cdot 13\,000 \text{ kWh} \cdot \text{home} \cdot 4\,500 \text{ persons} \cdot \text{d} \cdot \text{year}}{0.08 \cdot 1 \text{ kW} \cdot \text{year-home} \cdot 3 \text{ persons} \cdot 8 \text{ h} \cdot 365 \text{ d}}$$

$$= 84\,000 \text{ m}^2$$

$$= 8.4 \text{ hm}^2$$

12.6.5 Nuclear power

As mentioned earlier in this chapter, nuclear power may be generated from two processes, fission and fusion. Fission is the splitting of an atom of nuclear fuel, usually uranium-235 (U-235). The splitting process releases a great quantity of heat, which can then be converted to other energy forms. Nearly all reactors on line today producing electric energy are fission reactors using U-235. However, U-235 is limited in supply and would soon be exhausted if there were total energy dependence on it. In addition, radioactive waste material has caused some disposal problems that must be solved. Some 200 000 m³/year of nuclear waste are expected by the year 2000.

The development of the fast breeder reactor promises to be the power reactor of the future. It uses U-238, a plentiful element, to

Fig. 12.15 The Duane Arnold Energy Center, Palo, Iowa, is a nuclear-powered electric generating station. Nuclear power is another of the potential sources for additional energy. (*Iowa Electric Light and Power Company.*)

generate heat and also produces the fissionable material plutonium-239. In essence then, it is manufacturing fuel as it produces energy. But again the problem of nuclear waste must be overcome before fast breeder reactors are constructed on a large scale.

Example problem 12.14 If 1 g of U-235 is capable of producing the same amount of heat as 3 500 kg of coal, how much water would have to be stored $1.0(10^3)$ m above sea level to provide the same energy as 1 kg of U-235? Assume that coal has a heating value of 30.0 MJ/kg.

Solution A kilogram of U-235 would possess the same energy as $3.5(10^6)$ kg of coal and therefore would have a total energy value of

$$(30.0\ \text{MJ/kg})(3.5 \times 10^6\ \text{kg}) = 105(10^6)\ \text{MJ}$$

Setting this equivalent to the potential energy of the water

$$mgh = 105(10^{12})\ \text{J}$$

$$m = \frac{105(10^{12})\ \text{J}}{(9.81\ \text{m/s}^2)(10^3\ \text{m})}$$

$$= 10.7(10^9)(\text{N}\cdot\text{m}\cdot\text{s}^2)/\text{m}^2$$

$$= 10.7(10^9)(\text{kg}\cdot\text{m/s}^2)(\text{m}\cdot\text{s}^2)/\text{m}^2$$

$$= 10.7(10^9) \text{ kg}$$
$$= 10.7 \text{ Tg}$$

Problems

12.1 What is the kinetic energy, in joules, of a $3.0(10^3)$-lb automobile traveling at 55 mi/h?

12.2 What force (newtons) would be required to stop the automobile in Prob. 12.1 in $1.0(10^2)$ m?

12.3 If the automobile in Prob. 12.1 were allowed to coast to a stop up an 11 percent grade, how far up the grade would it travel, assuming no friction?

12.4 An automobile weighing 1 750 kg has a kinetic energy of 788 kJ. What is the speed in kilometers per hour?

12.5 An 1 800-kg automobile starting from rest coasts $3.0(10^2)$ m down an 8.0 percent grade. If air and road friction are neglected, what speed will be attained?

12.6 Water flows over a falls and drops freely for 150 m. What velocity is attained just before it hits the bottom? What is its kinetic energy (per kilogram) at this point?

12.7 (*a*) A manmade lake of area 10.0 mi^2, average depth of 50.0 ft, and surface elevation of $2.00(10^2)$ ft above sea level has what potential energy with respect to sea level? Express in joules.
(*b*) What volume rate of flow (m^3/s) is required to provide energy at the rate of $1.0(10^4)$ kW if the water is allowed to fall to sea level?

12.8 An object is moving at a velocity of 65 ft/s. From what height (m) would it have to fall to achieve this velocity?

12.9 A force of 225 N at an angle of 25° with the horizontal is required to move a block along a level surface.
(*a*) How much work is done in moving the block 35 m?
(*b*) If the block is moved the 35 m in 12 s, what is the power output? Express in horsepower.

12.10 A mass of 15 kg is elevated 12 m above the surface of the moon (one-sixth the gravitational field of that on earth). How much work is done? Express in joules.

12.11 The pressure of air inside a piston-cylinder combination is 1.4 MPa. If a force of $1.0(10^4)$ N is required to hold the piston, determine the diameter of the piston.

12.12 A piston is 9.5 cm in diameter. If the pressure on the piston is 1.6 MPa, what force is required to hold the piston?

12.13 Determine the total energy of a $7.00(10^3)$ kg airplane traveling at $5.00(10^2)$ km/h at an altitude of $2.00(10^3)$ m.

12.14 A closed system receives 15 Btu of heat and 8 500 ft · lb of work. What is the change in internal energy of the system? Express as joules.

12.15 A closed system consists of 5.0 kg of gas. During a process, 530 kJ of heat is added to the gas and the internal energy is decreased by 26 kJ. How much work was done by the gas?

12.16 A closed system process is carried out adiabatically. If $5.0(10^4)$ ft · lb of work is removed from 12 lb of fluid in this process, what is the change in internal energy per pound of fluid?

12.17 A system performing a cyclic operation receives heat at the rate of $1.1(10^2)$ Btu/min and delivers work at the rate of 0.85 hp.

(*a*) How much heat leaves the system per minute? Express as joules.

(*b*) What is the overall efficiency of the system?

12.18 A tractor burns fuel at the rate of 28 L/h. If the fuel has a heating value of 36 MJ/L, what power is output if the overall efficiency is 35 percent?

12.19 What is the Carnot efficiency of a heat engine operating between temperatures of 200°C and 20°C?

12.20 Is it possible to produce 75 kW of power from a Carnot engine if it receives $5.3(10^5)$ W at 540°C and rejects heat at 260°C?

12.21 A Carnot engine develops 25 hp while rejecting $2.0(10^1)$ Btu/s to a low-temperature reservoir. Determine the temperature of the heat source. Reservoir temperature is 40°C.

12.22 An internal combustion engine is used to drive an electric generator as a standby power plant. The engine output is a steady 40 kW. The generator provides 160 A at 240 V dc. The engine burns 8.0 L/h of fuel having a heating value of 36 MJ/L. Calculate

(*a*) Efficiency of the engine

(*b*) Efficiency of the generator

(*c*) Overall efficiency of the system.

12.23 It is desired to use a pump driven by an electric motor to pump 1 200 gal of oil to a height of 36 ft in $\frac{1}{2}$ h. The oil has a specific gravity of 0.81. If the combination of motor and pump has an efficiency of 62 percent, determine the power requirements to the motor in kilowatts.

12.24 A 35-hp centrifugal pump is used to pump a coal and water slurry. If the pump is 65 percent efficient, what input horsepower is required? What power (watts) must an electric motor draw if it is 85 percent efficient and must drive the centifugal pump?

12.25 A 0.50-hp electric motor drives a pump that has an efficiency of 71 percent. If the electric motor is 87 percent efficient,

(*a*) How much power (watts) does the electric motor draw?

(*b*) How much water could be pumped in 1 h to a height of 3.0 m? Give answer in liters.

12.26 An automobile traveling at 92 km/h has a force of 820 N acting on it due to air and road friction. What power is being developed by the engine?

12.27 If electricity costs 4.5 ¢/kWh, what does it cost per year to keep four 100-W bulbs lit continuously?

12.28 How many homes, each using 13 000 kWh of electricity each year, could be furnished with electric energy each year from 11 kg of U-235? Assume 75 percent conversion efficiency. U-235 energy equivalent is 10^8 MJ/kg.

12.29 A refrigeration unit for a cold storage facility is rated at 12 tons with an EER of 7.5. If electricity costs 4.5 ¢/kWh, how much does it cost to run this unit continuously for a year?

12.30 In Fig. 12.7 let the piston diameter *d* be 0.070 m and the cylinder length be 0.090 m. If a force of 830 N is required to hold the piston in this position, determine the pressure of the air.

12.31 A home is cooled by a 3-ton air-conditioning unit. What percent of capacity of the unit is required to dissipate the heat produced from four 100-W bulbs, six 75-W bulbs, four 40-W bulbs, a 500-W heat coil for a coffee maker, and eight persons playing bridge? The average heat output of a person at rest is about 480 Btu/h.

12.32 A barrel of oil contains approximately $5.8(10^6)$ Btu of energy. Energy demand in the United States is about 38 million barrels per day of oil equivalent (MB/DOE). Express this number of MB/DOE in terms of

(*a*) tons, coal equivalent (13 000 Btu/lb)
(*b*) ft^3, natural-gas equivalent [$1.0(10^3)$ Btu/ft^3]
(*c*) liters, gasoline equivalent (160 000 Btu/gal)
(*d*) kg, uranium equivalent (1.0×10^8 MJ/kg).

Engineering economy

CHAPTER 13

13.1 Introduction

With the increase in the importance of technology has come a greater role for the engineer in management. Engineers in many cases have become managers or executive officers of businesses and must be involved in decision making. When the managers are not engineers, they most often will have engineering advisors who provide reports and analyses that influence their decisions. Steadily over the past decades, the amount of capital investment (money spent for equipment, expansion, etc.) has risen in American business and industry to the point that the average investment now exceeds the annual salary cost. The amount of the investment obviously fluctuates with the type of industry and the degree of automation. This statistic underlines the importance of knowledge of the technical aspects of the enterprise and the increasingly important role of the engineer in financial decision making.

It is common for engineers to be concerned about efficiency. In the discussion of engineering mechanics, Chap. 8, it was stated that

Fig. 13.1 The economic services section of a large consulting engineering firm. (Black and Veatch Consulting Engineers.)

the engineer designs so that material stresses will be below yield values but also with reasonable safety factors. It was implied that a good design is not one that uses excessive materials; in other words, the materials should be used efficiently. In mechanical and electrical designs, efficiency is defined as output divided by input with a maximum value of 1.00, or 100 percent. Such a relationship is obvious when dealing with energy or common materials. When considering the use of capital, it is equally obvious that anything less than 100 percent efficiency is unacceptable; after all, who wants to spend more on a venture than will be returned.

Few businesses have ever been started with the sole goal of producing a product; the aim of business is to produce a profit. In like manner, an established business is interested only in new ventures or products that will produce a favorable rate of return on the investment. Rate of return is defined as profit divided by investment, or

$$\text{Annual rate of return} = \frac{\text{annual profit}}{\text{investment}}$$

Since any venture has at least some risk involved, few investors will become interested unless there is a promise of a much greater return than could be realized by deposits in banks and savings and loans or through the purchase of treasury notes or government bonds. These considerations are part of the engineer's concern, either when having to be the decision maker or in attempting to convince the decision maker of the worth of an idea or invention.

13.2 Simple interest

The idea of interest on an investment is certainly not new. The New Testament refers to banks, interest, and return. History records business dealings involving interest at least 40 centuries ago. Early business was largely barter in nature with repayment in kind. It was common during the early years of the development of the United States for people to borrow grain, salt, sugar, animal skins, etc., from each other to be repaid when the commodity was again available. Since most of these items depended on the harvest, annual repayment was the normal process. Likewise, since the lender expected to be repaid after no more than a year, simple annual interest was the usual transaction. When it became impossible to repay the loan after a year, the interest was calculated by multiplying the principal amount by the product of the interest rate and the number of periods (years).

$$I = Pni \qquad 13.1$$

where I = interest accrued

P = principal amount

n = number of interest periods

i = interest rate per period (as a decimal, not in percent)

Therefore, if $1 000 were to be loaned at 7 percent annual interest for 5 years, the interest would be:

$$\begin{aligned} I &= Pni \\ &= (1\,000)(5)(0.07) \\ &= \$350 \end{aligned}$$

and the amount, S, to be repaid is

$$\begin{aligned} S &= P + I \\ &= 1\,000 + 350 \\ &= \$1\,350 \end{aligned} \tag{13.2}$$

It can be seen that

$$\begin{aligned} S &= P + I = P + Pni \\ &= P(1 + ni) \end{aligned} \tag{13.3}$$

13.3 Compound interest

As time progressed and business developed, the practice of borrowing became more common, and the use of money replaced the barter system. It also became increasingly more common that money was loaned for longer periods of time. Simple interest was relegated to the single-interest period, and the practice of compounding developed. It can be shown using Eq. 13.3, $n = 1$, that the amount owed at the end of one period is

$$P + Pi = P(1 + i)$$

The interest generated during the second period is then $(P + Pi)i$. It can be seen that interest is being calculated not only on the principal but upon the previous interest as well. The sum S at the end of two periods becomes

P	Principal amount
$+ Pi$	Interest during first period
$+ Pi + Pi^2$	Interest during second period
$P + 2Pi + Pi^2$	Sum after two periods

This can be factored as follows:

$$P(1 + 2i + i^2) = P(1 + i)^2$$

The interest during the third period is

$$(P + 2Pi + Pi^2)i = Pi + 2Pi^2 + Pi^3$$

and the sum after three periods is

$P + 2Pi + Pi^2$	Sum after second period
$+ \quad Pi + 2Pi^2 + Pi^3$	Interest during third period
$P + 3Pi + 3Pi^2 + Pi^3 = P(1 + i)^3$	Sum after three periods

This procedure can be generalized to n periods of time and will result in

$$S_n = P(1 + i)^n \qquad 13.4$$

where S_n is the sum generated after n periods.

Therefore, based on interest compounded annually, the preceding example becomes (5 years at 7 percent):

$$S_n = P(1 + i)^n$$

$$S_5 = (1\,000)(1.07^5)$$

$$S_5 = \$1\,402.55$$

The same result can be obtained by the following arithmetic:

Principal amount	=	\$1 000.00
Interest during first year: (1 000)(0.07)	=	70.00
Sum after 1 year	=	1 070.00
Interest during second year: (1 070)(0.07)	=	74.90
Sum after 2 years	=	1 144.90
Interest during third year: (1 144.90)(0.07)	=	80.14
Sum after 3 years	=	1 225.04
Interest during fourth year: (1 225.04)(0.07)	=	85.75
Sum after 4 years	=	1 310.79
Interest during fifth year: (1 310.79)(0.07)	=	91.76
Sum after 5 years	=	\$1 402.55

Care must be exercised in using interest rates and payment periods to make sure that the interest rate used is the rate for the period selected. If the annual rate is 12 percent, the semiannual rate is 6 percent, the monthly rate is 1 percent, and the daily rate is $\frac{12}{365}$ percent.

Example problem 13.1 What sum must be repaid if \$8 000 is borrowed at 12 percent interest for 4 years?

Solution If compounded annually,

$S_4 = (8\,000)(1.12^4) = \$12\,588.15$

If compounded semi-annually,

$S_8 = (8\,000)(1.06^8) = \$12\,750.78$

If compounded quarterly,

$S_{16} = (8\,000)(1.03^{16}) = \$12\,837.65$

If compounded monthly,

$S_{48} = (8\,000)(1.01^{48}) = \$12\,897.81$

If compounded daily,

$$S_{1\,460} = (8\,000)\left(1 + \frac{0.12}{365}\right)^{1\,460} = \$12\,927.57$$

13.4 Present worth

The transaction given in Example Problem 13.1 can be presented as a free-body diagram in the form of a *time line*.

The principal is placed at i interest for n periods (five in this figure) and S_5 is drawn out at the end of five periods. See Fig. 13.2.

When the arrow is directed toward the time line, it indicates an amount placed into the account. Conversely, the arrow (S_5) directed away from the time line indicates an amount withdrawn from the account.

It should be clear that if you were the banker (lender), the direction of the arrows would be reversed. This leads us to the term present worth.

Present worth is the amount that must be invested now to produce a prescribed sum at another date. Thus, in Fig. 13.2, P is the present worth of S_5 if the interest is i and $n = 5$. Algebraically, if $S_n = P\,(1 + i)^n$, then

$$P = S_n\,(1 + i)^{-n}$$

Therefore, the present worth of \$1 469.33 to be collected (or paid) 5 years from now at 8 percent annual interest is \$1 000.00.

$$P = \frac{1\,469.33}{1.08^5} = \$1\,000.00$$

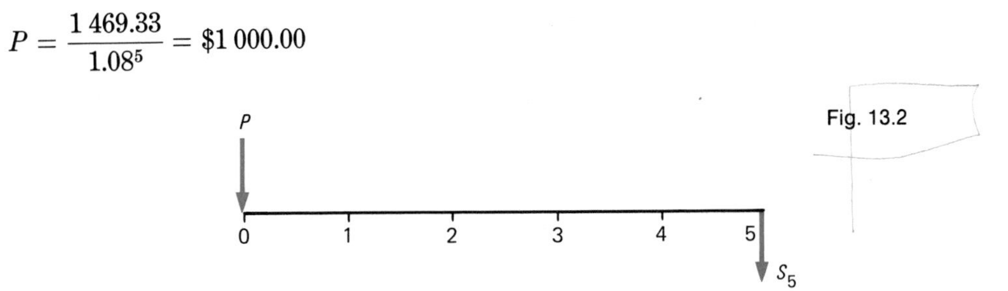

Fig. 13.2

In situations that involve economic decisions, the following questions may arise:

1. Does it pay to make an investment now?
2. What is the benefit of a payment now that will be made at some other date?
3. If a series of payments is made over specific intervals throughout a designated time span, what is this worth now?

In such cases, the answer is the present worth of the transaction. It follows then that competing proposals can be compared by determining the present worth of each.

Present worth is, in other words, the inverse of compound interest: $P = S_n(1 + i)^{-n}$

Example problem 13.2 If money is worth 5 percent annual interest compounded annually, what is the present net cash equivalent of each of the following items (with no interest payments having been withdrawn or paid in the meantime):

$1 000 deposited 2 years ago (Fig. 13.3*a*)

$1 000 deposited 1 year ago (Fig. 13.3*b*)

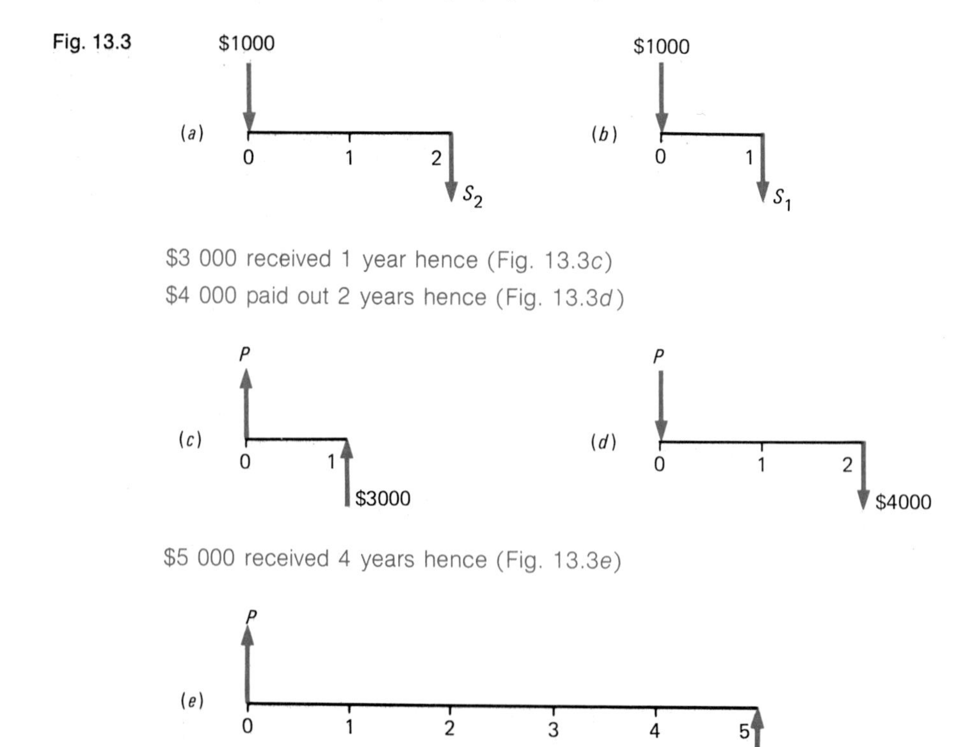

Fig. 13.3

$3 000 received 1 year hence (Fig. 13.3*c*)

$4 000 paid out 2 years hence (Fig. 13.3*d*)

$5 000 received 4 years hence (Fig. 13.3*e*)

Solution

$S_2 = (1\,000)(1.05^2) \quad = \$1\,102.50$

$S_1 = (1\,000)(1.05^1) \quad = \$1\,050.00$

$P = (3\,000)(1.05^{-1}) \quad = \quad 2\,857.14$

$P = -(4\,000)(1.05^{-2}) \quad = -3\,628.12$

$P = (5\,000)(1.05^{-4}) \quad = \quad 4\,113.51$

$5 495.03 Present worth of the five transactions

Many businesses calculate their present worth annually, so the change in present worth is a measure of the growth of the business.

Example problem 13.3 On December 31, 1970, $4 212 was available. Find the sum of money at December 31, 1977, if the interest rate is 6 percent compounded annually.

Solution See Fig. 13.4.

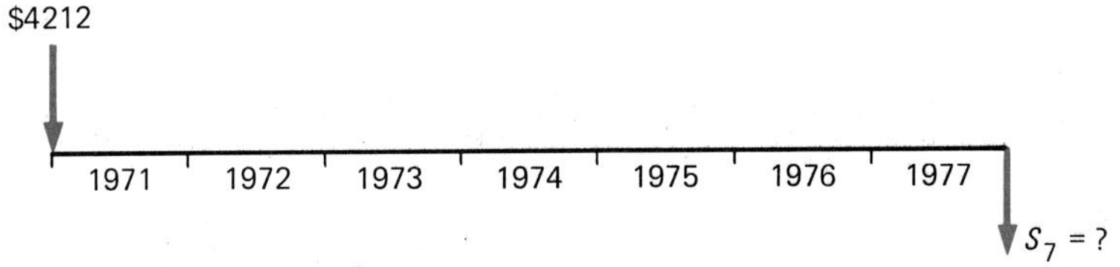

$4212

1971 1972 1973 1974 1975 1976 1977

$S_7 = ?$

$S_7 = (4\,212)[1.06]^7 = \$6333.29$

Fig. 13.4

13.5 Annuities

A series of equal payments at regular intervals accruing compound interest is known as an *annuity*. Annuities are used by business in several ways, each of which has been given a name to distinguish it from the others.

13.5.1 Sinking fund

A *sinking fund* is an annuity that is established in order to produce an amount of money at some future time. It might be used to develop cash for an expenditure that one knows is going to occur—for instance, a Christmas fund to pay for presents. If you wish to trade cars at regular intervals, it is much more economical to pay in advance. A time line for the sinking fund is shown in Fig. 13.5.

If A is deposited at the end of each period and interest is compounded at i, the sum S_n will be produced after n periods.

It can be seen that the last payment will produce no interest, the payment at period $n-1$ will produce interest equal to A times i,

Fig. 13.5

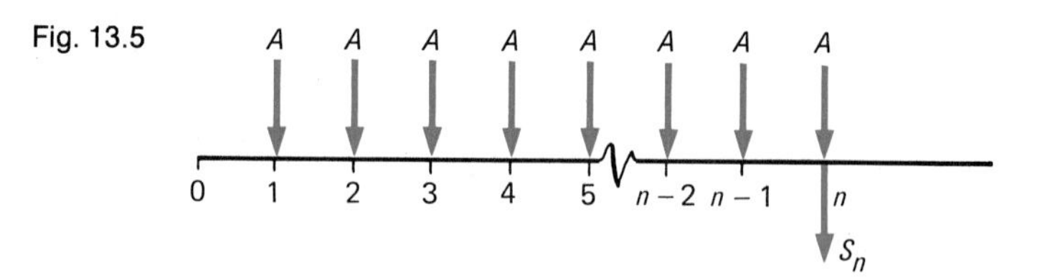

and the payment at period $n-2$ will produce $A(1 + i)^2$, and so forth. Hence, the sum produced will be as follows:

Deposit at end of period	Interest generated	Sum due to this payment
n	None	$A(1)$
$n-1$	$A(i)$	$A(1 + i)$
$n-2$	$A(1 + i)i$	$A(1 + i)^2$
$n-3$	$A(1 + i)^2 i$	$A(1 + i)^3$

$$S_4 = \text{Sum for four payments} = A(4 + 6i + 4i^2 + i^3)$$

and

$$S_4 = A\left[\frac{(4i + 6i^2 + 4i^3 + i^4 + 1) - 1}{i}\right]$$

$$= A\left[\frac{(1 + i)^4 - 1}{i}\right]$$

It can be shown that

$$S_n = A\left[\frac{(1 + i)^n - 1}{i}\right] \qquad 13.5$$

Therefore, if you want to accumulate S_n during n periods, A must be deposited at the end of each period at i interest rate, compounded at each period.

Sinking funds are often used to accumulate sufficient money to replace worn-out or obsolete equipment.

Example problem 13.4 How much will be produced by a sinking fund if $90 is deposited each month for 3 years with an interest rate of 6 percent per year?

Solution

$$S_{36} = 90\left[\frac{(1 + 0.06/12)^{36} - 1}{0.06/12}\right]$$

$$= 90\left[\frac{1.005^{36} - 1}{0.005}\right] = \$3\,540.25$$

Example problem 13.5 It is desired to accumulate $10 000 to replace a piece of equipment after 8 years. How much money must be placed annually into a sinking fund that earns 7 percent interest? Assume that the first payment is made today and the last one 8 years from today with interest compounded annually.

Solution See Fig. 13.6.

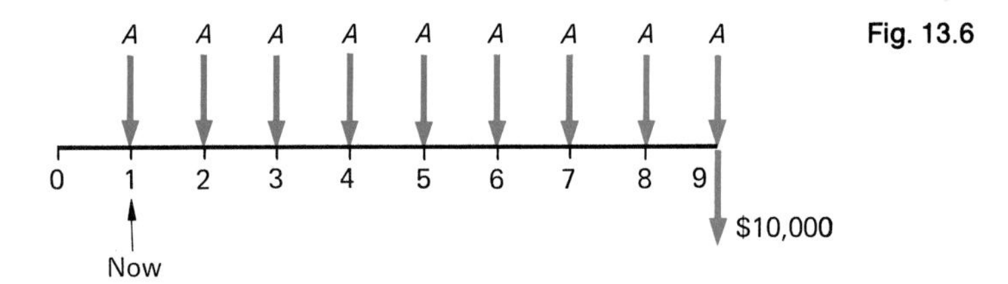

Fig. 13.6

$$S_9 = A\left[\frac{(1 + 0.07)^9 - 1}{0.07}\right] = 10\,000$$

$$A = \left[\frac{0.07(10\,000)}{1.07^9 - 1}\right] = \$834.86$$

Note that n equals 9 in this case because the first payment must occur at the end of the first period for the sinking fund formula to be valid.

13.5.2 Installment loan

A second way that annuities are used is to retire a current debt by making periodic payments instead of a single larger sum. This is the time payment plan used by most retail businesses. It is also used to amortize bond issues. The time line is shown in Fig. 13.7.

In this case the principal amount P is the size of the debt and A is the amount of the periodic payment that must be made with interest compounded at the end of each period. It can be seen that if P is removed from the time line and S_n placed at the end of the nth period, it would be a sinking fund. Furthermore, it can be shown that S_n would also be the value of P placed at compound interest for n periods $[S_n = P(1 + i)^n]$. Likewise, P can be termed the present

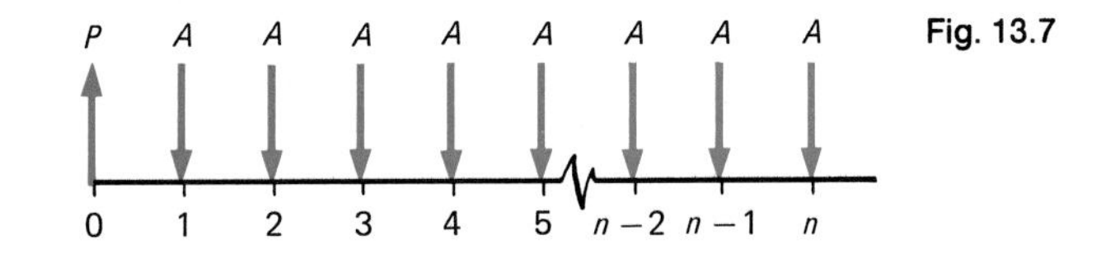

Fig. 13.7

worth of the sinking fund that would be accumulated by the deposits. Therefore, since

$$S_n = P(1 + i)^n \qquad \text{and} \qquad S_n = A\left[\frac{(1 + i)^n - 1}{i}\right]$$

the present worth becomes

$$P = A\left[\frac{(1 + i)^n - 1}{i(1 + i)^n}\right] \tag{13.6}$$

The term within the brackets is known as the present worth of a sinking fund, or the uniform annual payment present worth factor.

It follows that

$$A = P\left[\frac{i(1 + i)^n}{(1 + i)^n - 1}\right] \tag{13.7}$$

where the term in brackets is most commonly called the capital recovery factor or the uniform annual payment annuity factor and is the reciprocal of the uniform annual payment present worth factor.

Example problem 13.6 An automobile that has a total cost of \$5 760 is to be purchased by trading in an older car for which \$1 320 is allowed. If the interest rate is 9 percent a year and the payments will be made monthly beginning at the end of 1 month for 3 years, what are the monthly payments?

Solution

$$A = (5\ 760 - 1\ 320)\left[\frac{(0.007\ 5)(1.007\ 5^{36})}{1.007\ 5^{36} - 1}\right]$$

$$= \$141.19$$

If 36 monthly payments of \$141.19 had been placed in a sinking fund at 9 percent interest, they would generate

$$(141.19)\left(\frac{1.007\ 5^{36} - 1}{0.007\ 5}\right) = \$5\ 810.35$$

Another way of expressing the relationship is by saying that \$4 440 (that is, 5 760 − 1 320) is the present worth of 36 monthly payments of \$141.19, beginning in 1 month at $\frac{3}{4}$ percent per month.

Example problem 13.7 Assume that the auto-purchase arrangement in Example Problem 13.6 is the same except that no payments are to be made until 6 months after purchase and then a total of 36 payments are to be made. What is the amount of the monthly payment?

Solution See Fig. 13.8.

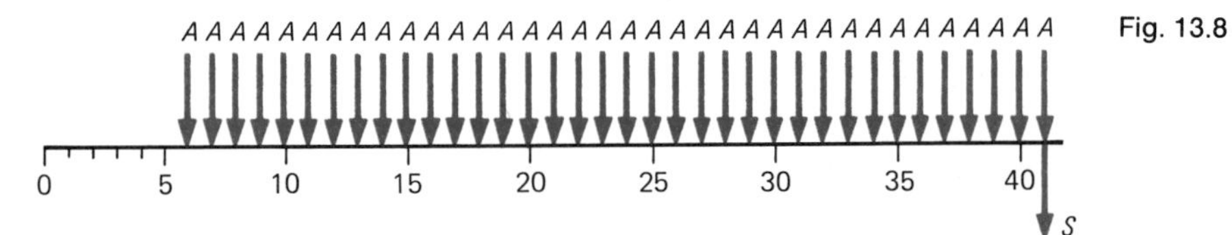

Fig. 13.8

$$\text{Balance due after trade-in} = 5\,760 - 1\,320 = \$4\,440$$

$$\text{Balance due after 5 months} = 4\,440(1.007\,5)^5 = \$4\,609.02$$

$$A = 4\,609.02\left[\frac{(0.007\,5)(1.007\,5^{36})}{1.007\,5^{36} - 1}\right] = \$146.57$$

Note that the unpaid balance is compounded for only 5 months because the first payment marks the end of the first period of the annuity. This is normally referred to as a deferred annuity.

13.5.3 Retirement plan

A third way to view annuities is the classic way in which most people think of annuities, that is, the time when periodic payments come to them, such as in retirement. The formula that applies is Eq. 13.6; and the time line is that shown in Fig. 13.9.

The problem might be stated as follows: How much money (P) must be available so that A can be received for n periods, assuming i interest rate? The example of the auto purchase and payments (Example Problem 13.6) is really of this type, except that the roles are changed and now the buyer becomes the banker. It can also be said that P is the present worth of the n future payments.

13.6 Depreciation

With the passage of time, the value of most physical property suffers a reduction, which is called depreciation. The price that a willing buyer will pay a willing seller for a property is called the market

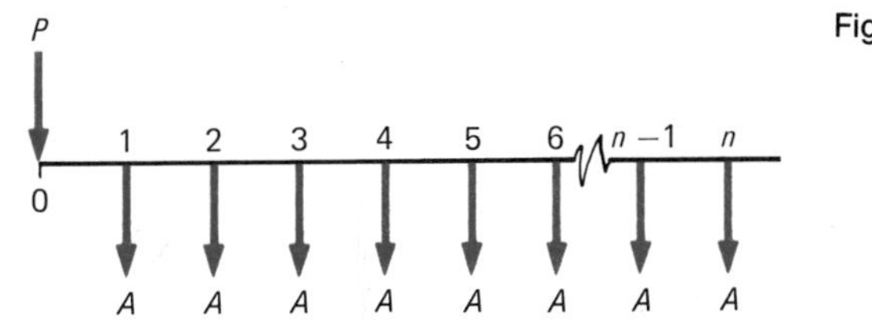

Fig. 13.9

value. The amount that the property can be sold for at a later date when it is no longer wanted by the owner is called the salvage value. Therefore, the market value at purchase minus the salvage value at discard is the amount of depreciation. The concept is simple and obvious; however, determination of depreciated value depends on an estimation of salvage value. The fact that most property does depreciate with time indicates that the amount of the depreciation is part of the cost of doing business. If the property in question is a machine that produces a given number of items per year and has a limited life expectancy, then a portion of the cost of the machine must be charged to each item produced and sold. A second reason for calculating depreciation can be to create a fund from which the cost of replacing the worn-out property can be taken. A third reason is that the Internal Revenue Service (IRS) accepts depreciation as a cost of doing business, and the taxable profits can legally be reduced by the amount of the calculated depreciation. There are several acceptable methods (by the IRS) of calculating depreciation, four of which will be examined. Many accepted depreciation methods fail to account for the time-value relationship of money.

13.6.1 Straight-line depreciation

One general requirement of methods of depreciation is that they not be overly complex. The least complex is the straight-line method. It assumes uniform depreciation over the life of the property and can be calculated from the formula

$$d = \frac{C - C_\ell}{\ell} \qquad 13.8$$

where d = annual amount of depreciation

C = cost, new

C_ℓ = value at age ℓ, years

ℓ = age, years

Example problem 13.8 If a machine costs \$8 000 new and can be traded in for \$1 600 after 16 years, what is the annual straight-line depreciation rate?

Solution

$$d = \frac{C - C_\ell}{\ell}$$

$$= \frac{8\,000 - 1\,600}{16} = \$400$$

This method has an added advantage in that during times of inflation, if the annual depreciation is invested, it will produce more than the original cost of the machine and will tend to overcome the increased cost of replacement.

13.6.2 Sinking-fund method

This method assumes that it is desirable to produce a fund to replace the property. Moreover, it also assumes that an amount equal to the new cost minus the salvage cost as shown below by the formula

$$d = (C - C_\ell)\left[\frac{i}{(1+i)^\ell - 1}\right] \qquad 13.9$$

will replace the property.

Example problem 13.9 Use the same data as Example Problem 13.8, assuming 8 percent annual interest.

Solution

$$d = (8\,000 - 1\,600)\left[\frac{0.08}{1.08^{16} - 1}\right]$$

$$= \$211.05$$

It can be seen that this method requires much less money to be put aside each year in order to accumulate the \$6 400 (8 000 − 1 600), which is an advantage if funds are limited and \$6 400 is truly the sum that is required. However, if the computation is to be used for tax purposes or to replace property whose cost will probably escalate, the sinking fund will not produce the most desirable results. It must be understood that the total depreciation at the end of any year is the amount of the sinking fund that is established. The total depreciation at the end of 2 years may be found, using Eq. 13.5, as follows:

$$211.05\left[\frac{1.08^2 - 1}{0.08}\right] = \$438.98$$

so that depreciation during the second year is \$227.93.

13.6.3 Sum of years' digits

A third method of calculating depreciation has been given the name sum of years' digits. It is calculated by assigning depreciation in decreasing amounts as follows: (Example is for an expected life of 8 years.)

Year of life	Reverse order of years of life	Fractional depreciation
1	8	8/36
2	7	7/36
3	6	6/36
4	5	5/36
5	4	4/36
6	3	3/36
7	2	2/36
8	1	1/36
	Sum = 36	Sum = 36/36 = 1

In this method, the analyst must first determine the total depreciation as before and then assign annual depreciation, the largest amount during the first year with decreasing amounts each year thereafter. The method produces a rapid depreciation in the early years and thereby has tax advantages.

Example problem 13.10 Calculate the depreciation for each year in the Example Problem 13.8.

Year of life	Reverse order of years of life	Fractional depreciation	Depreciation during the year
1	16	16/136	752.94
2	15	15/136	705.88
3	14	14/136	658.82
4	13	13/136	611.76
5	12	12/136	564.71
6	11	11/136	517.65
7	10	10/136	470.59
8	9	9/136	423.53
9	8	8/136	376.47
10	7	7/136	329.41
11	6	6/136	282.35
12	5	5/136	235.29
13	4	4/136	188.24
14	3	3/136	141.18
15	2	2/136	94.12
16	1	1/136	47.06
	Sum = 136		

13.6.4 The Matheson formula

This formula produces what is commonly called the declining balance method. It is sometimes called the constant percentage method in as much as it assumes that the depreciation is a percentage of the remaining value. This produces a greater depreciation the first year and smaller amounts in each succeeding year, similar to the sum of years' digits method. The declining balance method can never depreciate a property to zero—not a serious problem because the depreciation in later years becomes very small.

If k represents the rate of depreciation, the depreciation in the first year is Ck. The value at the start of the second year is $C - Ck$, or $C(1 - k)$. The depreciation during the second year is $C(1 - k)k$ and the total depreciation in 2 years is $2Ck - Ck^2$, so that $C(1 - k)^2$ is the value after 2 years.

It can be seen that the value remaining after n years is $C_n = C(1 - k)^n$, where C_n is the depreciated value.

$$\frac{C_n}{C} = (1 - k)^n \qquad \text{and} \left(\frac{C_n}{C}\right)^{1/n} = 1 - k$$

so that for ℓ years

$$k = 1 - \left(\frac{C_\ell}{C}\right)^{1/\ell}$$

Substitution of this expression for k into $C_n = C(1 - k)^n$ yields

$$C_n = C\left(\frac{C_\ell}{C}\right)^{n/\ell} \qquad 13.10$$

Example problem 13.11 What is the depreciated value at the end of 4 years of a property that costs \$15 000 new and has an estimated value of \$3 000 at the end of 9 years?

Solution

$$C_n = C\left(\frac{C_\ell}{C}\right)^{n/\ell}$$

$$= 15\,000\left(\frac{3\,000}{15\,000}\right)^{4/9}$$

$$= \$7\,335.64$$

The total depreciation during these 4 years is clearly 15 000 − 7 335.64 = \$7 664.36. This method produces a very rapid depreciation and is widely used for that reason.

Table 13.1 summarizes some of the data given above and compares the four methods of depreciation that have been discussed. (See also Fig. 13.10.) It can be seen that these methods produce values that vary considerably. The choice of the method employed will depend on the purpose of the calculation and the convention established by a particular industry or company.

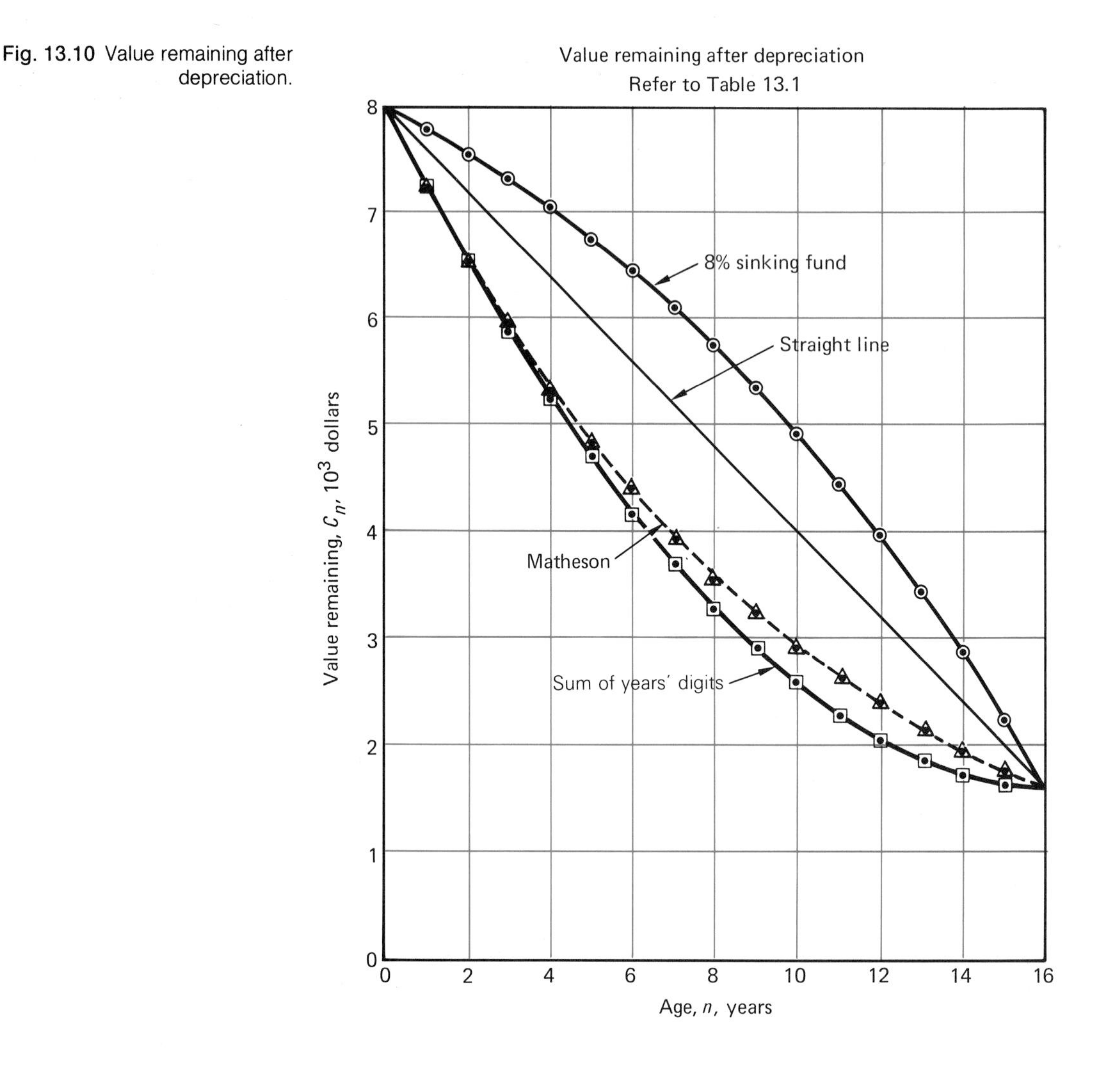

Fig. 13.10 Value remaining after depreciation.

13.7 Economic selection

The principal use that engineers can make of the material on interest just discussed is to analyze situations so that a decision can be made. In general, it is true that several alternatives are available, each having some obviously strong attributes. The task is to compare these alternatives and choose the one that appears to be superior, all things considered. A review of the formulas indicates that many estimations, predictions, and assignments of values must be made, such as new cost, salvage value, interest rates, operating costs, productivity, etc. Few, if any, of these values can be made with certainty, and the engineer must often work with data that are inexact. Even though the data are not perfect, an engineer

Table 13.1 Comparison of Depreciation Methods*

Age years	Depreciation during the year				Depreciated value remaining			
	Straight line	Sinking fund	Sum of years' digits	Matheson	Straight line	Sinking fund	Sum of years' digits	Matheson
1	400	211.05	752.94	765.57	7600.00	7788.95	7247.06	7234.43
2	400	227.93	705.88	692.31	7200.00	7561.02	6541.18	6542.12
3	400	246.17	658.82	626.05	6800.00	7314.85	5882.36	5916.07
4	400	265.86	611.76	566.15	6400.00	7048.99	5270.60	5349.92
5	400	287.14	564.71	511.96	6000.00	6761.85	4705.89	4837.96
6	400	310.10	517.65	462.98	5600.00	6451.75	4188.24	4374.98
7	400	334.91	470.59	418.67	5200.00	6116.84	3717.65	3956.31
8	400	361.70	423.53	378.60	4800.00	5755.14	3294.12	3577.71
9	400	390.64	376.47	342.37	4400.00	5364.50	2917.65	3235.34
10	400	421.89	329.41	309.61	4000.00	4942.61	2588.24	2925.73
11	400	455.64	282.35	279.98	3600.00	4486.97	2305.89	2645.75
12	400	492.05	235.29	253.19	3200.00	3994.92	2070.60	2392.56
13	400	531.56	188.24	228.96	2800.00	3463.36	1882.36	2163.60
14	400	573.98	141.18	207.05	2400.00	2889.38	1741.18	1956.55
15	400	619.90	94.12	187.23	2000.00	2269.49	1647.06	1769.32
16	400	669.50	47.06	169.32	1600.00	1600.00	1600.00	1600.00

*Table is based on an original cost of $8000.00, salvage value of $1600.00, life expectancy of 16 years and an interest rate of 8 percent.

can make better decisions with economic comparisons than without.

The most obvious method of comparing costs is to determine the total cost of each alternative. An immediate problem arises in that the various costs occur at intervals, and the total number of dollars spent is not a valid method. It has been shown that the present worth of an expenditure can be calculated. If this is done for all costs, the present worth of buying, operating, and maintaining two or more alternatives can then be compared. The present worth method is referred to as the capitalized cost. Simply stated, it is the sum of money necessary to buy, maintain, and operate a facility. The alternatives must obviously be compared for the same length

Fig. 13.11 Each machine has different characteristics. An economic analysis is required to choose the one to be purchased. (*Sperry New Holland.*)

of time, requiring that one or more be replaced due to short-life expectancies.

A second method, preferred by those who work with annual budgets, is to calculate the average annual cost of each. The approach is similar to the capitalized cost method, but the numerical value is in essence the annual contribution to a sinking fund that would produce a sum identical to the capitalized cost placed at compound interest.

Many investors approach decisions on the basis of the profit a venture will produce in terms of percent per year. The purchase of a piece of equipment, a parcel of land, a new product line, etc., is thus viewed favorably only if it appears that it will produce an annual profit greater than a certain percent. The amount of an acceptable percent return is not constant but fluctuates with the money market. Since there is doubt about the amount of the profit and certainly there is a chance of a loss, it would not be prudent to proceed if the prediction of return was not considerably above "safe" investments such as bonds.

Example problem 13.12 Compare two units, *A* and *B*. *A* has a new cost of $42 000, a life expectancy of 14 years, a salvage value of $4 000, and an annual operating cost (including taxes) of $3 000. *B* has a new cost of $21 000, a life expectancy of 7 years, a salvage value of $2 000, and an operating cost of $5 000. Assume an annual interest rate of 7 percent.

Solution

Present worth

	A	*B*
First cost	$42 000.00	$21 000.00
Operating cost (present worth of sinking fund) $\left[\frac{1.07^{14}-1}{(0.07)(1.07^{14})}\right]$(annual operating cost)	26 236.40	43 727.34
Present worth − replacement $(21\,000-2\,000)\div 1.07^{7}$		11 832.25
Present worth − salvage		
$4\,000 \div 1.07^{14}$	−1 551.27	
$2\,000 \div 1.07^{14}$		−775.63
Present worth 14 years' basis	$66 685.13	$75 783.96

Annual cost

Find present worth of initial cost minus salvage.

For *A*. Present worth − salvage = −$1 551.27.

For *B*. Present worth − salvage (14 years) = −$775.63.

Present worth of purchase at 7 years = (21 000 − 2 000)/(1.07^7) = \$11 832.25.
Present worth of A = 42 000 − 1 551.27 = 40 448.73.
Present worth of B = 21 000 + 11 832.25 − 775.63 = \$32 056.62.
Annual equivalent of present worth:

$$A = P\left[\frac{i(1+i)^n}{(1+i)^n - 1}\right]$$

For A:

$$A_A = 40\,448.73\left[\frac{0.07(1+0.07)^{14}}{(1+0.07)^{14} - 1}\right] = \$4\,625.11$$

$$A_B = 32\,056.62\left[\frac{0.07(1+0.07)^{14}}{(1+0.07)^{14} - 1}\right] = \$3\,665.51$$

Annual cost:

	A	*B*
Operation	3 000.00	5 000.00
Depreciation	4 625.11	3 665.51
	7 625.11	8 665.51

Therefore, A is less expensive than B.

Problems

13.1 What is the present worth of \$1 000 payable in 5 years if money is thought to be worth (*a*) 5 percent, (*b*) 15 percent, (*c*) 25 percent?

13.2 The taxes on a home are \$1 500.00 per year. The mortgage stipulates that the owner must pay one-twelfth of the annual taxes on the first day of each month (in advance) to the bank so that the taxes can be paid on April 1 of the next year. (1985 taxes are paid on April 1, 1986, for example.). Assuming 6 percent annual interest, compounded monthly, determine how much profit the bank makes on the owner's tax money?

13.3 A business receives \$300 000 that must be held for 2 years before it can be used for capital expenditures. The business can invest the money at 7.1 percent interest.

(*a*) If the money is compounded annually, what will the investment be worth in 2 years?
(*b*) What happens if it is compounded semiannually at the same interest rate (7.1 percent)?
(*c*) What happens if the bank suddenly decides to compound the money quarterly? What will the investment be worth in eight quarters?

13.4 How much must be invested now to grow to \$25 000 in 2 years if the interest rate is $5\frac{1}{4}$ percent compounded annually?

13.5 It is expected that by December 31 of each of 4 years beginning December 31, 1978, annual amounts of $1 828 will be available. What is the sum of money on December 31, 1981, equivalent to the $1 828 per year with interest rate of 6 percent annually?

13.6 Suppose that a company is to receive a $5 000 payment 10 years from today. If the interest rate is 7 percent annually, what actually is the worth of this money *today*?

13.7 A present loan of $1 000 is to be repaid by payments of $400 at the end of each of the next 3 years. Find the interest rate.

13.8 As of December 31, 1977, $6 335 is available. Find the annual sum of money that can be withdrawn for each of 4 years beginning December 31, 1978, if the interest rate is 6 percent annually.

13.9 By December 31 of each of 5 years beginning December 31, 1981, annual amounts of $1 000 will be available. Find the sum of money on December 31, 1980, equivalent to the $1 000 per year series if $i = 6$ percent.

13.10 A company has agreed to pay $5 000 to XYZ Electronics 1 year from today, 2 years from today, and 3 years from today. Assuming an annual interest rate of 8 percent, determine what the present worth for XYZ Electronics is.

13.11 After having thought the situation over, a company decides that it will change the terms of the transaction in Prob. 13.10 and instead start payment as of today. How will this change XYZ's present worth?

13.12 The ownership of automobiles is increasing at the rate of $3\frac{1}{2}$ percent per year in the United States. At this rate, when will the number of autos be double the present number?

13.13 In 1941, the United States issued War Bonds that cost $18.75 and could be cashed in for $25 at the end of 10 years. What was the annual compound interest rate on these bonds?

13.14 On October 17, 1977, one of a certain class of government bonds was purchased for $9 879. On October 17, 1981, it was redeemed for $10 000. Every 6 months the bondholder received $447.50 in interest (8.95 percent annual interest rate). He immediately invested the interest in a savings account that paid 6 percent annual interest, compounded semiannually. How much will he have earned during the 4-year period? Express his earnings in terms of annual interest rate on the original investment, compounded annually.

13.15 An automobile is purchased for $3 960. After a down payment of $500, the rest is financed. If payments are $101.29 per month for 42 months, what is the interest rate?

13.16 One of two machines, *A* and *B*, is to be purchased to provide for a certain production operation in a factory. Machine *A* costs $10 000 and machine *B*, $15 000. However, machine *B* will result in a net annual savings in operating costs of $800 over machine *A*. Which machine would you recommend purchasing and why if each has a useful life of 10 years and the money is worth 9 percent? Assume that both machines will be worthless at the end of 10 years.

13.17 A business can borrow $500 000 at 8 percent interest compounded annually and pay it back as follows:

At the end of 2 years	$100 000 plus all accrued interest
At the end of 4 years	$100 000 plus all accrued interest
At the end of 6 years	$100 000 plus all accrued interest
At the end of 8 years	$100 000 plus all accrued interest
At the end of 10 years	$100 000 plus all accrued interest

What is the average annual cost of this transaction?

13.18 Your supervisor requests that you estimate the average annual cost of maintenance and operation of a portion of the plant at which you are employed. Your studies indicate that during the next 5 years, the cost will be $6 000 per year; during the second 5 years, the cost will be $12 000 per year; and during the third 5 years, it will be $16 000 per year. If the prevailing interest rate is 9 percent per year, what will be the average annual cost over the next 15 years? Average annual cost means that the payments into a fund each year are equal and will produce the same dollar value at the end of 15 years as the payments stated in the problem.

13.19 Money will double in value in how many years at 9 percent annual interest, compounded quarterly.

13.20 How much should you invest monthly, beginning at age 25, to be able at age 65 to withdraw $1 500 per month for 20 years? Assume a uniform interest rate of 6 percent.

13.21 You plan to buy a home and need to mortgage it for $50 000. If the interest rate is $9\frac{1}{4}$ percent, what will your monthly payments be in order to retire the mortgage in 15 years?

13.22 For how long must you deposit $100 per month to a savings account that pays $6\frac{1}{2}$ percent interest in order to produce a large enough sum so that $100 per month can be withdrawn forever?

13.23 A piece of machinery is purchased today for $20 000. It is estimated that the selling price of this machinery will increase at the rate of 3 percent per year. If the trade-in value is estimated to be $1 000 at the end of 6 years, how much must be set aside annually to create a fund sufficient to buy a replacement if the interest rate is 6 percent?

13.24 Your real estate taxes are $1 800 per year with one-half due April 1 and the remainder due October 1 each year. What single sum of money must you place in an account earning $5\frac{1}{4}$ percent interest 15 months before the first payment is due in order to accumulate enough money to pay each tax bill? Assume monthly compounding.

13.25 Compare the following:

(*a*) $\frac{1}{4}$ percent interest per week compounded weekly
(*b*) 1 percent interest per month compounded monthly
(*c*) 3 percent interest per quarter compounded quarterly

13.26 What is the nominal interest rate if

(*a*) Payments of \$4 000 per year for 6 years will retire a loan of \$20 000?
(*b*) Payments of \$1 000 per quarter for 6 years will retire a loan of \$20 000?

13.27 What uniform annual payment is equivalent to the following payment schedule if the interest rate is $7\frac{1}{2}$ percent, compounded annually?

(*a*) \$600 at the end of the first year
(*b*) \$800 at the end of the second year
(*c*) \$1 200 at the end of the third year
(*d*) \$2 000 at the end of the fourth year
(*e*) \$2 400 at the end of the fifth year

13.28 You plan to develop a fund of \$500 to be available to you on the first Saturday of December of this year. How much must you deposit in a $5\frac{1}{2}$ percent savings account each Saturday beginning the first week in January? (Assume weekly compounding.)

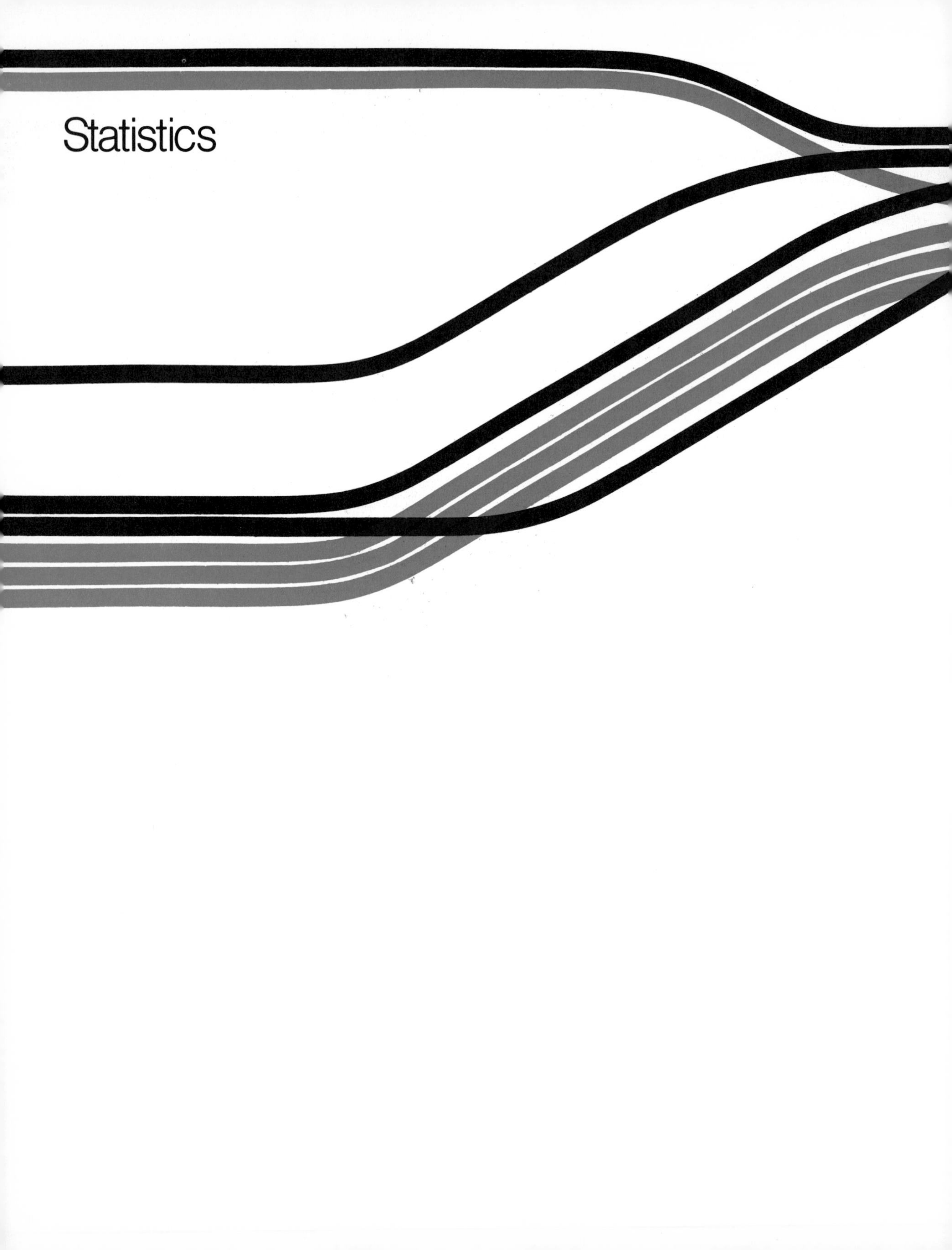

Statistics

CHAPTER 14

14.1 Introduction

Statistics, as used by the engineer, can most logically be called a branch of applied mathematics. It constitutes what some call the science of decision making in a world full of uncertainty. In fact, a degree of uncertainty exists in most day-to-day activities from something as simple as the tossing of a coin to the results of an election, the outcome of a ball game, or the comparison of the efficiency of two production processes.

There can be little doubt that it would be virtually impossible to understand a great deal of the work done in engineering without having a close association with the subject area of statistics. Numerical data derived from surveys and experiments constitute the raw material upon which interpretations, analyses, and decisions are based; and it is essential that engineers learn how to properly use information derived from such data.

Everything concerned even remotely with the collection, processing, analysis, interpretation, and presentation of numerical data belongs to the domain of statistics.

Fig. 14.1 Random sampling of a production coil. (*McGraw-Edison.*)

There exists today a number of different and interesting stories about the origin of statistics, but most historians believe it can be traced to two dissimilar areas: games of chance and political science.

During the eighteenth century, various games of chance involving the mathematical treatment of errors led to the study of probability and ultimately the foundation of statistics. At approximately the same time, an interest in the description and analysis of the voting of political parties led to the development of methods that today fall under the category of *descriptive statistics*, which is basically designed to summarize or describe important features of a set of data without attempting to infer conclusions that go beyond the data.

Descriptive statistics is an important part of the entire subject area; it is still used whenever a person wishes to represent data derived from observation.

In more recent years, however, statisticians have shifted their emphasis from methods that merely describe the data to methods that make generalizations about the data, called *statistical inference*.

To understand the distinction between descriptive statistics and statistical inference, consider the following example.

Suppose that two freshman engineers are enrolled in mathematics and each completes five quizzes. Student A receives grades of 94, 89, 92, 80, and 85; student B receives grades of 82, 61, 88, 78, and 81. On the basis of this information, it can be said that student A has an average of $(94 + 89 + 92 + 80 + 85)/5 = 88$ and that student B has an average of $(82 + 61 + 88 + 78 + 81)/5 = 78$.

Manipulating numbers belongs to the domain of descriptive statistics. Concluding that student A is a better student than B is a generalization, or a statistical inference.

From the information alone, it does not follow that student A is better than student B. Student B may have had an off day on the second quiz, may have been ill, or may have studied the wrong material for the quiz. On the other hand, student A may have studied the correct material for the quiz. There are always uncertainties, so that in this example, it may or may not be correct to conclude that one student is better than the other. The careful evaluation and analysis of all elements involving chance or risk that are normally taken when making such generalizations is an integral part of any statistical inference.

An important step to take when considering the generalization of data is that of carefully examining how the variables were controlled. For instance, if student A had been told which pages to study and student B had not, it is obvious that no reasonable or meaningful comparison can be made.

This last point is mentioned to stress the fact that in statistics, it is not enough to consider only sets of data and calculated results when arriving at conclusions. Items such as control and authenticity of collected data and how the experiment or survey was planned are of major importance. Unless proper care is taken in the planning and execution stages, it may be impossible to arrive at valid results or conclusions.

Although this chapter will not cover any of the methods or techniques associated with statistical inference, it plays such an important role in modern-day engineering that this brief introduction was considered appropriate.

14.2 Frequency distribution

Various ways of describing measurements and observations, such as the grouping and classifying of data, are a fundamental part of statistics. In fact, when dealing with a large set of collected numbers, a good overall picture of the data can often be conveyed by proper grouping into classes. The following example will serve to illustrate this point.

Consider the individual test scores received by students on the first major exam in their freshman computations course (see Table 14.1)

Table 14.2 is a type of numerical arrangement showing scores distributed among selected classes. Some information such as the highest and lowest values will be lost once the raw data has been sorted and grouped.

The construction of numerical distributions as in this example normally consists of the following steps: select classes into which the data are to be grouped; distribute data into appropriate classes; and count the number of items in each class.

Since the last two steps are essentially mechanical processes, our attention will be directed primarily toward the *classification* of data.

Two things must be considered when arranging data into classes: the number of classes into which the data are to be grouped, and the range of values each class is to cover. Both these areas are somewhat arbitrary, but they do depend on the nature of the data and the ultimate purpose the distribution is to serve.

Table 14.1

92	71	89	91	53	93	90	96	95
98	76	96	94	68	91	82	82	44
88	87	93	78	85	98	82	90	70
78	70	87	88	89	95	99	88	88
77	65	85	64	79	50	81	80	76

Table 14.2

Test scores	Tally	Frequency
41–50	||	2
51–60	|	1
61–70	~~||||~~	5
71–80	~~||||~~ |||	8
81–90	~~||||~~ ~~||||~~ ~~||||~~ |	16
91–100	~~||||~~ ~~||||~~ |||	13
	Total	45

The following are guidelines that should be followed when constructing a frequency distribution.

1. Use no less than 6 and no more than 15 classes.
2. Select classes that will accommodate all the data points.
3. Make sure that each data point fits into only one class.
4. Whenever possible, make the class intervals of equal length.

The numbers in the right-hand column of Table 14.2 are called the class frequencies which denote the number of items that are in each class.

Since frequency distributions are constructed primarily to condense large sets of data into more easily understood forms, it is logical to display or present the data graphically. The most common form of graphical presentation is called the *histogram.*

It is constructed by representing measurements or grouped observations on the horizontal axis and class frequencies along the graduated and calibrated vertical axis. This representation affords a graphical picture of the distribution with vertical bars whose bases equal the class intervals and whose heights are determined by the corresponding class frequencies. Figure 14.2 demonstrates a histogram of the test scores tabulated in Table 14.2.

Frequency distributions are occasionally approximated with a smooth line to indicate the general shape of the curve.

14.3 Measures of central tendency

The solution of many engineering problems in which a large set of data is collected can be somewhat facilitated by the determination of single numbers that describe unique characteristics about the data. The most popular measure of this type is called the arithmetic mean.

The arithmetic mean, or mean of a set of n numbers, is defined as the sum of the numbers divided by n. In order to develop a notation

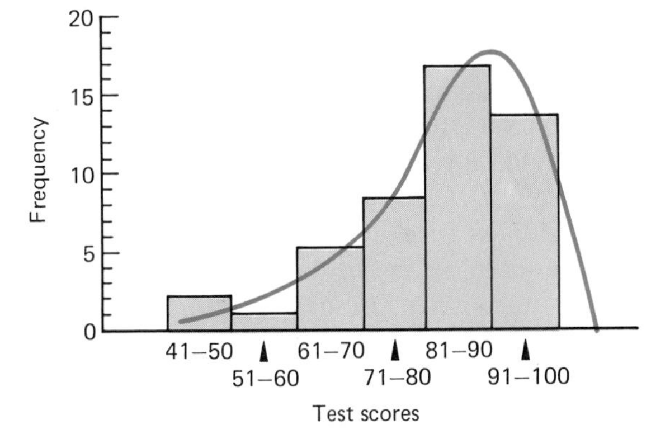

Fig. 14.2 Test scores.

and a simple formula for arithmetic mean, it is helpful to use an example.

Suppose that the average, or mean, height of a starting basketball team is to be determined. Let the height in general be represented by the letter x and the height of each individual player be represented by x_1, x_2, x_3, x_4, and x_5. More generally, there are n measurements that are designated $x_1, x_2, \ldots, x_n$. From this notation, the mean can be written as follows.

$$\text{Mean} = \frac{x_1 + x_2 + x_3 \cdots + x_n}{n}$$

A mathematical notation that indicates the summation of a series of numbers is normally written

$$\sum_{i=1}^{n} x_i$$

which represents $x_1 + x_2 + x_3 + \cdots + x_n$. This notation will be written in the remainder of the chapter as $\Sigma\, x_i$, but the intended summation will be from 1 to n.

When the x values are representative of a random sample and not an entire population, the notation for arithmetic mean will be given as $\bar{x}$. When a complete set of observations is used, it is referred to as the *population;* when a finite portion or subset of that population is used, it is referred to as a *sample.*

These standard notations provide the following common expression for the *arithmetic mean.*

$$\bar{x} = \frac{\Sigma x_i}{n} \qquad 14.1$$

The mean is a popular measure of central tendency because (1) it is familiar to most people, (2) it takes into account every item,

(3) it always exists, (4) it is always unique, (5) it lends itself to further statistical manipulations, and (6) it is reliable.

One disadvantage of the arithmetic mean, however, is that any gross error in a number can have a pronounced effect on the value of the mean. To avoid this difficulty, it is possible to describe the "center" of a set of data with other kinds of statistical descriptions. One of these is called the *median*, which can be defined as the value of the middle item of data arranged in increasing or decreasing order of magnitude. For example, the median of the five numbers 15, 27, 10, 18, and 22 can be determined by first arranging in increasing order: 10, 15, 18, 22, and 27. The median is 18.

If there are an even number of items, there is never a specific middle item, so the median is defined as the mean of the values of the two middle items. For example, the median of six numbers, 5, 9, 11, 14, 16, and 19, is $(11 + 14)/2 = 12.5$.

The mean and median of a set of data rarely coincide. Both terms describe the center of a set of data, but in different ways. The median divides the data so that half of all items is greater than or equal to the median; the mean is more correctly described as the center of gravity of the data.

The median, like the mean, has certain desirable properties. It always exists and is always unique. Unlike the mean, the median is not easily affected by extreme values.

In addition to the mean and the median, there is one other average, or center, of a set of data that we call the *mode*. It is simply the value that occurs with the highest frequency. In the following set of numbers, 18, 19, 15, 17, 18, 14, 17, 18, 20, 19, 21, and 14, the number 18 is the mode because it appears more often than any of the other values.

An important point for a practicing engineer to remember is that there are any number of ways to suggest the middle, center, or average value of a data set. If comparisons are to be made, it is essential that similar methods be compared. It is only logical to compare the mean of Brand A with the mean of Brand B, not the mean of one with the median of the other. If one particular item, brand, process, etc., is to be compared with another, the same measures must be used. If the average grade in one section of college calculus is to be compared with other sections, the mean of each section would be one important statistic.

14.4 Measures of variation

It would not be likely that the mean values of the course grades of different sections of college calculus would be of equal magnitude. The extent to which the means are dissimilar, however, is also of fundamental importance.

Measures of variation indicate the degree to which data are dispersed, spread out, or bunched together. Suppose by coincidence that two sections of a college calculus course have exactly the same mean grade values on the first-hour exam. It would be of interest to know how far individual scores varied from the mean. Perhaps one class was bunched very closely around the mean while the other class demonstrated a wide variation with some very high scores and some very low scores. This situation is typical and often of interest to the engineer.

It is reasonable to define this variation in terms of the distances by which numbers depart from the mean value. It is impossible, however, to determine the variation in a set of numbers, x_1, x_2, . . . , x_n, whose mean is $\bar{x}$, by considering only the difference between the individual number and the mean as $x_1 - \bar{x}, x_2 - \bar{x}, \ldots, x_n - \bar{x}$. These quantities are called deviations from the mean and their sum is always zero. Since it is the magnitude of these deviations that is important, the squares of the deviations from the mean are determined; then to get a number that is more representative of the original deviations, the square root is calculated. The formula obtained by this technique, called the *standard deviation s*, is

$$s = \left[\frac{\Sigma(x_i - \bar{x})^2}{n - 1}\right]^{0.5} \qquad 14.2$$

Statisticians in recent years have divided by $n - 1$ in preference to n when the data under consideration are a sample and not the entire population. This modification is significant only when n is small (i.e., less than 100). It is intended to compensate for the fact that estimates which use n tend to be small because there is usually less variability in a sample than in the entire population.

Another common measure of variation is called *sample variance*, which is the square of the standard deviation.

$$s^2 = \frac{\Sigma(x_i - \bar{x})^2}{n - 1} \qquad 14.3$$

When deviations from the mean $(x_i - \bar{x})$ result in whole numbers, then calculations by either Eq. 14.2 or 14.3 are convenient to make. When such is not the case, then it is better to use Eq. 14.4.

$$s = \left[\frac{n(\Sigma x_i^2) - (\Sigma x_i)^2}{n(n - 1)}\right]^{0.5} \qquad 14.4$$

Example problem 14.1 Calculate the mean, standard deviation, and sample variance of the mass of ten randomly selected female students. Table 14.3 is a record of the data.

Table 14.3

Mass, x_i, kg	x_i^2	$(x_i - \bar{x})$	$(x_i - \bar{x})^2$
45.6	2079.36	−5.32	28.30
48.9	2391.21	−2.02	4.08
52.2	2724.84	1.28	1.64
59.0	3481.00	8.08	65.29
51.7	2672.89	0.78	0.61
54.0	2916.00	3.08	9.49
44.5	1980.25	−6.42	41.22
46.7	2180.89	−4.22	17.81
52.2	2724.84	1.28	1.64
54.4	2959.36	3.48	12.11
Σ = 509.2	26 110.64	0.00	182.18

Solution By substituting into the correct formulas, you can determine mean, standard deviation, and sample variance.

From Eq. 14.1, you can obtain the mean as follows:

$$\bar{x} = \frac{\Sigma x_i}{n}$$

$$= \frac{509.2}{10}$$

$$= 50.9 \text{ kg}$$

From Eq. 14.2, you can calculate the standard deviation.

$$s = \left[\frac{\Sigma(x_i - \bar{x})^2}{n - 1}\right]^{0.5}$$

$$= \left(\frac{182.18}{9}\right)^{0.5}$$

$$= 4.50 \text{ kg}$$

Standard deviation can also be found from Eq. 14.4.

$$s = \left[\frac{n(\Sigma x_i^2) - (\Sigma x_i)^2}{n(n - 1)}\right]^{0.5}$$

$$= \left[\frac{10(26\,110.64) - 509.2^2}{10(9)}\right]^{0.5}$$

$$= 4.50 \text{ kg}$$

You can obtain sample variance from Eq. 14.3:

$$s^2 = \frac{\Sigma(x_i - \bar{x})^2}{n - 1}$$

$$= \frac{182.18}{9}$$

$$= 20.2 \text{ kg}^2$$

14.5 Normal curve

A normal distribution is a theoretical frequency distribution for a specific set of variable data. Its graphical representation is a bell-shaped curve that extends indefinitely in both directions. As can be seen by the drawing in Fig. 14.3, the curve comes closer and closer to the horizontal axis without ever reaching it, no matter how far the axis is extended from the mean (zero on the horizontal scale). The location and shape of a normal curve can be specified by two parameters: (1) the mean, which locates the center of the distribution; and (2) the standard deviation, which describes the amount of variability or dispersion of the data. The normal distribution can be represented by Eq. 14.5.

$$f(x) = \frac{1}{s\sqrt{2\pi}} e^{-\frac{1}{2}[(x - \bar{x})/s]^2} \qquad 14.5$$

A normal curve is symmetrical about the mean; however, the actual shape of the distribution depends on the deviation of the data about the mean. If the data are bunched around the mean, the curve will drop rapidly toward the x axis. If the data have a wide deviation about the mean, then the curve will approach the x axis more slowly. At one point, called the point of inflection, the curve changes from concave to convex. Mathematically this is exactly one standard deviation from the mean.

In any normal distribution, the following percentages of all data are included in the indicated intervals.

68.26 percent Mean $\pm$ one standard deviation

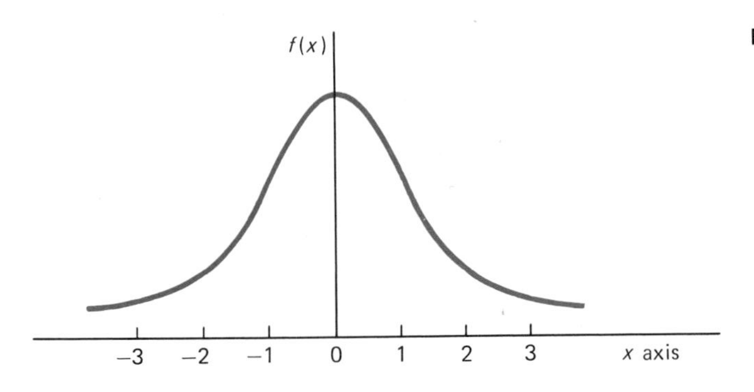

Fig. 14.3 Normal distribution.

94.40 percent	Mean $\pm$ two standard deviations
99.73 percent	Mean $\pm$ three standard deviations

For purposes of illustration, consider the following situation. Suppose that an auto manufacturer installs identical steel-belted radial tires on all 250 000 new Brand X automobiles produced in a single year. Approximately 10 percent, or 25 000, of these automobiles is observed until the tires can no longer meet certain minimum requirements. The exact mileage on each automobile is then recorded, at which time the mean and standard deviation are determined: $\bar{x} = 42\ 000$ mi and $s = 3\ 000$ mi.

If the data collected are plotted and the resulting distribution is normal, it is reasonable to suggest that 68 percent of all Brand X autos were driven between 39 000 and 45 000 mi (the mean 42 000 mi plus or minus one standard deviation, 3 000 mi) before new tires were required.

Although the normal distribution is the only continuous distribution that will be discussed in this chapter, it is not the only one encountered in the study of statistics. Other distributions such as the t distribution, the chi-square distribution, and the F distribution also play important roles in problems of statistical inference.

14.6 Linear regression

A main objective in the solution of many engineering problems is to predict or forecast the outcome of certain events. Of the many equations that can be used for purposes of prediction, the simplest and most widely used is the linear equation of the form

$$y = mx + b \qquad 14.6$$

where m and b are numerical constants. Once these constants are known, it is possible to calculate a predicted value of y for any value of x by simple substitution.

Example problem 14.2 Suppose that a class of 20 students is given a math test and the results are tabulated. The students' IQ scores are recorded with each test result. The scores are listed in Table 14.4.

Solution A number of interesting problems emerge from this example. For instance, can iQ be used to predict success in mathematics; and if so, how accurately? If the dependent variable (math score) is to be predicted in terms of the independent variable (IQ), then how can the best straight line be fit to the data?

Table 14.4

Student #	Math score	IQ	Student #	Math score	IQ
1	85	120	11	100	130
2	62	115	12	85	130
3	60	100	13	77	118
4	95	140	14	63	112
5	80	130	15	70	122
6	75	120	16	90	128
7	90	130	17	80	125
8	60	108	18	100	140
9	70	115	19	95	135
10	80	118	20	75	130

First, of course, the data must be plotted on rectangular coordinate paper to see if they can be represented by a straight line. (See Fig. 14.4.)

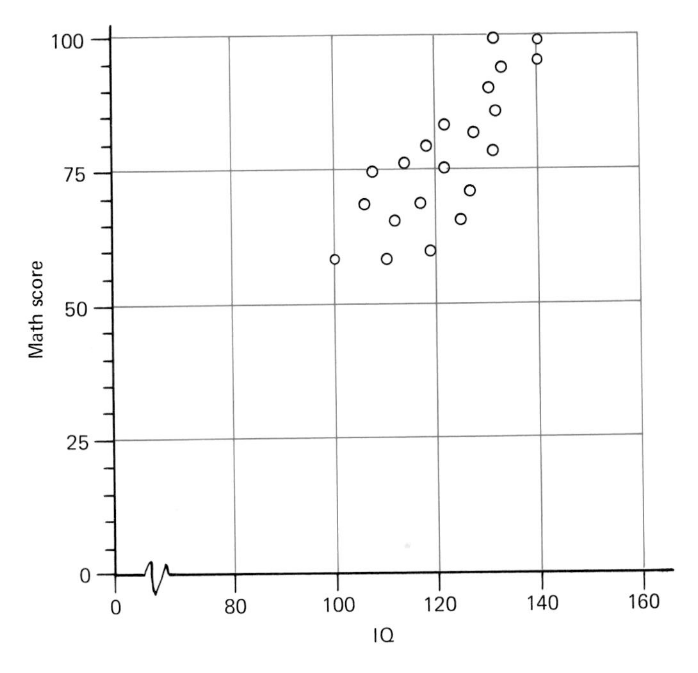

Fig. 14.4 Math and IQ scores.

In Chap. 3 it was pointed out that any time that data can be represented by a straight line, there are a number of methods for fitting the line to the data. Among the different techniques discussed, the most accurate technique available is called the method of least squares.

There is no limit to the number of straight lines that can be drawn through the data in Fig .14.4. In order to single out one line as the best fit, it is necessary to state what is meant by best. The method of least squares requires that the sum of the squares of the vertical deviations from the data points to the straight line be a minimum.

To demonstrate how a least-squared line is fit to data, let us again consider Example Problem 14.2. There are n pairs of numbers (x_1,y_1), (x_2,y_2), . . . , (x_n,y_n), where $n = 20$ with x and y being IQ and math scores, respectively.

Suppose that the equation of the line that best fits the data is of the form

$$y' = mx + b \tag{14.7}$$

where the symbol y' (y prime) is used to differentiate between the observed values of y and the corresponding values calculated by means of the equation of the line. In other words, for each value of x, there exists an observed value (y) and a calculated value (y') obtained by substituting x into the equation $y' = mx + b$.

The least-squares criterion requires that the sum of all $(y - y')^2$ terms, as illustrated in Fig. 14.5, be the smallest possible. One must determine the constants m and b so that the differences between the observed and the predicted values of y will be minimized.

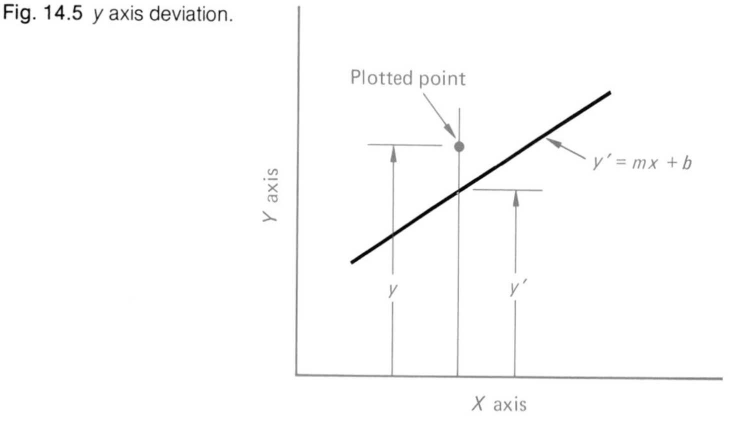

Fig. 14.5 y axis deviation.

The actual mathematical derivation is not included here, but when the sum of all the $(y - y')^2$'s is minimized, b and m may be determined, resulting in Eqs. 14.8 and 14.9:

$$m = \frac{n(\Sigma x_i y_i) - (\Sigma x_i)(\Sigma y_i)}{n(\Sigma x_i^2) - (\Sigma x_i)^2} \tag{14.8}$$

$$b = \frac{\Sigma y_i - m(\Sigma x_i)}{n} \tag{14.9}$$

Table 14.5 is a tabulation of the values necessary to determine the constants from these equations for Example Problem 14.2. The independent variable x is IQ and the dependent variable y is the math score.

Table 14.5

Dependent variable		Independent variable		
x (IQ)	x^2	y (Math score)	y^2	xy
120	14 400	85	7 225	10 200
115	13 225	62	3 844	7 130
100	10 000	60	3 600	6 000
140	19 600	95	9 025	13 300
130	16 900	80	6 400	10 400
120	14 400	75	5 625	9 000
130	16 900	90	8 100	11 700
108	11 664	60	3 600	6 480
115	13 225	70	4 900	8 050
118	13 924	80	6 400	9 440
130	16 900	100	10 000	13 000
130	16 900	85	7 225	11 050
118	13 924	77	5 929	9 086
112	12 544	63	3 969	7 056
122	14 884	70	4 900	8 540
128	16 384	90	8 100	11 520
125	15 625	80	6 400	10 000
140	19 600	100	10 000	14 000
135	18 225	95	9 025	12 825
130	16 900	75	5 625	9 750
2 466	306 124	1 592		198 527

Substituting these numbers into the appropriate equations, we get the following values for the two constants.

$$m = \frac{20(198\ 527) - (2\ 466)(1\ 592)}{20(306\ 124) - 2\ 466^2}$$

$$= 1.081$$

$$b = \frac{1\ 592 - (1.081)(2\ 466)}{20}$$

$$= -53.7$$

The equation of the line using the method of least squares is as follows:

$$y' = -53.7 + 1.08x$$

or

$$\text{Math score} = 1.08(\text{IQ}) - 53.7$$

Therefore,

Math score ≈ 1.1(IQ) − 54

Even though this equation is the best fit of the given data and could be used to predict math scores given any IQ, it might provide extremely misleading results based on the limited number of data points collected or it may indicate that there is no basis for assuming a correlation between math results and IQ. The exercise is intended solely as an example of the method of least squares.

14.7 Coefficient of correlation

The technique of finding the best possible straight line to fit experimentally collected data is certainly useful, as previously discussed. The next logical and interesting question is how well such a line actually fits. It stands to reason that if the differences between the observed y's and the calculated y''s are small, the sum of squares $\Sigma(y - y')^2$ will be small; and if the differences are large, the sum of squares will tend to be large.

Although $\Sigma(y - y')^2$ provides an indication of how well a least-squares line fits particular data, it has the disadvantage that it depends on the units of y. For example, if the units of y are changed from dollars to cents, it will be like multiplying $\Sigma(y - y')^2$ by a factor of 10 000. To avoid this difficulty, the magnitude of $\Sigma(y - y')^2$ is normally compared with $\Sigma(y - \bar{y})^2$. This allows the sum of the squares of the vertical deviations from the least-squares line to be compared with the sum of squares of the deviations of the y's from the mean.

To illustrate, Fig. 14.6*a* shows the vertical deviation of the y's from the least-squares line while Fig. 14.6*b* shows the deviations of the y's from their collective mean. It is apparent that where there is a close fit, $\Sigma(y - y')^2$ is smaller than $\Sigma(y - \bar{y})^2$.

Fig. 14.6 Deviation from y' and $\bar{y}$

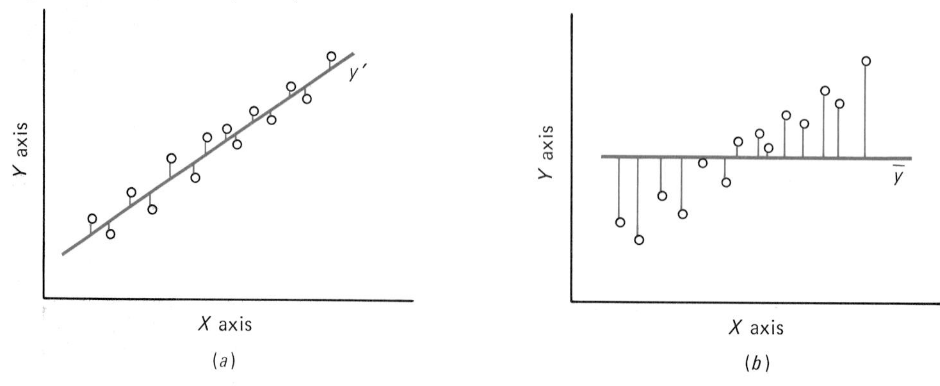

In contrast, consider Fig. 14.7. Again, Fig. 14.7*a* shows the vertical deviation of the y's from the least-squares line, and Fig. 14.7*b* shows the deviation of the y's from their mean. In the latter case, $\Sigma(y - y')^2$ is approximately the same as $\Sigma(y - \bar{y})^2$. This would seem to indicate that if the fit is good as in Fig. 14.6, $\Sigma(y - y')^2$ is much less than $\Sigma(y - \bar{y})^2$; and if the fit is as poor as in Fig. 14.7, the two sums of squares are approximately equal.

The coefficient of correlation puts this comparison on a precise basis.

$$r = \pm\sqrt{1 - \frac{\Sigma(y_i - y')^2}{\Sigma(y_i - \bar{y})^2}} \qquad 14.10$$

If the fit is poor, the ratio of the two sums is close to 1 and r is close to zero. However, if the fit is good, the ratio is close to zero and r is close to $+1$ or -1. From the equation, it is obvious that the ratio can never exceed 1. Hence, r cannot be less than -1 or greater than $+1$.

The statistic is undoubtedly the most widely used measure of the strength of a linear relationship between any two variables. It indicates the goodness of fit of a line determined by the method of least squares, and this in turn indicates whether there exists a relationship between x and y.

Although Eq. 14.10 serves to define the coefficient of correlation, it is seldom used in practice. A popular short-cut (but exact) formula is

$$r = \frac{n(\Sigma x_i y_i) - (\Sigma x_i)(\Sigma y_i)}{\sqrt{n(\Sigma x_i^2) - (\Sigma x_i)^2}\,\sqrt{n(\Sigma y_i^2) - (\Sigma y_i)^2}} \qquad 14.11$$

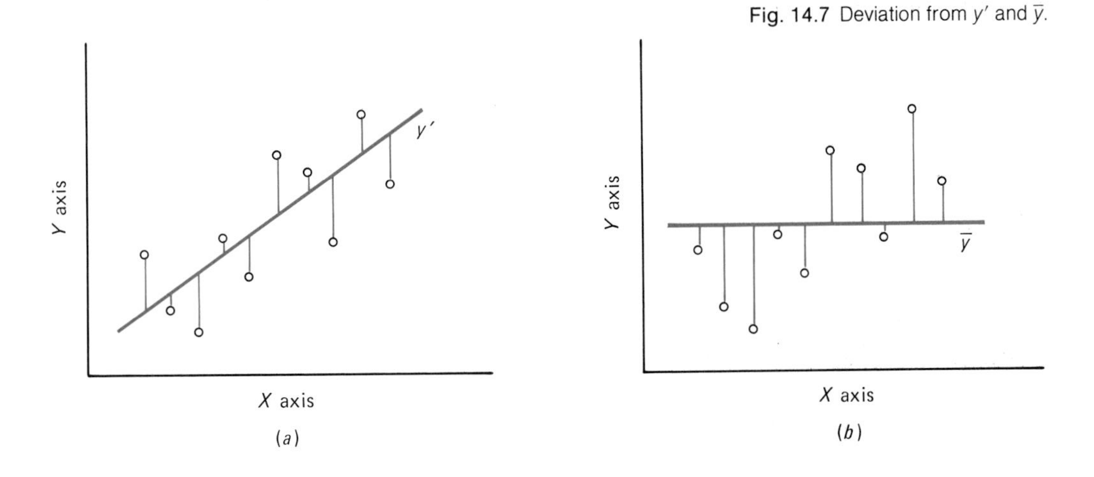

Fig. 14.7 Deviation from y' and $\bar{y}$.

The interpretation of r is not difficult if it is ± 1 or 0: when it is 0, the points are scattered and the fit of the regression line is so poor that a knowledge of x does not help in the prediction of y; when it is $+1$ or -1, all the points actually lie on the straight line, so an excellent prediction of y can be made by using x values. The problem arises when r falls between 0 and $+1$ or 0 and -1.

The simplest physical interpretation of r can be explained in the following manner. If the coefficient of correlation is known for a given set of data, then $100r^2$ percent of the variation of the y's can be attributed to differences in x, namely, to the relationship of y with x. If $r = 0.6$ in a given problem, then 36 percent, that is, $100(0.6^2)$ of the variation of the y's is accounted for (perhaps caused) by differences in x values.

Again consider Example Problem 14.2 on IQ and math scores, substituting values from Table 14.5 into the short-cut equation for correlation coefficient.

$$r = \frac{(20)(198\,527) - (2\,466)(1\,592)}{\sqrt{(20)(306\,124) - 2\,466^2}\,\sqrt{(20)(129\,892) - 1\,592^2}}$$

$$= 0.87$$

$$100r^2 = 76 \text{ percent}$$

This would indicate that 76 percent of the variations in math scores can be accounted for by differences in IQ.

One word of caution when using or considering results from linear regression and coefficients of correlation. There is a fallacy in interpreting high values of r as implying cause-effect relations. If the increase in television coverage of professional football is plotted against the increase in traffic accidents at a certain intersection over the past 3 years, an almost perfect positive correlation $(+1.0)$ can be shown to exist. This is obviously not a cause-effect relation, so it is wise to interpret correlation coefficient carefully. The variables must have a measure of association if the results are to be meaningful.

Problems

14.1 If utility bills paid by residents of a small town during the month of October varied from \$32.16 to \$63.20, construct a table with seven classes into which these amounts may be grouped.

14.2 A state highway department is interested in the variation of two-lane highways. By measuring the width of all two-lane concrete roads, the following data are collected. Each reading is recorded to the nearest tenth of a decimeter.

63.5	64.9	66.4	66.9	61.5
71.2	66.6	68.9	68.9	63.6
67.8	67.5	67.6	71.3	56.2
64.2	55.0	68.8	68.5	65.9
67.5	72.0	63.9	64.3	73.0
61.4	63.4	68.2	68.7	64.4
70.3	68.2	68.2	69.2	65.6

(*a*) Group these measurements into a frequency distribution table having five equal classes from 54.0 to 73.9.
(*b*) Construct a histogram of the distribution.
(*c*) Determine the median, mode, and mean of the data.

14.3 A farm-implement manufacturing company in the Midwest purchases steel castings from a Chicago-area foundry. Thirty castings were selected at random and weighed, and their masses were recorded to the nearest kilogram, as shown below.

235	232	228	228	240	231
225	220	218	230	222	229
217	233	222	221	228	228
238	232	230	226	236	226
227	227	229	229	224	227

(*a*) Group the measurements into a frequency distribution table having six equal classes from 215 to 244.
(*b*) Construct a histogram of the distribution.
(*c*) Determine the median, mode, and mean of the data.

14.4 An approximation of missile velocities were recorded over a predetermined fixed distance. Each value is rounded to the nearest 10 m/s.

980	960	950	1 010
930	880	870	960
850	1 020	970	940
970	900	1 030	950
1 000	940	970	600

(*a*) Group these measurements into a frequency distribution table having six equal classes that range from 500 to 1 100.
(*b*) Construct a histogram of the distribution.
(*c*) Determine the median, mode, and mean of the data.

14.5 The following test scores were recorded by a class of freshman engineering students on a chemistry test.

69	90	67	85	70	40	70	77	80	85
58	70	67	62	75	87	73	73	74	70
83	63	72	95	62	65	90	58	68	99
58	69	60	83	88	79	80	68	100	75
70	31	93	79	72	64	52	65	77	72

(*a*) Group these test scores into a frequency distribution table.
(*b*) Construct a histogram of the distribution.
(*c*) Determine the median, mode, and mean of the data.

14.6 Using the data given in Prob. 14.4, calculate

(*a*) The standard deviation using the short-cut formula.

(*b*) The standard deviation using a scientific calculator that has that feature.

(*c*) The variance.

14.7 Using the data given in Prob. 14.5, calculate

(*a*) Standard deviation using the traditional formula.

(*b*) Standard deviation using the short-cut formula.

(*c*) Standard deviation using a calculator with that feature.

(*d*) The range of grades that fall within one standard deviation, two standard deviations. Assume the data has a normal distribution.

14.8 The following are grades that 20 students obtained on the midterm and final examination in a freshman graphics course.

Midterm x	Final y
84	78
75	85
97	91
68	78
86	81
91	75
53	64
84	91
77	78
92	89
62	52
83	73
36	50
51	40
89	83
91	87
82	80
74	70
85	89
96	98

(*a*) Using the method of least squares, determine the best straight line through the data.

(*b*) Plot the data on linear graph paper.

(*c*) Represent the equation in part *a* on the graph from part *b*.

(*d*) Calculate the coefficient of correlation.

PART FIVE

Foundations of design

Engineering design—a process

CHAPTER 15

15.1 Introduction

In Chap. 1 we alluded to design in engineering as an iterative process involving decisions at each step (see Fig. 15.1). The purpose of this chapter is to expand on these introductory remarks by considering the design process in more detail. It is possible to subdivide the entire design process in many different ways. In this chapter, it has been broken down into nine separate steps, so that each step can be explained separately. To make each phase as understandable and practicable as possible, we relate the progress of an actual student design project at the end of each step. We recount, in other words, how a group of freshmen accomplished each phase of the process from start to finish.

Before we begin our discussion, we must first examine engineering design in general and its place in the activities of the technology team that was described in Chap. 1. Many practicing engineers hold the opinion that design is *the* distinguishing feature of engineering. They feel that most of our efforts are directed toward producing systems and devices that use our natural resources in the most effective, efficient manner to satisfy human needs. The real test of the systems and devices is found in their use: do they truly fit into our society and improve it. Design is what engineers do, distinct from other practitioners in their ability to use technology.

It is estimated that only about 20 percent of the products that are available today were produced 25 years ago (see Fig. 15.2). One

1. Identification of a need
2. Definition of the problem
3. Search
4. Criteria and constraints
5. Alternative solutions
6. Analysis
7. Decision
8. Specification
9. Communication

Fig. 15.1 The design process.

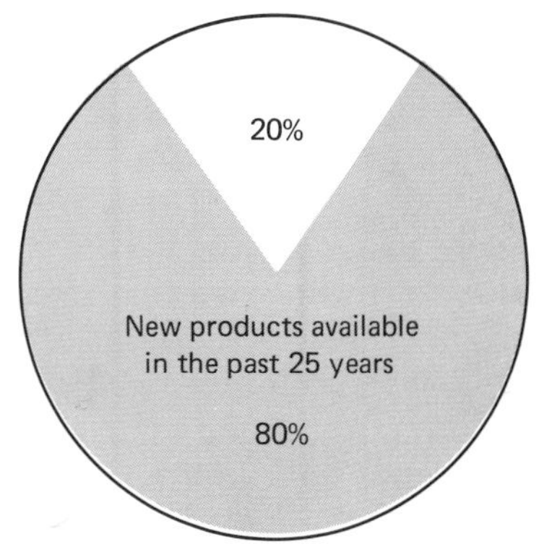

Fig. 15.2 Changing technology.

Fig. 15.3 A member of a design team explains how a component fits into the project. (*Allen Bradley.*)

can reasonably assume, therefore, that similar changes will be made between now and the beginning of the twenty-first century. Many of the items that human beings consider necessities today will have been superseded by innovations in the next two decades; and others will have undergone a series of changes that are still more evolutionary in nature. Perhaps the flea markets of the year 2000 will contain hand-held calculators because the graduates of the class of 1985 developed something that made them as obsolete as the slide rules of the class of 1970.

15.1.1 The design process

A simple definition of design is: to create according to a plan. A process, on the other hand, is a phenomenon identified through step-by-step changes that lead toward a required result. Both these definitions suggest the idea of an orderly, systematic approach to a desired end. Figure 15.4 shows the design process as continuous and cyclic in nature. This idea has validity in that many problems arise during the design process that generate subsequent designs. You should not assume that each of your design experiences will necessarily follow the sequential steps without deviation. Experienced designers will agree that the steps as shown are quite logical; but on many occasions, designers have had to repeat some steps or perhaps have been able to skip one or more.

Before beginning an overview of the entire design process, we must state that limits are always placed on the amount of time available. Normally we establish a time frame or a series of dead-

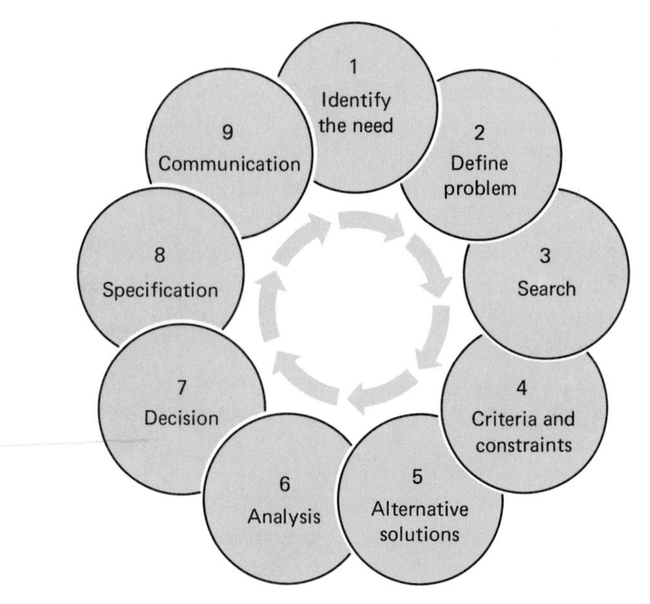

Fig. 15.4 Cyclic process.

lines for ourselves before we begin the process. It is almost impossible for us to tell you how much time should be allocated for each step, because the problems are so varied. A sample of a time frame is shown in Fig. 15.5. You may wish to try its form for your class project.

The whole process begins when a need is recognized: put simply, someone feels that something must be done. Oftentimes it is not the designers who are involved at step 1; but they usually assist in

Design steps	Activity time schedule — Percentage of total time									
	10	20	30	40	50	60	70	80	90	100
Identify need										
Define problem										
Search										
Criteria										
Alternatives										
Analysis										
Decision										
Specifications										
Communication										

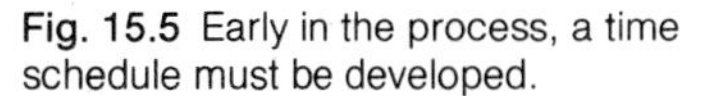
Fig. 15.5 Early in the process, a time schedule must be developed.

defining the problem (step 2) in terms that allow it to be scrutinized. Information is gathered at step 3, and then the nature of the solution to the problem is determined and boundary conditions (constraints) are established. At step 5, in which several possible solutions are entertained, the creative, innovative talents of the designer come into play. This step is followed by detailed analysis of the alternatives, after which a decision is reached regarding which one should be completely developed. Specifications of the chosen concept are prepared, and its merits are explained to the proper people or agencies so that implementation (construction, production, etc.) can be accomplished. A more complete explanation of each step as well as a reporting of the actions of the freshman design team constitute the remainder of this chapter.

15.2 Identification of a need—step 1

Before the process can begin, someone has to recognize that some constructive action needs to be initiated. This may sound vague, but understandably so, because such is the way the process normally begins. Engineers do not have supervisors who tell them to identify a need. You might be asked to do so in the classroom, because some professors may ask you to work on a project that you choose rather than one that is assigned. When most of us speak of a need, we generally refer to a lack or shortage of something we consider essential or highly desirable. Obviously, this is an extremely relative thing, for what may be a necessity to some could be a luxury to others.

More often than not, then, someone other than the engineer decides that a need exists. In private industry, it is essential that products sell for the company to survive. Most of the products have a life cycle that goes from the development stage, when the expenditures by the organization are high and sales are low, to the peak demand period, when profits are high, and eventually to the point where the product becomes obsolete. Even though a human need may still exist, the economic demand does not, because a more attractive alternative has become available. With obsolescence of a product, the company perceives a need to phase out the product and to develop one that is profitable. Inasmuch as most companies exist to make a profit, profit can be considered to be the basic need.

A bias toward profit and economic advantage should not be viewed as a selfish position, because products are purchased by people who feel that what they are buying will satisfy a need that they perceive as real. Society appreciates anyone who provides essential and desirable services, as well as goods that we use and enjoy. The consumers are ultimately the judges of whether there is truly a need. In like manner, the citizens of a community decide whether or not to have paved streets, parks, libraries, adequate police and fire protection, and scores of other things. City councils vote on the details of the programs. And during the period when citizens and decision makers are formulating their plans, engineers are involved in supplying factual information to assist them. After the policy decisions have been made, engineers conduct studies, surveys, tests, and computations that allow them to prepare the detailed design plans, drawings, etc., that shape the final project.

15.2.1 The chapter example—step 1

Throughout the remainder of the chapter, we will trace the steps five freshmen engineering students[1] took to produce a design for their class project. As a starting point, a professor may assign students the task of identifying a need. It usually is easier to approach such an assignment by beginning with a very broad area of technology, such as energy, for example.

The five students we are concerned with were informed that all the student teams would be involved in some area dealing with the energy problem. Their professor began with construction of a decision tree, shown as Fig. 15.6. The class discussed sources of energy and jointly added the first level of subproblems: fossil, wind, geothermal, solar, nuclear, and organic. The class was then divided into groups; and the groups began to further subdivide *one* of the energy sources listed above. There probably is no end to this procedure, but it does provide quickly a wide range of topics from which needs may be more easily recognized. Our student design team developed Fig. 15.6, as shown, and thus began their discussion of the general topic of firewood for use in fireplaces. They recorded statements such as the following:

1. More and more people are electing to have fireplaces in their homes and apartments.
2. Firewood is not as commercially available as it used to be.
3. The price of firewood has risen significantly.
4. People are now more willing to cut and split their own firewood than they were previously.
5. The small, inexpensive chain saw has made the cutting portion of the task more acceptable, but splitting the wood is still a major problem.

After a period of discussion involving these topics and others related to firewood, they agreed upon the following initial statement of need:

[1]The five students were John W. Benike, Douglas L. Carper, Patrick J. Grablin, Rick Sessions, and David L. White.

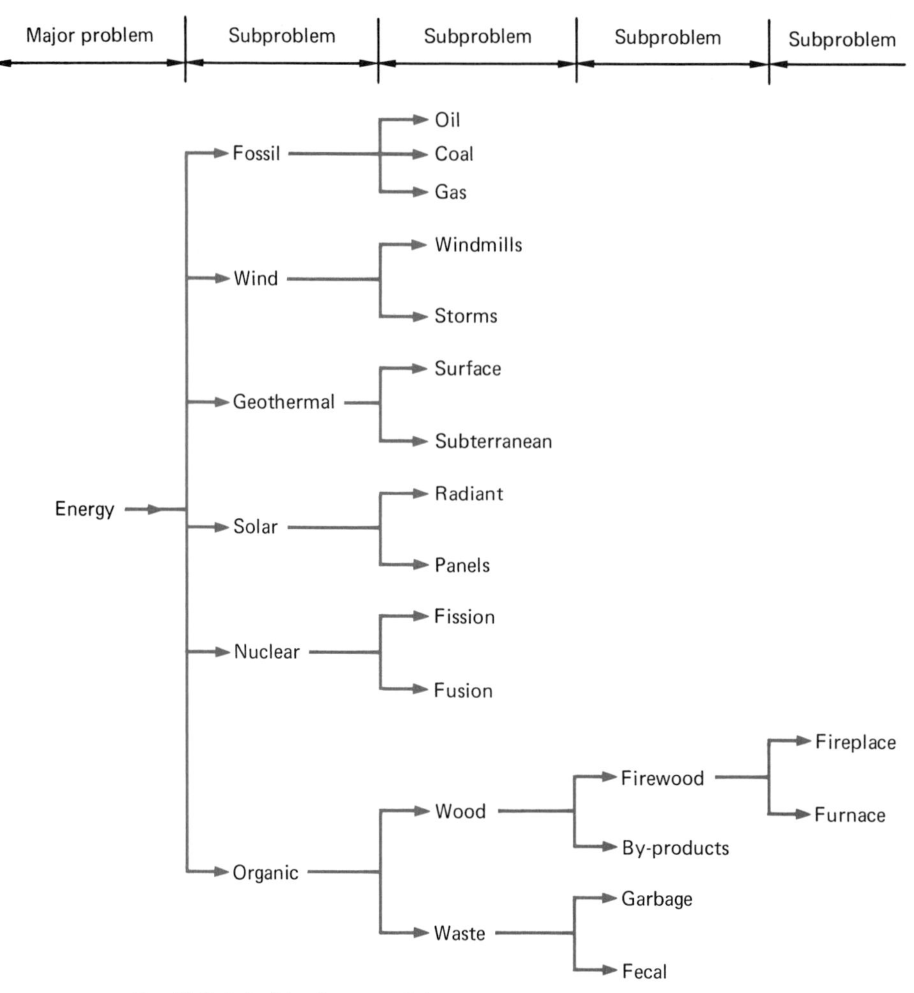

Fig. 15.6 A decision tree pertaining to energy.

There is a need for an inexpensive supply of firewood for use in the home.

(We will see later that this statement is changed by the students in much the same way that professional engineers refine and redefine problems during a design process.)

15.3 Problem definition—step 2

There is often a temptation to construct quickly a mental picture of a gadget that if properly designed and manufactured will satisfy a need. In the general case of a need for firewood, we obviously know that we can call a supplier and have firewood delivered, but the cost factor is equally obvious. We could get a bit facetious and decide to burn our furniture. There have been emergency situations during extreme storms when this was the best solution available. If some friendly neighbor will supply us with firewood at no cost or effort, a problem does not exist. An important point to realize at this time is had we allowed ourselves to focus on a specific piece of equipment or a single method of obtaining firewood in the very beginning, we would never have considered the statements mentioned above.

15.3.1 Broad definition first

The need as previously stated does not point to any particular solution and thereby leaves us with the opportunity to consider a wide range of alternatives before we agree on a specific problem statement. Consider for a moment a partial array of possibilities that would satisfy the original statement that there is a need for an inexpensive supply of firewood for use in the home.

1. Purchase firewood from a supplier.
2. Use something other than firewood (coal or rolled newspaper).
3. Make use of existing equipment.
4. Hire a portion of the work done (probably the splitting of logs).
5. Design improved equipment.

This is not an exhaustive list. You may want to take a few minutes to add to it. The first item is the current solution for many people. The other items show some promise of being an improvement over simply purchasing the firewood. You may think that item 1 should not be listed because it offers no change. If so, you are mistaken, because the status quo is the solution that is selected most often, at least as a temporary measure.

15.3.2 Symptom versus cause

If you cough and do nothing but suck on a cough drop, you may be treating the symptom (the tickle in your throat) but doing little to alleviate the cause of the tickle. This approach may be expedient; but it can many times result in a repetition of the problem if the tickle is caused by a virus or a foreign object of some sort. Engineers seldom tell a client to take two aspirins and call back tomorrow, but they can sometimes be guilty of failing to see the real problem.

For many years residential subdivisions were designed so that the rainfall would drain away quickly, and expensive storm sewer systems were constructed to accomplish the task. Not only were the sewers expensive, but they also resulted in transporting the water problem downstream for someone else to handle. In recent years, perceptive engineers have designed land developments so that the rainfall is temporarily held and released over a longer period of time. This approach employs smaller, less expensive sewers and reduces the likelihood of flooding downstream. The problem was not how to get rid of the rainfall as originally assumed, but what should be done with the water.

15.3.3 Solving the wrong problem

In the 1970s, the problem of increasing fatalities in auto accidents was clearly recognized. It was shown that the fatality rate could be significantly reduced if the driver and front-seat passenger used lap and shoulder belts. The solution technique that was implemented was to build in an interlock system that required the belts to be latched before the auto could be started. That solution certainly should have solved the problem but it didn't. It attacked the problem of requiring that the belts be physically used but it did nothing to solve the real problem—that of driver attitude. The driver and passenger still did not wish to use belts and did everything possible to avoid it even to having the interlock system removed.

15.3.4 The chapter example—step 2

The students whose progress we are following considered the possibilities outlined in Sec. 15.3.1. Their discussions covered a range of topics; and from their notebook we have listed a few of their pertinent recorded thoughts.

1. The range of possible solutions must be reduced before the problem can be solved.
2. People really like the smell of burning wood (in preference to paper or coal).
3. There already exists an adequate supply of wood in most areas.
4. Chain saws are already well developed; hence, cutting the wood into proper lengths is not a pertinent problem.
5. Time is very short, so a problem must be chosen that we (the students) can solve.

Most assuredly these young people had other thoughts, many of which were not recorded. The result of their considerations was a slightly revised problem definition:

There needs to be available to the average household an inexpensive, efficient method of splitting a small quantity of firewood.

In practice, you and other engineers will face restrictions that will affect the quality of your solutions. Many times your solution will have to meet governmental regulations in order to qualify for grants of money; or perhaps safety requirements by some agency cannot be met if certain materials are used. In almost all your projects, there will be cost and time constraints that force you to make decisions that are not what you really want to do. Such decisions, once made, then control many of your subsequent actions on that project.

This situation was faced by our freshman design students. By limiting the range of possible solutions and accepting the present method of cutting wood, they eliminate even the consideration of other burning materials. We are not being critical of their decisions because we have experienced similar time and resource constraints.

15.4 Search—step 3

Most of your productive professional time will be spent locating, applying, and transferring information—all sorts of information. This is not the popular opinion of what engineers do, but it is the way it will be for you. Engineers are problem solvers, skilled in applied mathematics and science, but they seldom, if ever, have enough information about a problem to begin solving it without first gathering more data. This search for information may reveal facts about the situation that result in redefinition of the problem.

15.4.1 Types of information

The problem usually dictates what types of data are going to be required. The one who recognizes that something was needed (step 1) probably listed some things that are known and some things that need to be known. The one or ones who defined the problem had to have knowledge of the topic or they could not have done their part (step 2). Generally, there are several things that we look for in beginning to solve most problems. For example,

1. What has been written about it?
2. Is something already on the market that may solve the problem?
3. What is wrong with the way it is being done?
4. What is right with the way it is being done?
5. Who manufactures the current "solution"?
6. How much does it cost?
7. Will people pay for a better one if it costs more?
8. How much will they pay (or how bad is the problem)?

Fig. 15.7 Research occurs throughout the design process and manufacturers' catalogs are often a source of information. (*Stanley Consultants.*)

15.4.2 Sources of information

If anything can be said about the last half of the twentieth century, it is that we have had an explosion of information. The amount of data that can be uncovered on most subjects is overwhelming. People in the upper levels of most organizations have assistants who condense most of the things that they must read, hear, or watch. When you begin a search for information, be prepared to scan many of your sources and catalog them so that you can find them easily if the data subsequently appears to be important.

Some of the sources that are readily available include the following:

1. Your library. Many universities have courses that teach you how to use your library. Such courses are easy when you compare them with those in chemistry and calculus, but their importance should not be underestimated. There are many sources in the library that can lead you to the information that you are seeking. You may find what you need in an index such as the *Engineering Index,* but don't overlook the possibility that a general index, such as *The Reader's Guide* or *Business Periodicals Index,* may also be useful. The *Thomas Register of American Manufacturers* may direct you to a company that makes a product that you need to know more about. *Sweets Catalog* is a compilation of manufacturer's information sheets and advertising material about a wide range of products. There are many other indexes that provide specialized information. The nature of your problem will direct which ones may be helpful to you. Don't hesitate to ask for assistance from the librarian.

2. Government documents. Many of these are housed in special sections of your library, but others are kept in centers of government—city hall, county

court houses, state capitols, and Washington, D.C. The agencies of government that regulate, such as Interstate Commerce Commission, Environmental Protection Agency, regional planning agencies, etc., make rules and police them. The nature of the problem will dictate which of the myriad of agencies can fill your needs.

3. Professional organizations. The American Society of Civil Engineers is a technical society that will be of interest to students majoring in civil engineering. Each major in your college is associated with not one but often several such societies. The National Society of Professional Engineers is an organization that most engineering students will eventually join, as well as at least one technical society such as the American Society for Mechanical Engineers (ASME), the Institute of Electrical and Electronics Engineers (IEEE), or any one of dozens that serve the technical interests of the myriad of specialties with which professional practices seem most closely associated. Many engineers are members of several associations and societies. Other organizations, such as the American Medical Association and the American Bar Association, serve various professions, and all have publications and referral services.

4. Trade journals. They are published by the hundreds, usually specializing in certain classes of products and services.

5. Vendor catalogs. Perhaps your college subscribes to one of the several information services that gather and index journals and catalogs. These data banks may have tens of thousands of such items available to you on microfilm. You need only learn how to use them.

6. Individuals that you have reason to believe are somewhat expert in the field. Your college faculty has at least several, maybe many. There are, no doubt, some practicing engineers in your city.

15.4.3 Recording your findings

The purpose of a bibliography is to direct you to more information than is included in the article you are reading. The form of the bibliography makes it easy to find the reference. So it seems reasonable for you to record your information sources in proper form so that if that reference is to be cited in your report, you are ready to do it properly. By so doing, you are ensuring that it can be found again quickly and easily. Few things are more disgusting than to be unable to locate an article that you found once and know will be helpful if you could locate it again.

It is usually a good procedure to record each reference on a card or sheet of paper. English teachers usually recommend the use of file cards, but engineers seem to prefer information put in a bound notebook. Whatever your choice, Fig. 15.8 is recommended as a reasonable form of record.

As we are looking at something like a piece of equipment we oftentimes have thoughts and ideas that should be recorded for

Fig. 15.8 Documentation of information is essential if it is to be useful later. This card will permit easy retrieval of the book.

TA 152.17 Inganere, M. E.
G273 *Heat, Air, and Gas Power*. McGraw-Hill, New York, 1973.

Has good tables in appendix and formulas on pages 52–55 covering heat transfer cases that may occur on project.

future reference. At such times, our ability to sketch is an invaluable tool because so many details can be graphically shown but are very difficult to describe in words.

15.4.4 The chapter example—step 3

The team of engineering students whose project we are following realized the importance of the research phase but were also aware of the overall time constraints on the design process. They decided that a detailed research plan was needed so that specific assignments could be made to avoid conflict or overlap of effort. After considerable discussion, specific research areas were assigned to each team member. They consulted home builders about the demand for fireplaces, suppliers of firewood, several manufacturers of chain saws, companies that sell chain saws, a landscape architect, city government, the library, a county extension service, an engineer, and a company that sells commercial log splitters.

One of the team members was assigned the task of checking the library to determine what products were currently available. The librarian explained about the various indexes, so the student selected the *Thomas Register*. After failing to find anything listed under "Logs," he tried "Splitters." Figure 15.9 is a reproduction of his notes. (The appendix of the student report includes copies of letters sent by the design team along with the responses received.)

With this information the student went to the *Yellow Pages* of the local telephone book, wherein he learned that one of the products was sold locally; so a team member was assigned to visit the dealer. The dealer was temporarily out of advertising pamphlets, so the team member sketched the floor model and recorded pertinent data about it, the notes for which are shown as Fig. 15.10. Their procedure is to be commended and highly recommended in that they adequately documented the information in sufficient detail so that it could be used or the manufacturer could be contacted for additional data.

The initial research stage has provided us with added information about the problem, so that we are now ready to begin to de-

Source: Thomas Register of American Manufacturers
Thomas Register Catalog File

Listing: Splitters: Wood, Firewood, Kindling, etc.

Location: Engineering Library – Reference Tables

Manufacturers

1) Gordon Corporation
P.O. Box 244-TR
Farmington, CT
(Hydraulic)

2) H. L. Diehl Co.
South Windham, CT

3) Vermeer Mfg. Co.
3804 New Sharon Rd.
Pella, IA
(Powered, log, hydraulic, trailer)
(515) 628-3141

4) Tree King Mfg. & Engineering, Inc.
North St.
Showhegan, ME
(Hydraulic)

5) Lindig Mfg. Corp.
1831 West County Rd.
St. Paul, MN
(612) 633-3072

6) Carthage Machine Co., Inc.
571 West 3rd Ave.
Carthage, NY
(Wood for pulp mills)

7) Equipment Design & Fabrication, Inc.
722 N. Smith St.
Charlotte, N.C.
(Log)

8) Pabco Fluid Power Co.
5752 Hillside Ave.
Cincinnati, OH
(Log)
(513) 941-6200

9) Piqua Engineering, Inc.
234-52 First Street
Piqua, OH
(513) 773-2464

10) Henke Mfg. Co., Inc.
433 W. Florida St.
Milwaukee, WI
(Log)

11) Didier Mfg. Co.
1652 Phillips Ave.
P.O. Box 806
Racine, WI
(Hydraulic, log)

12) Murray Machinery, Inc.
104 Murray Road
Wausau, WI
(Hydraulic, paper roll)

Fig. 15.9 A record of manufacturers who produce log-splitters.

scribe the design in terms of things it must be or must have and what attributes are most important.

15.5 Criteria and constraints—step 4

When we approach a problem, we always know some things about the answer that tend to describe what the final solution will be like. Please don't misread that statement; we did not say that we know what the best solution is, rather that we know some things *about* it. We may know, for instance, that it has to fit into a certain place, a fact that gives us an idea about its size. We may know that it can't

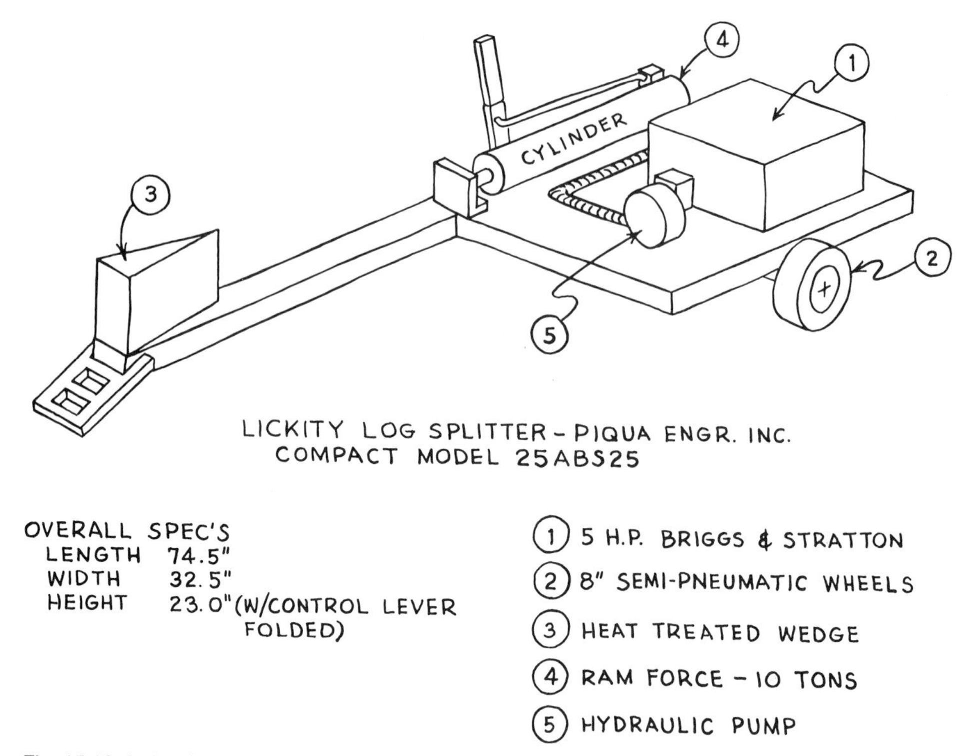

Fig. 15.10 A sketch is often the best way to record certain information.

cost more than so many dollars or that certain colors simply will not work. These three items tend to put limits or restrictions—what we commonly call constraints—on the solution. Such an idea is not new to you and it certainly is a routine experience for the practicing engineer. We face such a situation in almost every decision we make, even those that are not really important. When you got up this morning, you had to choose some clothes to wear. You probably limited the choice to those hanging in your closet (or maybe in your roommate's closet). This was a constraint *you* placed on the solution, not one that really existed until you made it so. In most fields of engineering, formulas have been developed and are used in designs of various kinds. Many, and probably most, of them are valid in a certain range of physical conditions. For instance, the hydraulic conditions of the flow of water are not valid below 0°C or above 100°C and are restricted to certain normal pressure ranges. Figure 8.21 shows a stress-strain curve for a particular type of steel. You can easily see that the relationship above and below the yield point is quite different. We normally refer to these constraints or limits as *boundary conditions* and they occur in many different ways.

Similar to constraints but clearly different are criteria. As we view a problem, we may know that some attributes of a solution will be good and some will be detrimental. Perhaps lower cost, less weight, or maybe tallest, will be desirable. Usually a considerable list of attributes can be made that help us evaluate the ideas that will come to us as we analyze the problem and consider alternatives. The criteria will be based on your background knowledge and the research that you have conducted, but they should not be based on ideas about what the solution should be. *Always* establish the criteria before you attempt to generate alternative solutions.

15.5.1 Design criteria

Whereas each project or problem has a personality all its own, there are certain characteristics that occur in one form or another in a great many projects. We should ask ourselves, "What characteristics are most desirable and which are not applicable?" Typical design criteria are listed below.

1. Cost—almost always a heavily weighted factor
2. Reliability
3. Weight
4. Ease of operation and maintenance
5. Appearance
6. Compatibility
7. Safety features
8. Noise level
9. Effectiveness
10. Durability
11. Feasibility
12. Acceptance

There will be other criteria and perhaps some of those given are of little or no importance in some projects, so the team members in industry or in the classroom must decide which ones are retained on their list. Since value judgments have to be made later, it probably makes little sense to include those which will be given relatively low weights of less than 5 percent. There are oftentimes mild disagreements at this point, not about which criteria are valid, but rather about how much weight should be assigned to each. It is often better if the team members make their assignments of weight independently and then compile all the results. This tends to dampen the effect of the more persuasive members at the same time that it forces all team members to contribute consciously.

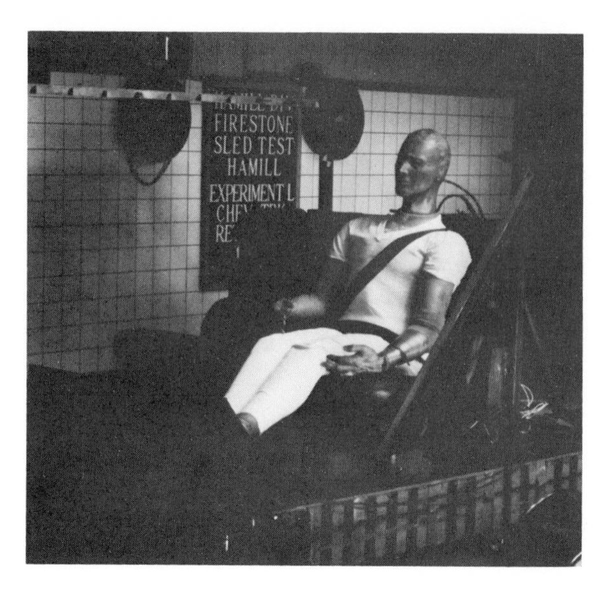

Fig. 15.11 The safety criterion may be evaluated by model testing. (*Firestone Tire & Rubber Co.*)

Usually there are not many instances where one of the members strongly disagrees with the mean value of the weight assigned to each criterion. Some negotiation may be required, but it is seldom a difficult situation to resolve.

15.5.2 The chapter example—step 4

Most people feel comfortable when they talk in general terms about a great many things, because as long as they do not have to get specific, an avenue of escape from their position is left open. The students whose progress we are following were not so fortunate, because they were facing a real problem. A review of their activity time schedule indicated that it was time to make some decisions. They therefore agreed on the following assigned weights.

1. Cost: 30 percent
2. Portability: 20 percent
3. Ease of operation: 15 percent
4. Safety: 15 percent
5. Durability: 10 percent
6. Use of standard parts: 10 percent

What areas of agreement and disagreement do you see between our list of 12 criteria and their list of 6?

The most obvious difference is that they included the use of standard parts as an important criterion, but it was not listed at all in Sec. 15.5.1. Just why this is important is not clear unless you try to place yourself in their position. If the young people have plans to manufacture their log-splitter, then it will be much easier and less expensive to begin operations if many of the components can be purchased rather than manufactured in their own plant.

They agreed that cost, weight (portability), ease of operation, and safety were important. The others on our list of 12 were either not considered or were considered to be of low importance (less than 5 percent).

They did not list any specific constraints, but we must assume that they are practical individuals who would assign a very low rating to a concept that exceeded some level of performance. For instance, if one of their ideas had an estimated cost of $500, it would no doubt be rated at near zero on a 10-point scale. They did not, however, tell us at what cost the zero rating begins; is it $200 or $100 or $75? In like manner they did not say at what weight they would consider a concept to be too heavy to be portable. They give us a little help in that they restrict the projected users to adults (excluding children).

It is our conclusion that at this point they have decided that the solution to their problem will be a portable log-splitter that can be operated by a single adult. If this is true, then we can conclude that a loop has been installed in the design process. This is not unusual. Figure 15.4 might very well have a number of arrows that show that problem solvers do return to steps in the design process that have supposedly been completed. In this particular case, we can assume that our students have redefined their problem even though they don't say so and don't report having undertaken any additional research. Again, this is not unusual.

15.6 Alternative solutions—step 5

We are now ready to see if we can think of a good solution. The best way that we can be reasonably sure that we arrive at a good (hopefully the best possible) solution is to examine a large number of possibilities.

Suppose that you are chief engineer for a manufacturing company and are faced with appointing someone to the position of director of product testing. This is an important position, because all the company's products are given rigorous testing under this person's direction before they are approved. You must compare all the candidates with the job description (criteria) to see who would do the best job. This seems to be a ridiculously simple procedure, doesn't it? Well, we think it does too, but many times such a process is not followed and poor appointments are made. In the same way that the list of candidates for the position has to be made, so must we produce a list of possible answers to our problem before we can go about the job of selecting the best one.

15.6.1 The nature of invention

The word "invention" strikes fear into the minds of many people. They say, "Me, an inventor?" The answer is, "Why not?" One reason why we don't fashion ourselves as inventors is that some of our earliest teaching directed us to be like the other boys and girls. Since much of our learning was by watching others, we learned how to conform. We also learned that if we were like the other kids, no one laughed at us. We can recall in our early days, even preschool, that the worst thing that could happen would be if people laughed at us. We'll bet that most of you have a similar feeling when you say something that is not too astute and it is followed by smiles and polite laughs. Moreover, we don't like to experiment because most experiments fail. It is a very secure person who never has to try something that he or she hasn't already done well. Think about it: when you were in the first few grades at school, didn't you feel great when you were called on by the teacher and you knew the answer? Don't you, even today, try to avoid asking your professors a question because you don't want the professor or your classmates to know that you don't know the answer? Most of us like to be in the majority. Please don't assume that we are saying that the majority of the people *are* wrong or that it is bad to be like other people. However, if we dwell on such behavior, then we will never do anything new. A degree of inventiveness or creativity is essential if we are to arrive at solutions to problems that are better than the way things are being done now. If we can remove the blocks to creativity, then we have a good chance of being inventive.

The father of one of the authors had a motto above his desk that read as follows:

Life's greatest art,
Learned through its hardest knocks,
Is to make
Stepping stones of stumbling blocks.

He doesn't know whether this was original with his father or not, but he remembers it after not having seen it for over 20 years, and it surely applies to the process of developing ideas.

15.6.2 Building the list

There are a great number of techniques that have been used to assist us in developing a list of possible solutions. Three of the more effective methods will be briefly discussed.

Checkoff lists, designed to direct your thinking, have been developed by a number of people. Generally the lists suggest possible ways that an existing solution to your problem might be changed and used. Can it be made: a different color, a different shape, stronger or weaker, larger or smaller, longer or shorter, of a different material, reversed or combined with something else? It is suggested that you write your list down on paper and try to conceive of how the present solution to the problem might be if you changed it according to each of the words on your list. Ask yourself: why is it like it is; will change make it better or worse; did the original designers have good reason for doing what they did or did they simply follow the lead of their predecessors.

Morphological listing gives a visual conception of the possible combinations that might be generated. These listings are usually shown as grids or diagrams. It is easy to visualize a cube, as shown in Fig. 15.12. This example indicates that we are considering the log-splitting problem as composed of three subdivisions: power source, power delivery, and splitting principles. These are then subdivided, as shown in Fig. 15.12. The cube shown produces 72 different combinations such as the one indicated by the shaded volume. Here we would have the logs split by torsion, applied by a reciprocating motion generated by an electric power source. Surely you can think of more than three major subdivisions and additional ideas for each of them than those shown in Fig. 15.12.

Brainstorming is a technique that has received wide discussion and support. The mechanics of a brainstorming session are rather simple. The leader states the problem clearly and ideas about its solution are invited. The length of productive sessions varies, but

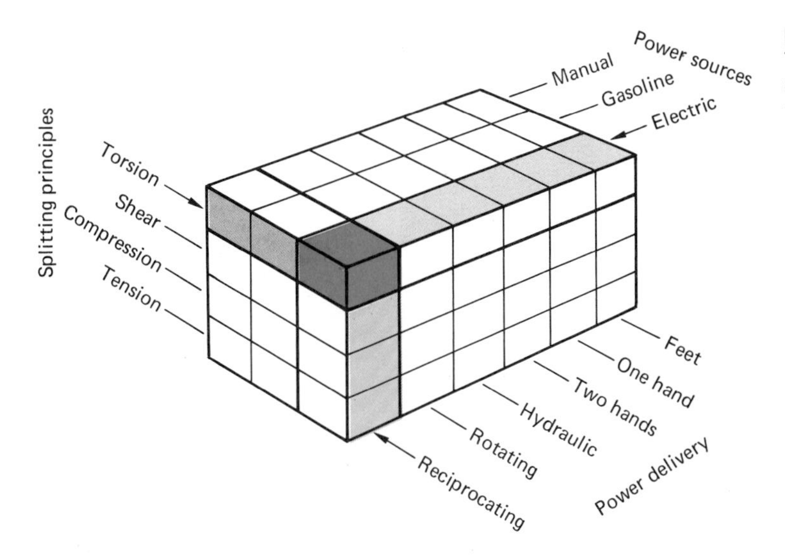

Fig. 15.12 A morphological chart. There are 72 combinations of these 3 attributes. The shaded example would employ an electric, reciprocating device that splits by torsion.

it is usually in the half-hour range. Often it takes a few minutes for a group to rid itself of its natural reserved attitude. But brainstorming can be fun, so choose a problem area and try it with some friends. Be prepared for a surprise at the number of ideas that will develop.

There are many descriptions of this process, most of which can be summed up as follows:

1. The size of the group is important. We have read of successful groups that range from three to fifteen; however, it is generally agreed that six or eight is an optimum size for brainstorming.

2. Free expression is essential. This is what brainstorming is all about. Any evaluation of the exposed ideas is to be avoided. Nothing should be said to discourage a group member from speaking out.

3. The leader is a key figure, even though free expression is the hallmark. The leader sets the tone and tempo of the session and provides a stimulus when things begin to drag.

4. The members of the group should be equals. No one should feel any reason to impress or support any other member. If your supervisor is also a member, you must steer clear of concern for his or her feelings or support for his or her ideas.

5. Recorders are necessary. Everything that is said should be recorded, mechanically or manually. Evaluation comes later.

We have discussed a few techniques that are recommended to stimulate your thought processes. You may choose one of the freewheeling techniques or perhaps a well-defined method. Regardless of your preferences, we think you will be pleased and even surprised at the large list of ideas that you can develop in a short period of time.

Fig. 15.13 The design team wrestles with a problem sketched by a team member. (*Bourns, Inc.*)

15.6.3 The chapter example—step 5

Our team of engineering students approaches their task of generating ideas by setting up a brainstorming session. Their minds are already tuned in on the problem; they have produced a list of candidates for the ultimate decision. They admit that they erred by giving preliminary evaluation to some of the ideas, which is a hard rule not to violate. The following is a list of ideas as it appears in their written report. It is not, however, their total list.

Ideas for splitting wood

1. Hydraulic cylinder (vertical or horizontal) used as a method to apply force.
2. Auto-jack or fence-tightener concept in order to apply pressure through a mechanical advantage. (See Fig. 15.14.)
3. Use of compressed air to force wedge through log.
4. Adaptations of conventional hand tools such as the axe, mall, or wedge.
5. Power or manual saws.
6. Heavy pile driver with block and tackle for raising weight.
7. High voltage arc between electrodes; similar to lightning.
8. Spring-powered wedge using either compression or tension.
9. Sliding mass that drives wedge into wood.
10. Drop wedge from elevated position onto the log.
11. Electronic sound that produces compression waves strong enough to split logs.
12. Wedge driven by explosive charge.
13. Spinning hammermill that breaks by shearing and concussion like a rock crusher.
14. Separate or split with intensive concentrated high energy such as laser beam.
15. Force a conical wedge into log and apply a torsional force.
16. Use a large mechanical vice with one jaw acting as a wedge.
17. Drill core (hole) in wood, fill with water, tap, and freeze.
18. Cut wood into slabs rather than across the grain.

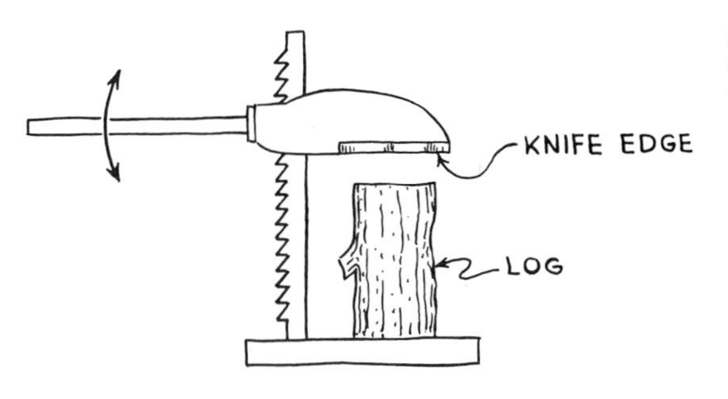

Fig. 15.14 Thumbnail sketches are oftentimes helpful in describing your ideas.

19. Apply couple to ends of log causing a shearing action.
20. Drop log from elevated position onto fixed wedge.

15.7 Analysis—step 6

At this point in the design process we have gathered information, set criteria and restrictions for a solution, and generated alternative solutions. In order to select the best solution in light of the available knowledge and criteria, the alternative solutions must be tested against various conditions. This testing procedure is called analysis, possibly a pivotal point in the design process. Often analysis leads to redefinition of the problem or continued research for more specific information about the problem.

A great deal of your engineering education will involve analysis. In fact, if analysis was not an integral part of the design process, we would have little need for engineering colleges. Analysis in the design process involves use of mathematical and engineering principles to verify the performance of an alternative solution. Another way to understand analysis is to consider a system, the laws of nature, input to the system, and output of the system. Analysis is conducted to determine system output when the system, laws, and input are given. For example, a simply supported beam, as shown in Fig. 15.15, may be considered as a system, the applied load as the input, and the equations of equilibrium as the laws. From this information we can determine several outputs, namely, the reactions at the supports A and B, the maximum shear stress, maximum tensile stress, and deflection at any point along the span of the beam. This information (output) may then be compared with criteria and constraints to determine if this particular length and beam configuration are satisfactory.

If any aspect of the analysis of the beam yields unsatisfactory results, we must go back to a previous point in the design process and make the appropriate alterations.

Fig. 15.15 A simple sketch can be a big help in the analysis of many systems.

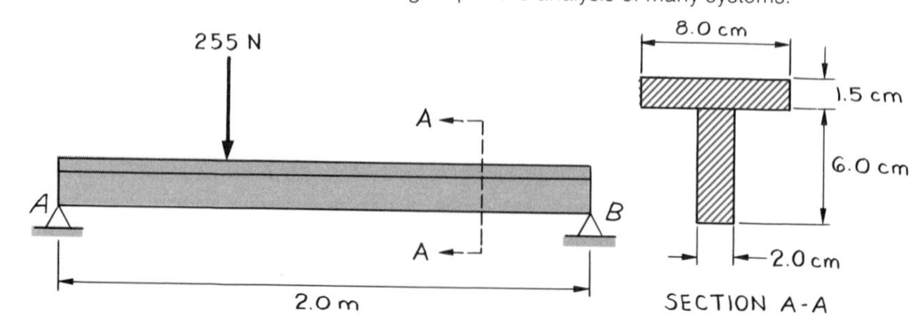

The analysis conducted by engineers in most design projects involves three areas: the laws of nature, the laws of economics, and common sense.

15.7.1 The laws of nature

You have already come into contact with many of the laws of nature and you will no doubt be exposed to many more. At this point in your education, you may have been exposed to the conservation principles: the conservation of mass, conservation of energy, conservation of momentum, and conservation of charge. From chemistry you are familiar with the laws of Charles, Boyle, and Guy-Lussac. In mechanics of materials, Hooke's law is a statement of the relationship between load and deformation. Newton's three principles serve as the basis of analysis of forces and the resulting motion and reactions.

Many methods exist to test the validity of an idea against the laws of nature. We might test the validity of an idea by constructing a mathematical model, for example. A good model will allow us many times to vary one parameter and examine the behavior of the other parameters. We may very well determine the limits within which we can work. Other times we will find that our boundary conditions have been violated and, therefore, the idea must be discarded.

Many mathematical models can be used to best advantage by plotting them as graphs. Very often the slopes of tangents to curves, points of intersection of curves, areas under or over or between curves, or other characteristics provide us with data that can be used directly in our designs.

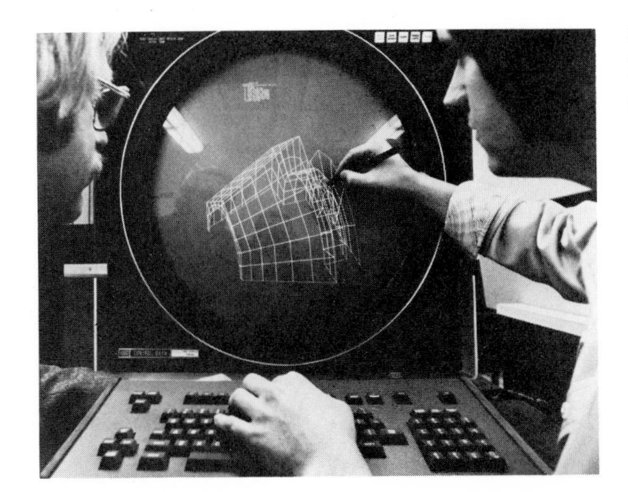

Fig. 15.16 A graphic display is part of most computer-aided designs—in this case, a complicated structural frame. (*Control Data Corporation.*)

The preparation of scale models of proposed designs is often a necessary step (see Fig. 15.17). This can be a simple cardboard cutout or it can involve the expenditure of great sums of money to test the model under simulated conditions that will predict how the real thing will perform under actual use. A prototype or pilot plant is sometimes justified because the cost of a failure is too great to chance. Such a decision usually comes only after other less-expensive alternatives have been shown to be inadequate.

You probably have surmised that the more time and money that you allot to your model, the more reliable is the data that you receive. This fact is often distressing because we want and need good data but have to balance our needs against the available time and money.

15.7.2 The laws of economics

Section 15.7.1 introduced the idea that money and economics are part of engineering design and decision making. We live in a society that is based on economics and competition. It is no doubt true that many good ideas never get tried because they are deemed to be economically infeasible. Most of us have been aware of this condition in our daily lives. We started with our parents explaining why we could not have some item that we wanted because it cost too much. Likewise, we will not put some very desirable component into our designs because the value gained will not return enough profit in relation to its cost.

Industry is continually looking for new products of all types. Some are desired because the present one is not competing well in the marketplace. Others are tried simply because it appears that people will buy them. How does a manufacturer know that a new product will be popular? They seldom know with certainty. Chapter 14 dealt with statistics, an important consideration in market

Fig. 15.17 Engineers study a model of a hydraulic structure before the real one is built. (*Stanley Consultants.*)

analysis. Some of you may find that probability and statistics are truly fascinating and get involved with sampling popular opinion. The techniques of this area of mathematics allow us to make inferences about how large groups of people will react based on the reactions of a few. It is beyond our study at this time to discuss the techniques, but industry routinely employs such studies and gambles millions of dollars on the results.

15.7.3 Common sense

We recall a story told by a college professor many years ago. Two young engineers from the same class go to work for the same company. One had been a good student and the other barely graduated. Their supervisor wants to test them early, so he asks them both to calculate the mass of a large casting that was on a railroad flatcar in the company's yards. Two hours later, the good student reports that his calculations indicated an approximate mass of 92 500 kg; the less-gifted student says, "That's not right." The second student admits that he didn't know how to calculate the correct amount, but he did know that one ordinary flatcar couldn't carry that much. Both students reveal shortcomings but both probably succeeded as engineers. However, we must not allow ourselves the luxury of failing to check our work.

During the 1930s, the Depression years, a national magazine conducted a survey of voters and predicted a Republican victory. They were wrong and the public lost confidence in them to the point that the magazine went out of business. They sampled the population by taking all of the telephone books in the United States and, by a system of random numbers, selected people to be called. They then applied good statistical analysis and made their prediction. Why did they miss so badly? It is a bit hard for us today to imagine the Depression years, but the facts are that large percentages of the voters did not have telephones, so this economic class of people was not included in the analysis. This group of people did vote and largely Democratic. Nothing was wrong with the analytical method, only the basic premise. The message is rather obvious: no matter how advanced our mathematical analysis, the results cannot be better than our basic assumptions. Likewise, we must always test our answers to see if they are reasonable.

15.7.4 The chapter example—step 6

Our design team generated many ideas for splitting wood, 20 of which are listed in Sec. 15.6.3. In addition, the criteria had been previously determined (Sec. 15.5.2) and value decisions made with regard to evaluating the importance of each criterion. At this point, decisions must be made to reduce the number of alternative solutions. The time available for completing the design project does not allow the team the luxury of thorough analysis of each of the ideas. Therefore, a decision is made to reduce the number of alternative solutions to five. These five will then be investigated and developed in more detail. The following is the result of the team's analysis.

Analysis of alternative solutions

*(Items marked with asterisks were kept by the team for further development.)

1. Hydraulic cylinder (vertical or horizontal)
   (*a*) Extreme cost for materials and manufacturing
   (*b*) High operational and maintenance costs
   (*c*) Nonportable for one person
   (*d*) Lack of standard parts

*2. Auto-jack principle or fence tightener (force by creating a mechanical advantage)
   (*a*) Reasonably portable
   (*b*) Minimum manual labor required

3. Use of compressed air (pneumatic)
   (*a*) Minimum portability
   (*b*) Extensive material, manufacturing, and operational costs

4. Adaptations of conventional hand tools such as the axe, mall, or wedge
   (*a*) Inefficient operation
   (*b*) Is the current solution
   (*c*) Unsafe for an inexperienced user

5. Power or manual saws
   (*a*) High cost of materials and manufacturing
   (*b*) Not a low-volume solution

6. Heavy pile driver with block and tackle used to raise weight
   (*a*) Not portable
   (*b*) Expensive

7. High-voltage arc between electrodes, similar to lightning bolt
   (*a*) Inefficient
   (*b*) Expensive
   (*c*) Impractical to use

*8. Spring-powered wedge using either compression or tension
   (*a*) Relatively easy to use
   (*b*) Portable
   (*c*) Low manufacturing cost

*9. Sliding mass that drives wedge into wood
   (*a*) Good for low-volume usage
   (*b*) Portable
   (*c*) Low initial cost and operational costs

*10. Drop a wedge from an elevated position onto the log
   (*a*) Uses a mechanical advantage

(b) Simple construction
(c) Good for low-volume production

11. Electronically produced sound that produces compression waves strong enough to split logs
(a) Impractical because of other damage that could be done
(b) Dangerous for average person to use

*12. Wedge driven by explosive charge
(a) Minimum work
(b) Low cost
(c) Easily portable

13. Hammermill that would chop wood much like a coal or rock crusher
(a) Expensive
(b) Much waste material (chips)
(c) Not easily portable

14. Separate or split with concentrated high energy from a laser beam
(a) Expensive
(b) Potentially dangerous

15. Force a conical wedge into log and apply a torsional force
(a) Complicated mechanical design for a single piece of equipment
(b) Expensive

16. Use a large vise with one jaw acting as a wedge
(a) Slow operating if powered by human
(b) Probably not easily portable

17. Drill core (hole) in wood, fill with water, cap, and freeze
(a) Time consuming
(b) Inefficient

18. Cut wood into slabs rather than splitting
(a) Inefficient
(b) Not suitable for ordinary fireplace

19. Apply couple to ends of log causing a shearing action
(a) Expensive
(b) Inefficient
(c) Destructive to wood

20. Drop log from elevated position onto fixed wedge
(a) High amount of manual labor required
(b) Inefficient

You will note that the analysis is based solely on the stated criteria and at best is very general. Many of the comments made show that no computations involving the mechanics of wood splitting were made. No testing (or test data) was used for any of the potential solutions. [A professional engineer would almost surely have spent considerable time and money on models, simulations, prototypes, and tests to verify (or disprove) his or her ideas about many aspects of the concepts before attempting a decision.] You should decide for yourself whether the team should be criticized at this point in their design effort. Remember, the team had only 9 weeks together and limited experience in college analysis courses.

We will list at this time a few analyses that may have been made of the alternative solutions by beginning engineering students. You may agree or disagree with some of them and perhaps add other items.

1. Determine the force necessary to split a log of a given type of wood. This may be done experimentally or analytically. Some data on this exists in the literature.

2. Determine the stamina of a human being with regard to lifting a specified weight a number of times in a given time period. This would be valuable information for the manually operated splitters.

3. For impact-type splitters, make a preliminary estimation of impulse required to split logs. That is, ascertain what combinations of mass and velocity are needed to cleave the log.

4. Find out if there is a wedge angle that would be more efficient than others.

5. Decide what masses we are talking about with respect to the alternative solutions. Since portability is a major consideration, locate or generate some data to be able to make comparisons based on total mass of the various devices.

15.8 Decision—step 7

As we mentally review our own professional practices, we can honestly say that the most difficult times for us have not been when the analysis of a problem was difficult, but when it required a "tough" decision. We have known many engineers who are technically knowledgeable but unable to make a final decision. They may be happy to suggest several possible solutions and outline the strong and weak points of each—indeed they may feel that their function is to do just that—but let someone else decide which course is to be followed. The truth of the matter is that most engineering assignments require both: providing information and making decisions.

What makes reaching a decision so difficult? The answer is *trade-offs*. If we can be certain about anything in the future, it is that with your decisions will come the necessity to compromise. Review the criteria in Sec. 15.5.3 that our team selected. In order to sell the gizmo that they are to design, it must be competitive in cost; so the lower the cost, the better. If they wish to make it more durable, chances are that they will use materials that are expensive or they will use more materials (heavier construction perhaps). If they use more materials, they are adding weight, which limits portability.

Each time one criterion is optimized, another moves away from its optimum position. If the relationships are complicated, one may have to go through very complex processes to reach a decision. You can be certain that no one idea will be better in all respects than all of the others; hence, you may have to choose a concept that you know is inferior to others in one or more of the decision criteria.

15.8.1 Conceptualization

If you are to decide among your several ideas, you need some detailed information about each. You need at least enough information so that you can evaluate each concept against each of the criteria. Otherwise, how can you feel that you have selected the best one? If time and money are available, you may want to build a prototype of each one to inspect and test. Such a luxury is not possible for most engineers, even once in their careers. In most cases, they have to settle for drawings or a model of their concepts. They most often can't even afford to prepare complete, detailed drawings of each concept. *But* you must have enough information to clearly explain what the object looks like, how it works, what its size, materials, and weight are, and what are some clearances, etc. Just how many drawings are needed is another decision that you must make. The trade-off is, the more drawings prepared, the more time and money expended. If you try to get by with a very limited amount of drawings, you will probably have very inaccurate and incomplete information with which to make a decision. Yet you must decide.

15.8.2 The chapter example—step 7

The analysis process (Sec. 15.7.4) reduced our student team's list of ideas to five, which are repeated here for reference.

1. Auto-jack principle—the pressure wedge (item 2)
2. Spring-powered wedge (item 8)
3. Sliding mass (item 9)
4. Dropping a wedge from an elevation (item 10)
5. Wedge driven by explosive charge (item 12)

Before a final decision can be made, each remaining idea must be looked at in greater depth. This is the basis for concept development.

Each member of the team was assigned the difficult and creative task of taking a very general idea and beginning the process of shaping it into a physical form. Figure 15.18 illustrates three of many possible ways to begin development of one general idea, i.e.,

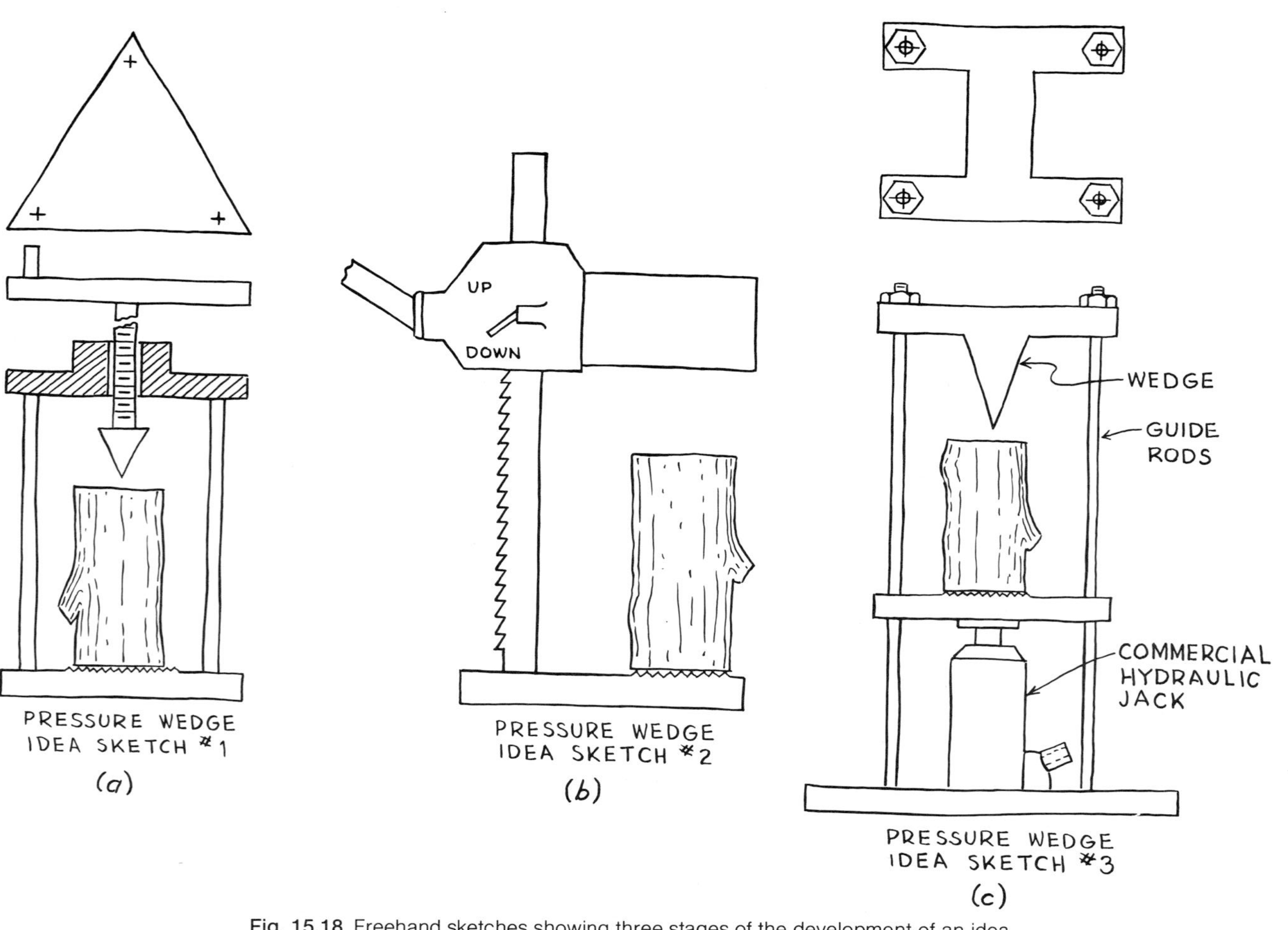

Fig. 15.18 Freehand sketches showing three stages of the development of an idea.

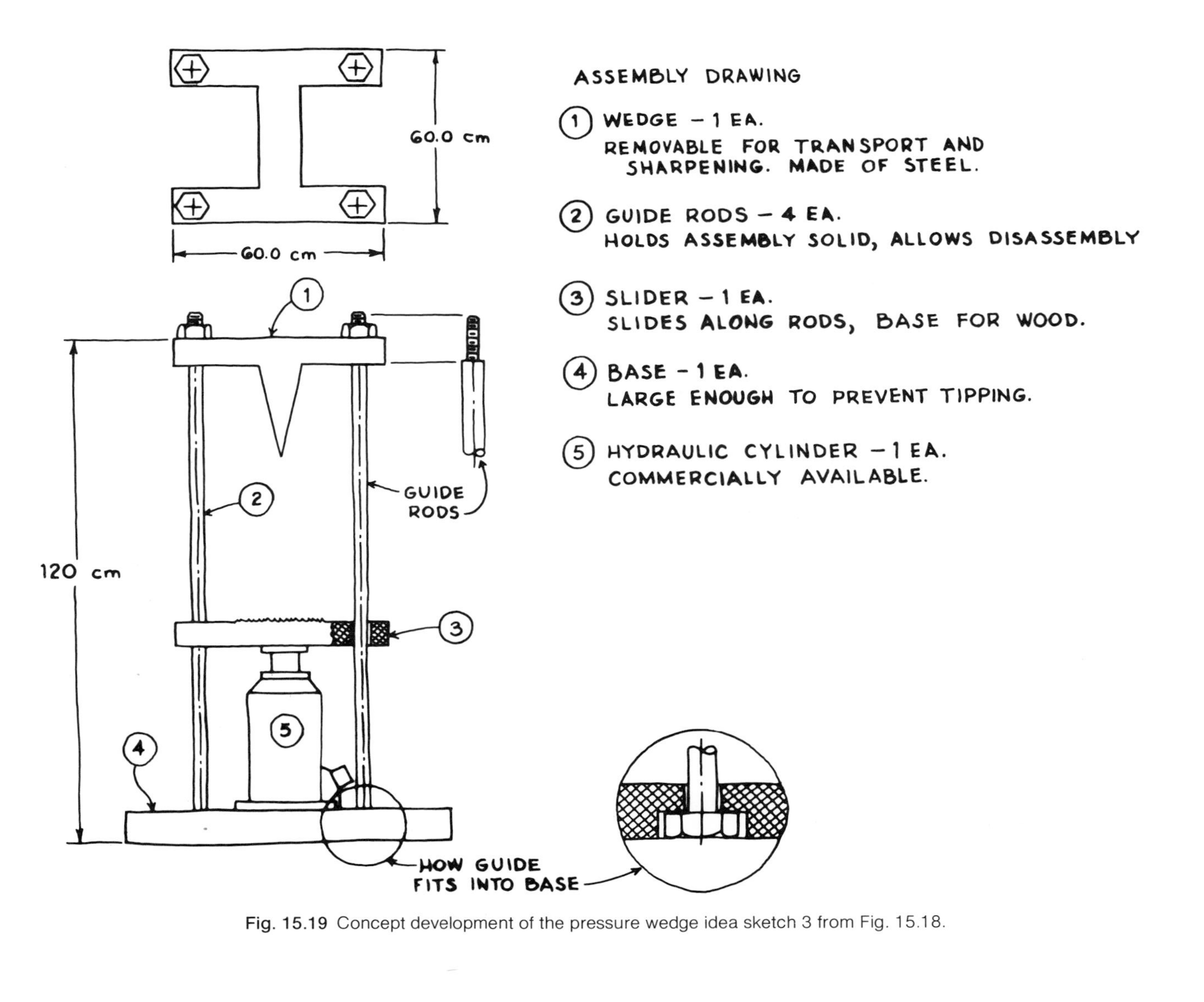

Fig. 15.19 Concept development of the pressure wedge idea sketch 3 from Fig. 15.18.

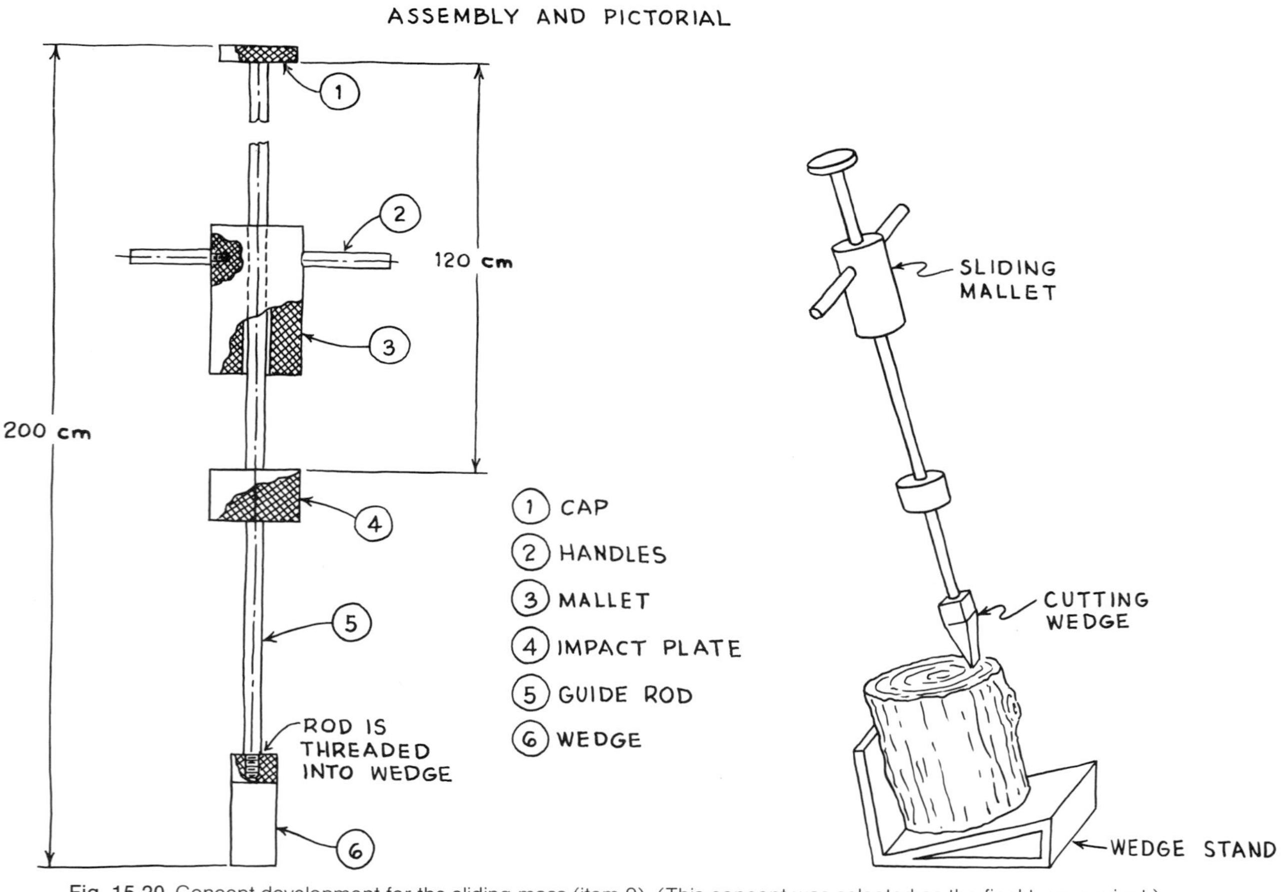

Fig. 15.20 Concept development for the sliding mass (item 9). (This concept was selected as the final team project.)

the pressure wedge. The team realized that this stage, or step, in the design sequence could result in many more idea sketches, but as always, time requires that decisions be made and the process be continued if prescribed deadlines are to be met.

Figures 15.19 and 15.20 are more complete concept developments that resulted from decisions made regarding which idea sketches were to be more completely laid out and which were to be forgotten.

Similar idea sketches followed by a single more-complete concept sketch were developed for each of the five areas listed above.

15.8.3 Criteria in decision

The objective of the entire design process is to choose the best solution for a problem. The steps that precede the decision phase are designed to give information that leads to the best decision. It should be quite obvious by now that poor research, a less-than-adequate list of alternatives, or inept analysis would reduce one's chances of selecting a good, much less the best, solution. Decision making, like engineering itself, is both an art and a science. There have been significant changes during the past few decades that have changed decision making from being primarily an art to what it is at the present with probability, statistics, optimization, and utility theory all routinely used. It is not our purpose to explore these topics, but simply to note their influence and to consider for a moment our task of selecting the best of the proposed solutions to our problem. The term "optimization" is almost self-explanatory in that it emphasizes that what we seek is the best, or optimum, value

Fig. 15.21 Three members of a design team are trying to reach a decision on the side slopes of an earth-filled dam. (*Stanley Consultants.*)

in light of a criterion. As you study more mathematics, you will acquire more powerful tools through calculus and numerical methods for optimization.

For now, to get a clearer idea of optimization, let's look at an example from hydraulics. An empirical equation was developed by an engineer named Manning to determine the velocity of flow of a fluid in a pipe not flowing full. The formula is

$$V = \frac{1.486}{n} R^{2/3} S^{1/2}$$

where V = velocity, ft/s

n = a roughness factor

R = hydraulic radius, ft

S = slope of the conduit as a decimal fraction

The hydraulic radius is defined as the area of the fluid divided by the wetted surface of the conduit. Suppose that you are required to find the maximum value of the hydraulic radius for a circular conduit. Two methods of finding the maximum value are readily apparent. First, you could write a mathematical expression of area divided by wetted perimeter, and by calculator determine a maximum. Second, you could calculate by trigonometry the values at several points, say 10, and plot them. By the geometry of the problem, you can safely assume that a smooth curve should go through all the points.

If you choose the second method, you will have a graph as shown in Fig. 15.22. The maximum value of the hydraulic radius occurs when the pipe is approximately 83 percent full. The Manning formula reveals that for given conditions of slope and pipe smoothness, the maximum velocity also occurs when the pipe is 83 percent full. Likewise, since the flow capacity of the conduit is equal to the velocity times the cross-sectional area of the fluid, the maximum capacity of the conduit occurs at the same point, 83 percent full.

We would be remiss at this point if we did not refer back to Sec. 15.7.3 and suggest applying a little common sense. What would happen if you designed your conduit on the basis of maximum capacity and all conditions were correct (very unlikely) except that someone attempted to increase the flow through the pipe? Immediately, it would literally jump from 83 percent full to completely full and be forced to operate under pressure. The seriousness of this depends on the situation. Common sense directs that we assume that some kind of safety factor is required or that other means must be developed to handle the overflow situation.

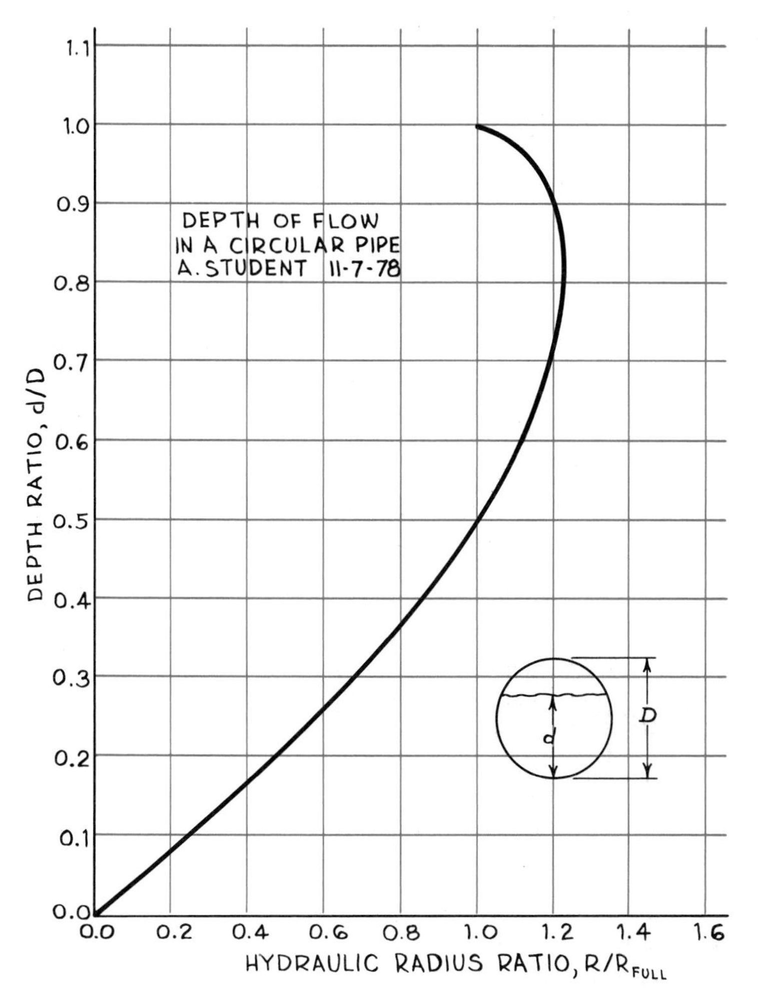

Fig. 15.22 Hydraulic characteristics of a circular pipe. Such a graph can speed calculations by showing relationships.

15.8.4 The chapter example—step 7 completed

The method of decision employed by the student design team is one that has considerable merit, and one that you may find simple to use. As mentioned in Sec. 15.5.3, they established six criteria and assigned weights to them. They later examined each of the five surviving concepts and graded them on a 0 to 10 scale. The grade was multiplied by the weight in percent and the points were recorded. A total of 1 000 points would be perfect. The concept with the greatest number of points is considered the best alternative. The results of their evaluation are shown in matrix form in Fig. 15.23.

Note that the winning alternative did not receive the highest rating for safety, ease of operation, and durability, and it tied for highest in portability. So, our team must report that this alternative has some shortcomings but that it is the best they can find.

Decision Matrix

Criteria	Weight W, percent	Selected concepts (see below) 1	2	3	4	5	6
Cost	30	6 / 180	7 / 210	7 / 210	7 / 210	9 / 270	
Ease of operation	20	10 / 200	7 / 140	9 / 180	10 / 200	7 / 140	
Safety	15	9 / 135	7 / 105	6 / 90	5 / 75	8 / 120	
Portability	15	6 / 90	5 / 75	4 / 60	10 / 150	10 / 150	
Durability	10	8 / 80	9 / 90	10 / 100	9 / 90	9 / 90	
Use of standard parts	10	7 / 70	8 / 80	8 / 80	6 / 60	9 / 90	
Total	100	755	700	720	785	860	

Rating scale R	
Excellent	9–10
Good	7– 8
Fair	5– 6
Poor	3– 4
Unsatisfactory	0– 2

Rating
6
180
$R \times W$

Selected concepts
1. Auto-jack principle (item #2)
2. Drop wedge from elevation (item #10)
3. Spring-powered wedge (item #8)
4. Wedge driven by explosion (item #12)
5. Sliding mass (item #9)
6. Additional concepts

Fig.15.23 Each concept was rated on a scale of 0 to 10 for each criterion. The rating was multiplied by the criterion weight and then summed. Concept 5 was chosen as the optimum even though it did not receive the highest rating on three of the six criteria.

15.9 Specification—step 8

After progressing through the design process up to the point of reaching a decision, many feel that the romance has gone out of the project. The suspense and uncertainty of the solution are over, but much work still lies ahead. Even if the new idea is not a breakthrough in technology but simply an improvement in existing technology, it must be clearly defined to others. Many very creative people are ill-equipped to convey to others just exactly what their proposed solution is. It is not the time to use vague generalities about the general scope and approximate size and shape of the chosen concept. One must be extremely specific about all of the details, regardless of the apparent minor role that each may play in the finished product.

15.9.1 Specification by words

One medium of communication that the successful engineer must master is that of language, written and spoken. Your problems in professional practice would be considerably less complicated if you were required to defend and explain your ideas only to other engineers. Few engineers have such luxury; most must be able to write and speak clearly and concisely to people who do not have comparable technical competence and experience. They may be officials of government who are bound so tightly by budgets and the need for public acceptance that only the best explanation is good enough to pierce their protective armor. They may be people in business who know that capital is limited and that they cannot defend another poor report to stockholders. Without appropriate documentation, they will not be nearly as certain as you are that your idea is a good one. Therefore, this phase of specification—communication—is so important that we have assigned it as the final step in the design process. Before discussing it, however, we will discuss another means of communication.

15.9.2 Graphical specification

Appendix C gives a brief overview of some of the basic principles of graphical procedures. As an engineer, you will not be expected to be an accomplished draftsman, but you will be expected to understand graphical techniques. You will probably have many occasions in which to work closely with technicians and draftsmen as they prepare the countless drawings essential to the manufacture of your design. You will not be able to do your job properly if you cannot sketch well enough to portray your idea or to read drawings well enough to know whether the plans that you must approve will actually result in your idea being constructed as you desire.

A lathe operator in the shop, an electronics technician, a contractor, or someone else, must produce your design. How is the person to know what the finished product is to look like, what materials are to be used, what thicknesses are required, how it is to work, what clearances and tolerances are demanded, how it is to be assembled, how it is to be taken apart for maintenance and repair, what connectors are to be used, etc.?

Typical drawings that are normally required include

1. A sufficient number of dimensioned drawings describing the size and shape of each part

Fig. 15.24 Detail drawings of each component must be prepared unless standard (stock) parts are used. (*Stanley Consultants.*)

2. Layouts to delineate clearances and operational characteristics
3. Assembly and subassemblies to clarify relationship of parts
4. Written notes, standards, specifications, etc., concerning workmanship and tolerances
5. A complete bill of materials

Included with the drawings are almost always written specifications, although certain classes of engineering work refer simply to documented standard specifications. Most cities have adopted one of several national building codes, so all structures constructed in that city must conform to the code. It is quite common for engineers and architects to refer simply to the building code as part of the written specifications and to write detailed specifications for only items that are not covered in the code. This procedure saves time and money for all by providing uniformity in bidding procedures. Many groups have produced standards that are widely recognized. For instance, there are standards for welds and fasteners for the obvious reasons—ease of specification and economy of manufacture. Moreover, there are such standards for each discipline of engineering.

15.9.3 The chapter example—step 8

Reproductions of two drawings prepared by the student team are given as Figs. 15.25 and 15.26. The following drawings were part of their completed report:

1. Detail drawings of all eight parts of the log-splitter
2. Exploded pictorial of the log-splitter
3. Detail drawing of the wedge stand (see Fig. 15.20 for pictorial)

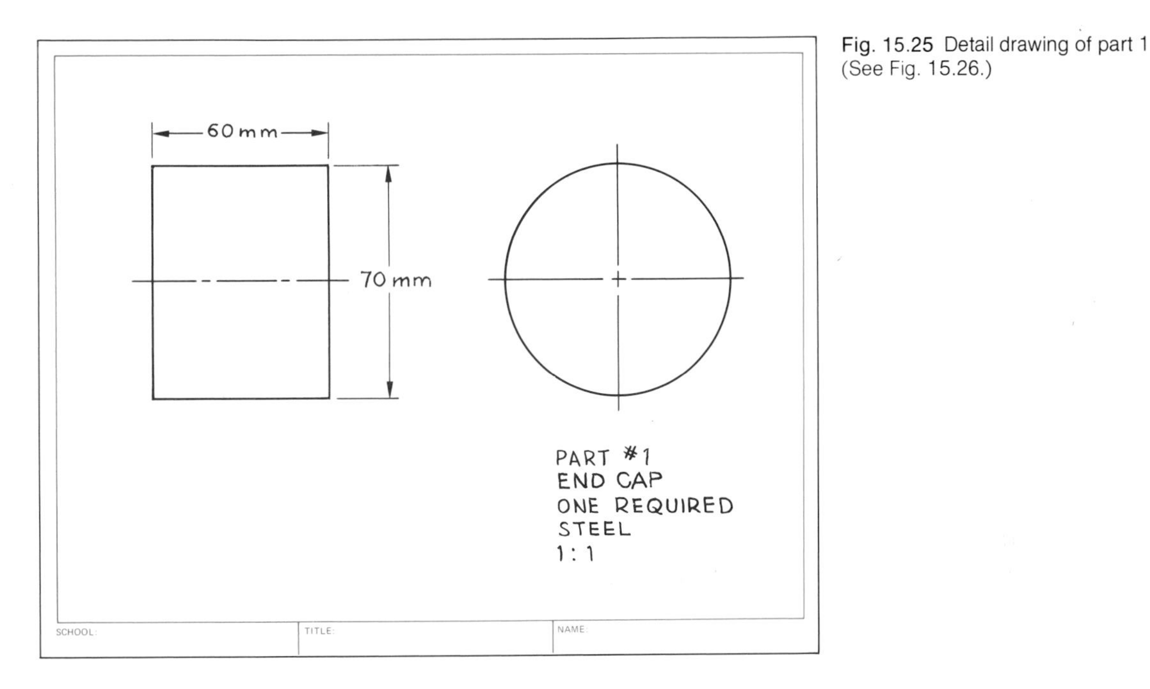

Fig. 15.25 Detail drawing of part 1 (See Fig. 15.26.)

4. Welding assembly of the log-splitter
5. Welding assembly of the wedge stand
6. Complete parts list

The drawings were accompanied by a cost analysis, weight summary, and description of the operating characteristics.

None of the team worked in a plant that produces such items, so no doubt an experienced detailer or draftsman would find reason to be critical of their drawings, but we feel that this phase of the process was performed quite well and that there will be few misunderstandings as a result of omissions on their part.

15.10 Communication—step 9

During the 1960s, the word "communication" seemed to take on a very high priority at conferences and in the professional journals. The need for conveying information and ideas had not changed, but there was an awareness of too much incomplete and inaccurate rendering of information. At most of the conferences the authors have attended over the past 20 years, one or more papers either discussed the need for engineers to develop greater skills in communication or demonstrated a technique for improving the skills. Students at most colleges are required to complete freshman English courses and, at some schools, a technical writing course

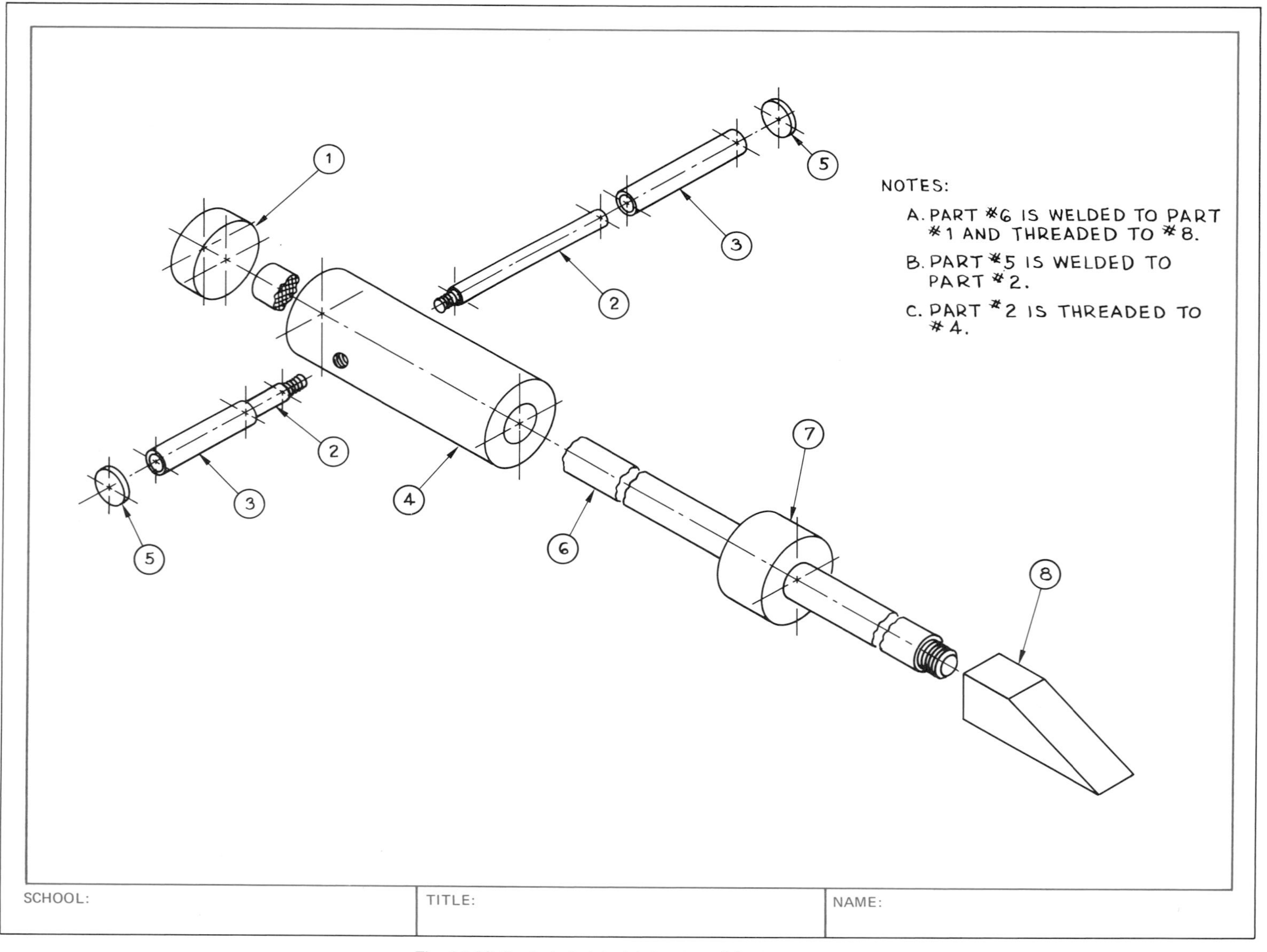

Fig. 15.26 Exploded pictorial drawing of the log-splitter.

in their junior or senior year; but many professors and employers feel that not enough of these types of courses are required.

For our purposes here, however, we will discuss only the salient points involved in design step 9.

15.10.1 Selling the design

It is certainly the responsibility of any profession to inform people of findings and developments. Engineering is no exception in this regard. Our emphasis here will be on a second type of communication, however: selling, explaining, or persuading.

Selling takes place all the way through the design process. Individuals who are the most skillful at it will see many of their ideas develop into realities. Those who are not so good at it will no doubt become frustrated with their supervisors for not exploring what they feel is a perfectly good idea in more depth.

If you are working as a design engineer for an industry, you cannot simply decide on your own that you will try to improve Model X of the gizmo they produce. Industry is anxious for their engineers to initiate ideas, but they won't necessarily approve all of them. As an engineer with a company, you must convince those who decide what assignments you get that the idea is worth the time and money required to develop it. Later, after the design has developed to the point where it can be produced and tested, you must again persuade management to place it into production.

It is a natural reaction to feel that your design has so many clear advantages that selling it should not be necessary. Such may be the case; but in actual situations, things seldom work so smoothly and simply. You will be selling or persuading or convincing others almost daily in a variety of ways. Among the many forms of communication are written and oral reports.

15.10.2 Written reports

The types of reports that you will write as an engineer will be varied, so a precise outline that will serve for all of the reports cannot be supplied. The two major divisions of reports are those used by individuals within the organization and those used primarily by clients or customers. Many times the in-house reports follow a strict form prescribed by the organization, whereas those intended for the client are usually designed for the particular situation. The nature of the project and the client usually determine the degree of formality employed in the report. Clients often state that they wish to use the report in some particular manner, which may direct you to the style of report to use. For instance, if you are a consultant for a city and have studied the needs for expansion of a power plant, the report may be very technical, brief, and full of equations, computations, etc., if it is intended for the use of the city's engineer and public utilities director. However, if the report is to be presented to citizens in an effort to convince them to vote for a bond issue to finance the expansion, the report will take a different flavor.

Reports generally will have the following divisions or sections:

1. Letter (memo) of transmittal
2. Introduction
3. Body
4. Summary
5. Conclusions and recommendations
6. Appendixes

These will be discussed in more detail below.

Letter of transmittal

The letter or memo should be no more than one page. It contains a statement of the agreement under which the assignment was undertaken, a word of thanks for assistance given by the client, and a statement that the report is attached or being sent by mail, by messenger, etc.

Introduction

This section identifies the problem, gives some background material, discloses the approach to the problem that will be employed, and briefly outlines the plan of the report.

Body

This is the primary section of the report. It usually follows a logical sequence of the operations of the study. If tests were conducted, research completed, or surveys undertaken, they are recounted and their significance is underscored. In essence, it is a narrative of what the engineer on the project did.

Summary

If the report is lengthy, few people will ever read all of it; and many people will read it in reverse order. The highest-level decision makers may read only the conclusions and recommendations. Others will also read the summary and then refer to the body and appendixes for clarification and/or amplification. The summary says what is in the body but in much less space. (It could be placed at the beginning of the report.)

Conclusions and recommendations

This section tells why the study was done and explains the purpose of the report. Herein you explain what you now believe to be true as a result of the work discussed in the body and what you recommend be done about it. You must lay the groundwork earlier in the report, and at this point you must sell your idea. If you have done the job carefully and fully, you may make a sale; but don't be discouraged if you don't. There will be other days and other projects.

Appendixes

Appendixes can be used to avoid interrupting your narrative so that it can flow more smoothly. Those who don't want to know everything about your study can read it without digression. What is in the appendix completes the story by showing all that was done. But it should not contain information that is essential to one's understanding of the report.

Some reports are quite formal; some include an abstract; and most have a table of contents with a list of figures and tables, if any. Each report must be designed to accomplish a specific goal.

Student reports oftentimes must follow an instructor's directions regarding form and topics that must be included. If you are asked to write a report in an introductory design course, refer to Fig. 15.1 to make sure that all of the steps in the design process have been successfully completed. You might also study the bias of your instructor and make sure that you have done especially the steps that he or she considers most important. This may sound as though pleasing the instructor and getting a good grade is all that is important. But perhaps in this respect the academic situation is something like that in industry or private practice: Your report must take into account the audience—its biases and its expectations whether professor in the classroom or supervisor in the business world.

15.10.3 The oral presentation

The objective of the oral presentation is the same as that of the written report—to furnish information and convince the listener. However, the methods and techniques are quite different. The written report is designed to be glanced at, read, and then studied. The oral presentation is a one-shot deal that must be done quickly. So it must be simple. There is no time to go into detail, to show complicated graphs and tables of data or many of the things that are given in a written report. What can you do to make a good presentation?

First, you must be prepared. No audience listens to people who have not bothered to prepare themselves. So you should rehearse with a timer, a mirror, and a tape recorder.

Stand in such a way that you don't detract from what you are saying or showing.

Look at your audience and maintain eye contact. You will be receiving cues from those who are listening, so be prepared to react to these cues.

Project your voice by consciously speaking to the back row. The audience quickly loses interest if it has to struggle to hear.

Speak clearly. We all have some problems with our voices—they are either too high, too low, or too accented and certain words or sounds are hard for us; but always be concerned for the listener.

Preparation obviously includes being thoroughly familiar with the material. It should also include determining the nature, size, and technical competence of the audience. You must know how much time will be allotted to your presentation and what else, if anything, is to be presented before or after your speech. It is essential that you know what the room is like, because the physical conditions of the room—its size, lighting, acoustics and seating arrangements—may very well control your use of slides, transparencies, records, and microphones.

The quality of your graphic displays can often influence the opinion of your audience. Again, consider to whom you are speaking carefully as you choose which and how many graphics to use. Be certain that they can be read and understood or don't use them at all. Don't clutter your displays with so many details that the message is obscured; and don't try to make a single visual aid accomplish too many tasks: it is good to change the center of emphasis. By all means, test your visual aids before the meeting and never apologize for their quality. (If they aren't good, don't use them.)

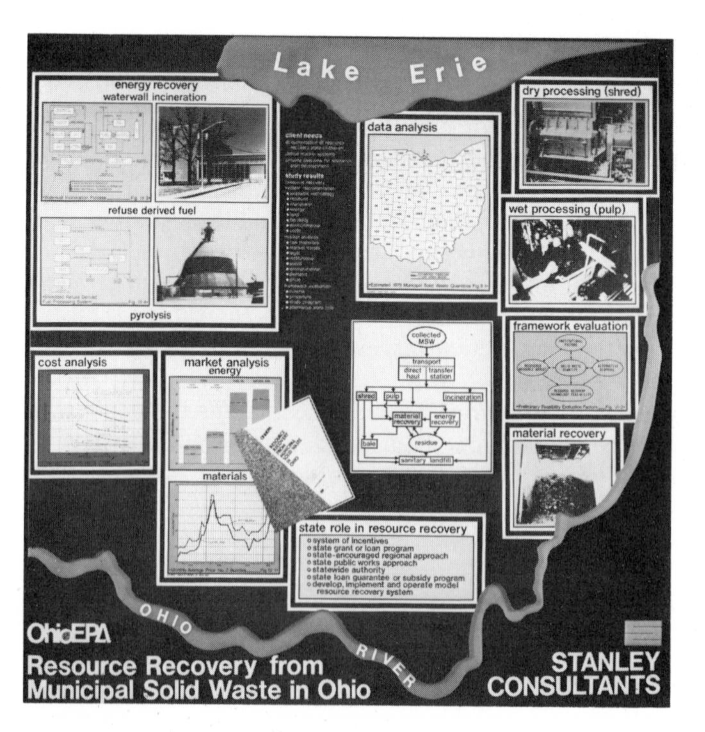

Fig. 15.27 Visual aid used by a consultant during an oral presentation to a client. (*Stanley Consultants.*)

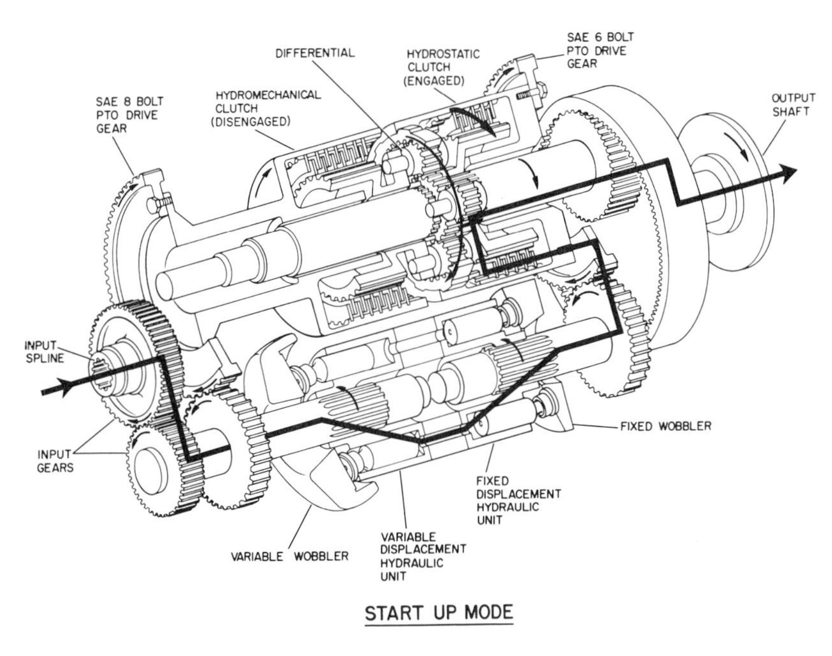

Fig. 15.28 A cutaway sectional view of a transmission. Visual aids are essential in presenting a complicated device such as this. (*Sundstrand Hydro-Transmission Div.*)

Figure 15.27 shows a number of visual aids that were used by an engineering consultant in presenting views on resource recovery. Included are photographs, flow diagrams, and several types of graphs and tables.

Figure 15.28 is a good, clear cutaway drawing that is excellent for use as a visual aid during an oral presentation. Figure 15.29 is

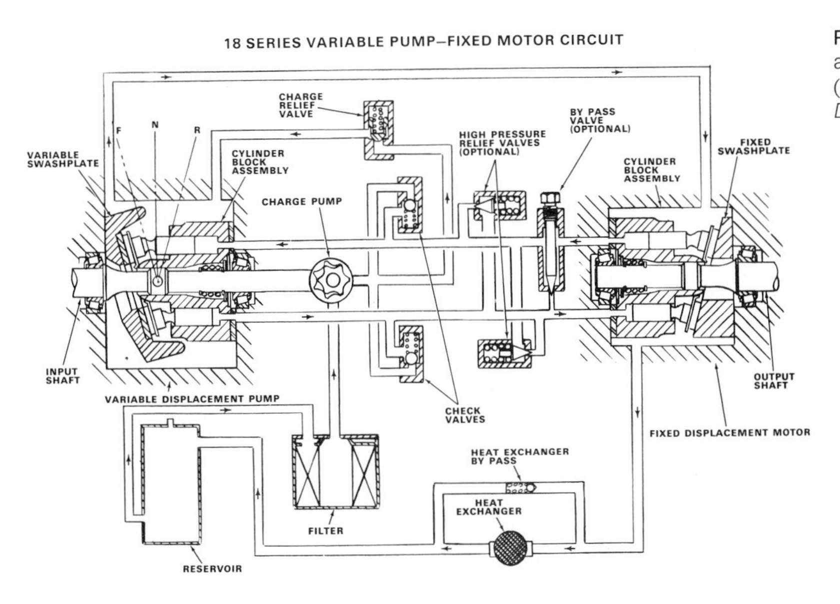

Fig. 15.29 A schematic used to augment an oral presentation. (*Sundstrand Hydro-Transmission Div.*)

a schematic drawing that can be used to explain a process. The quality of your visual aids can influence many people for you or against you before they hear all you have to say.

Have a good finish. Save something important for the last and make sure everyone knows when the end has come. By all means, don't end with, "Well, I guess that's about all I have to say." You have much more to say, you just don't have the time to say it.

15.10.4 The chapter example—step 9

The written and oral reports prepared and presented by the students were significant parts of their design experience. Both reports were regulated somewhat by their professor in much the same way that reports are in industry. They were told who would be reading the report and who would judge the oral presentation. They were given copies of the written report grading sheet and the oral presentation judging card. They correctly accepted these constraints as real (not imagined) and they were given high ratings by their evaluators.

15.11 Epilogue

It would be unfair to the reader and to the students who were the central figures of our story if we ended our report as they completed the course. The school term did end, and they went ahead with their studies in pursuit of engineering degrees. One of the team members found it necessary to interrupt his education and return to his home. As he sought work to earn funds for his eventual return to college, a friend asked him to construct a log-splitter similar to his class project design. In producing the unit for a customer (client), he learned that

Fig. 15.30 Two examples (right and on top of page 371) of devices that have progressed through several model changes (redesigns). (*Beech Aircraft and FMC Corporation.*)

1. It was a good design but not ready to be manufactured without change

2. Other details would have to be changed if it were ever to be mass-produced

The log-splitter has, in fact, become a reality, with more than 50 having been made and sold. The most recently produced units are different from the first one, as should be expected. Most items that have been on the market for years are modified.

To illustrate, consider the evolution of the two products shown in Fig. 15.30. The potato harvestor is the latest improvement in a machine that did not exist when the authors of this book were freshmen. It all began when the labor cost of plowing and gathering potatoes became too great for the competition it faced in the form of grain combines. The light airplane shown is not as sophisticated as the Boeing 747 or the Concorde SST, but neither does it have much in common with the original flying machine of the Wright brothers.

Figure 15.31 shows a mockup of the log-splitter, very much like the ones that have been produced and sold. The profit from the sale of the log-splitters will provide the means for our young "designer" to return and complete his formal education.

Suggested projects

The following list of project topics is included to stimulate your thinking and to show the variety of projects that freshman students have undertaken.

Headlights that follow the wheels' direction

Protective "garage" that can be stored in trunk

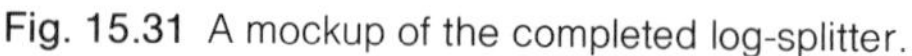

Fig. 15.31 A mockup of the completed log-splitter.

Means of preventing body rust

A device to prevent theft of helmets left on motorcycles

Conversion kit for winter operation of motorcycles

An improved rack for carrying packages or books on a motorcycle or bicycle

Child's seat for motorcycle or bicycle

Tray for eating, writing, and playing games in the back seat of a car

A system for improving traction on ice without studs or chains

An inexpensive built-in jack for raising a car

Auto-engine warmer

A better way of informing motorists of speed limits, road conditions, hazards, etc.

Theft- and vibration-proof hub caps

Better way to check engine oil level

Device to permit easier draining of the oil pan by weekend mechanics

A heated steering wheel for cold weather

Less-expensive replacement for auto air-cleaner elements

Overdrive for trail bike

Sun shield for automobile

Well-engineered, efficient automobile instrument panel

SOS sign for cars stalled on freeways

Remote car-starting system for warmup

Car-door positioner for windy days
Bicycle trailer
Automatic rate-sensitive windshield wipers
Corn detasseler
Improved wall outlet
Beverage holder for card table
Car wash for pickups
Better rural mailbox
Home safe
Improved automobile traffic pattern on campus
An alert for drowsy or sleeping drivers
Improved automobile headlight
Improved bicycle for recreation or racing
Improved bicycle brakes
Transit system for campus
Pleasure boat with retractable trailer wheels
Improved pedestrian crossings at busy intersections
Transportation within large airports
Improved baggage-handling system at airports
Improved parking facilities in and around campus
A simple but effective device to assist in cleaning clogged drains
Device to attach to a paint can for pouring
Improved soap dispenser
A better method of locking weights to a barbell shaft
Shoestring fastener to replace knot
Automatic moisture-sensitive lawn waterer
Better harness for seeing-eye dogs
Better jar opener
System or device to improve efficiency of limited closet space
Shoe transporter and storer
Pen and pencil holder for college students
Acceptable rack for mounting electric fans in dormitory windows
Device to pit fruit without damage
Riot-quelling device to subdue participants without injury
An automatic device for selectively admitting and releasing an auxiliary door for pets
Device to permit a person to open a door when loaded with packages
More efficient toothpaste tube
Fingernail catcher for fingernail clippers

A more effective alarm clock for reluctant students

An alarm clock with a display to show it has been set to go off

A device to help a mother monitor small children's presence and activity in and around the house

A chair that can rotate, swivel, rock, or stay stationary

A simple pocket alarm that is difficult to shut off, for discouraging muggers

Improved storage for luggage, books, etc., in dormitories

A lampshade designed to permit one to study while his or her roommate is asleep

A device that would permit blind people to vote in an otherwise conventional voting booth

One-cup coffee maker

Solar greenhouse

Quick-connect garden-hose coupling

Device for recycling household water

Silent wakeup alarm

Home aids for blind (or deaf)

Safer, more efficient, and quieter air mover for room use

Lock that can be opened by secret method without a key

Can crusher

Rain-sensitive house window

Better grass catchers for (riding) lawnmowers

Winch for hunters of large game

Gauges for water, transmission, etc., in auto

Built-in auto refrigerator

Better camp cooler

Dormitory-room cooler

Device for raising and lowering TV racks in classroom

Overdrive for trail bike

Impact hammer adapter for electric drills

Improved method to detect and control level position of bucket on bucket loader

Shields to prevent corn spillage where drag line dumps into sheller elevator (angle varies)

Automatic tractor-trailer hitch aligning device

Jack designed expressly for motorcycle use (special problems involved)

A motorbike using available (junk) materials

Means of improving road traction without using chains or studs

Improved road signs for speed limits, curves, deer crossings, etc.

More effective windshield wipers
Windshield deicer
Shock-absorbing bumpers for minor accidents
Home fire alarm device
Means of evacuating buildings in case of fire
Automatic light switches for rooms
Carbon monoxide detector
Indicator to report need for oil change
Collector for dust (smoke) particles from stacks
Means of disposing of or recycling soft-drink containers
A way to stop dust storms, resultant soil loss, and air entrainment
Attractive system for handling trash on campus
A self-decaying disposable container
Device for dealing with oil slicks
Means of preventing heat loss from greenhouses
Way of creating energy from waste
A bookshelf with horizontally and vertically adjustable shelves and dividers
A device that would make the working surface of graphics desks adjustable in height and slope, retaining the existing top and pedestal
Egg container (light, strong, compact) for camping and canoeing
Ramps or other facilities for handicapped students
Multifunctional (suitcase/chair/bookshelf, etc.) packing device for students
Self-sharpening pencil for drafting
Adapter to provide tilt and elevation control on existing graphics tables
Compact and inexpensive camp stove for backwoods hiking
Road trailer operable from inside car
Hood lock for cars to prevent vandalism
System to prevent car thefts
Keyless lock

Appendixes

Selected topics from algebra

APPENDIX A

A.1 Introduction

This appendix includes material on exponents and logarithms, simultaneous equations, and the solution of equations by approximation methods. The material can be used for reference or review. The reader should consult an algebra textbook for more detailed explanations or additional topics for study.

A.2 Exponents and radicals

The basic laws of exponents are stated below along with an illustrative example.

Law	Example
$a^m a^n = a^{m+n}$	$x^5 x^{-2} = x^3$
$\dfrac{a^m}{a^n} = a^{m-n} \quad a \neq 0$	$\dfrac{x^5}{x^3} = x^2$
$(a^m)^n = a^{mn}$	$(x^{-2})^3 = x^{-6}$
$(ab)^m = a^m b^m$	$(xy)^2 = x^2 y^2$
$\left(\dfrac{a}{b}\right)^m = \dfrac{a^m}{b^m} \quad b \neq 0$	$\left(\dfrac{x}{y}\right)^2 = \dfrac{x^2}{y^2}$
$a^{-m} = \dfrac{1}{a^m} \quad a \neq 0$	$x^{-3} = \dfrac{1}{x^3}$
$a^0 = 1 \quad a \neq 0$	$2(3x^2)^0 = 2(1) = 2$
$a^1 = a$	$(3x^2)^1 = 3x^2$

These laws are valid for positive and negative integer exponents and for a zero exponent, and can be shown to be valid for rational exponents. Some examples of fractional exponents are illustrated here. Note the use of radical ($\sqrt{\ }$) notation as an alternative to fractional exponents.

Law	Example
$a^{m/n} = \sqrt[n]{a^m}$	$x^{2/3} = \sqrt[3]{x^2}$
$\dfrac{\sqrt[n]{a}}{\sqrt[n]{b}} = \sqrt[n]{\dfrac{a}{b}} \quad b \neq 0$	$\dfrac{\sqrt[3]{16}}{\sqrt[3]{2}} = \sqrt[3]{8} = 2$
$a^{1/2} = \sqrt[2]{a^1} = \sqrt{a} \quad a \geq 0$	$\sqrt{25} = 5 \quad (\text{not } \pm 5)$

A.3 Exponential and power functions

Functions involving exponents occur in two forms—power and exponential. The power function contains the base as the variable and the exponent is a rational number. An exponential function has a fixed base and variable exponent.

The simplest exponential function is of the form

$$y = b^x \qquad b \geq 0$$

where b is a constant. Note that this function involves a power but is fundamentally different from the power function $y = x^b$.

The inverse of a function is an important concept for the development of logarithmic functions from exponential functions. Consider a function $y = f(x)$. If this function could be solved for x, the result would be expressed as $x = g(y)$. For example, the power function $y = x^2$ has as its inverse $x = \pm\sqrt{y}$. Note that in $y = x^2$, y is a single-valued function of x whereas the inverse is a double-valued function. For $y = x^2$, x can take on any real value whereas the inverse $x = \pm\sqrt{y}$ restricts y to only positive values or zero. This result is important in the study and application of logarithmic functions.

A.4 The logarithmic function

The definition of a logarithm may be stated as follows.

A number L *is said to be the logarithm of a positive real number* N *to the base* b *(where* b *is real, positive, and different from 1), if* L *is the exponent to which* b *must be raised to obtain* N.

Symbolically, the logarithm function is expressed as

$$L = \log_b N$$

for which the inverse is

$$N = b^L$$

For instance:

$$\log_2 8 = 3 \qquad \text{since } 8 = 2^3$$

$$\log_{10} 0.01 = -2 \qquad \text{since } 0.01 = 10^{-2}$$

$$\log_5 5 = 1 \qquad \text{since } 5 = 5^1$$

$$\log_b 1 = 0 \qquad \text{since } 1 = b^0$$

Several properties of logarithms and exponential functions can be identified when plotted on a graph.

Example problem A.1 Plot graphs of $y = \log_2 x$ and $x = 2^y$ that are inverse functions.

Solution

Since $y = \log_2 x$ and $x = 2^y$ are equivalent by definition, they will graph into the same line. Choosing values of y and computing x from $x = 2^y$ yields Fig. A.1.

Fig. A.1 The logarithmic function.

Some properties of logarithms that can be generalized from Fig. A.1 are

1. $\log_b x$ is not defined for negative or zero values of x.
2. $\log_b 1 = 0$.
3. If $x > 1$, then $\log_b x > 0$.
4. If $0 < x < 1$, then $\log_b x < 0$.

Other properties of logarithms that can be proven as a direct consequence of the laws of exponents are, with P and Q being real and positive numbers,

1. $\log_b PQ = \log_b P + \log_b Q$.
2. $\log_b \dfrac{P}{Q} = \log_b P - \log_b Q$.
3. $\log_b (P)^m = m \log_b P$.
4. $\log_b \sqrt[n]{P} = \dfrac{1}{n} \log_b P$.

The base b, as stated in the definition of a logarithm, can be any real number greater than 0 but not equal to 1, since 1 to any power remains 1. When using logarithmic notation, the base is always indicated, with the exception of base 10, in which case the base is frequently omitted. In the expression $y = \log x$, the base is understood to be 10. A somewhat different notation is used for the natural (Naperian) logarithms discussed in the Sec. A.5.

Sometimes it is desirable to change the base of logarithms. The procedure is shown by the following example.

Example problem A.2 Given that $y = \log_a N$, find $\log_b N$.

Solution

$$y = \log_a N$$

$$N = a^y \qquad \text{(inverse function)}$$

$$\log_b N = y \log_b a \qquad \text{(taking logs to base } b\text{)}$$

$$\log_b N = (\log_a N)(\log_b a) \qquad \text{(substitution for } y\text{)}$$

$$= \frac{\log_a N}{\log_a b} \qquad \left(\text{since } \log_b a = \frac{1}{\log_a b}\right)$$

A.5 Natural logarithms and e

In advanced mathematics, the base e is usually chosen for logarithms to achieve simpler expressions. Logarithms to the base e are called natural, or Naperian, logarithms. The constant e is defined in the calculus as

$$e = \lim_{n \to 0} (1 + n)^{1/n} = 2.7182818284 \ldots$$

For purposes of calculating e to a desired accuracy, an infinite series is used.

$$e = \sum_{n=0}^{\infty} \frac{1}{n!}$$

The required accuracy is obtained by summing sufficient terms. For example,

$$\sum_{n=0}^{6} \frac{1}{n!} = 1 + 1 + \frac{1}{2} + \frac{1}{6} + \frac{1}{24} + \frac{1}{124} + \frac{1}{720}$$

$$= 2.718055$$

which is accurate to four significant figures.

Natural logarithms are denoted by the symbol ln, and all the properties defined previously for logarithms apply to natural logarithms. The inverse of $y = \ln x$ is $x = e^y$. The following examples illustrate applications of natural logarithms.

Example problem A.3

$\ln 1 = 0 \qquad \text{since } e^0 = 1$

$\ln e = 1 \qquad \text{since } e^1 = e$

Example problem A.4 Solve for x:

$2^x = 3^{x-1}$

Specify answer to four significant figures.

Solution Taking natural logarithms of both sides of the equation and using a calculator for evaluation of numerical quantities,

$$x \ln 2 = (x - 1)\ln 3$$

$$\frac{x}{x-1} = \frac{\ln 3}{\ln 2} = 1.5850$$

$$x = 2.709 \qquad \text{(four significant figures)}$$

This problem could have been solved by choosing any base for taking logarithms. However, in general, base e or 10 should be chosen so that a scientific calculator can be used for numerical work.

A.6 Simultaneous equations

Several techniques exist for finding the common solution to a set of n algebraic equations in n unknowns. A formal method for solution of a system of linear equations is known as Cramer's rule, which requires a knowledge of determinants.

A second-order determinant is defined and evaluated as

$$\begin{vmatrix} a_1 b_1 \\ a_2 b_2 \end{vmatrix} = a_1 b_2 - a_2 b_1$$

A third-order determinant is defined and evaluated as

$$\begin{vmatrix} a_1 b_1 c_1 \\ a_2 b_2 c_2 \\ a_3 b_3 c_3 \end{vmatrix} = a_1 \begin{vmatrix} b_2 c_2 \\ b_3 c_3 \end{vmatrix} - a_2 \begin{vmatrix} b_1 c_1 \\ b_3 c_3 \end{vmatrix} + a_3 \begin{vmatrix} b_1 c_1 \\ b_2 c_2 \end{vmatrix}$$

where the second-order determinants are evaluated as indicated above. The procedure may be extended to higher-order determinants.

Cramer's rule for a system of n equations in n unknowns can be stated as follows:

1. Arrange the equations to be solved so that the unknowns x, y, z, and so forth, appear in the same order in each equation; if any unknown is missing from an equation, it is to be considered as having a coefficient of zero in that equation.

2. Place all terms that do not involve the unknowns in the right member of each equation.

3. Designate by D the determinant of the coefficients of the unknowns in the same order as they appear in the equations. Designate by D_i the determinant obtained by replacing the elements of the ith column of D by the terms in the right member of the equations.

4. Then, if $D \neq 0$, the values of the unknowns x, y, z, and so forth, given by

$$x = \frac{D_1}{D}, \; y = \frac{D_2}{D}, \; z = \frac{D_3}{D}, \; \ldots$$

Example problem A.5 Solve the following system of equations that have already been written in proper form for application of Cramer's rule.

$$3x + y - z = 2$$

$$x - 2y + z = 0$$

$$4x - y + z = 3$$

Solution

$$x = \frac{\begin{vmatrix} 2 & 1 & -1 \\ 0 & -2 & 1 \\ 3 & -1 & 1 \end{vmatrix}}{\begin{vmatrix} 3 & 1 & -1 \\ 1 & -2 & 1 \\ 4 & -1 & 1 \end{vmatrix}} = \frac{2\begin{vmatrix} -2 & 1 \\ -1 & 1 \end{vmatrix} - 0\begin{vmatrix} 1 & -1 \\ -1 & 1 \end{vmatrix} + 3\begin{vmatrix} 1 & -1 \\ -2 & 1 \end{vmatrix}}{3\begin{vmatrix} -2 & 1 \\ -1 & 1 \end{vmatrix} - 1\begin{vmatrix} 1 & -1 \\ -1 & 1 \end{vmatrix} + 4\begin{vmatrix} 1 & -1 \\ -2 & 1 \end{vmatrix}}$$

$$= \frac{2(-2 + 1) - 0(1 - 1) + 3(1 - 2)}{3(-2 + 1) - 1(1 - 1) + 4(1 - 2)}$$

$$= \frac{-5}{-7}$$

$$= \frac{5}{7}$$

The reader may verify the solutions $y = 6/7$ and $z = 1$.

There are several other methods of solution for systems of equations that are illustrated by the following examples.

Example problem A.6 Solve the system of equations

$$9x^2 - 16y^2 = 144$$

$$x - 2y = 4$$

Solution The common solution represents the intersection of a hyperbola and straight line. The method used is substitution. Solving the linear equation for x yields

$$x = 2y + 4$$

Substitution into the second-order equation gives

$$9(2y + 4)^2 - 16y^2 = 144$$

which reduces to

$$20y^2 + 144y = 0$$

Factoring gives

$$4y(5y + 36) = 0$$

which yields

$$y = 0, \frac{-36}{5}$$

Substitution into the linear equation $x = 2y + 4$ gives the corresponding values of x

$$x = 4, -\frac{52}{5}$$

The solutions are thus the coordinates of intersection of the line and the hyperbola

$$(4,0), \left(-\frac{52}{5}, -\frac{36}{5}\right)$$

which can be verified by graphical construction.

Example problem A.7 Solve the system of equations

(*a*) $3x + y = 7$

(*b*) $x + z = 4$

(*c*) $y - z = -1$

Solution Systems of equations similar to these arise frequently in engineering applications. Obviously, they can be solved by Cramer's rule. However, a more rapid solution can be obtained directly by elimination.

From Eq. c,

$$y = z - 1$$

From Eq. a,

$$y = 7 - 3x$$

From Eq. b,

$$x = 4 - z$$

Successive substitution yields

$$z - 1 = 7 - 3x$$

$$z - 1 = 7 - 3(4 - z)$$

$$-2z = -4$$

$$z = +2$$

Continued substitution gives

$$y = 1$$

$$x = 2$$

Every system of equations should first be carefully investigated before a method of solution is chosen so that the most direct method, requiring the minimum amount of time, is used.

A.7 Approximate solutions

Many equations developed in engineering applications do not lend themselves to direct solution by standard methods. These equations must be solved by approximation methods to the accuracy dictated by the problem conditions. Experience is helpful in choosing the numerical technique for solution.

Example problem A.8 Find to three significant figures the solution to the equation

$$2 - x = \ln x$$

Solution One method of solution is graphical. If the equations $y = 2 - x$ and $y = \ln x$ are plotted, the common solution would be the intersection of the two lines. This would not likely give three-significant-figure accuracy, however. A more accurate method requires use of a scientific calculator.

Inspection of the equation reveals that the desired solution must lie between 1 and 2. It is then a matter of setting up a routine that will continue to bracket the solution between two increasingly accurate numbers. Table A.1 shows the intermediate steps and indicates that the solution is $x = 1.56$ to three significant figures.

For greater accuracy and/or more complex problems, a computer or programmable calculator could easily be used to determine a solution by the method just described. The time available and equipment on hand will always influence the numerical technique to be used.

Table A.1 Solution of $2 - x = 1n\ x$

x	1	2	1.5	1.6	1.55	1.56	1.557
$2 - x$	1	0	0.500	0.400	0.450	0.440	0.443
$\ln x$	0	0.693	0.405	0.470	0.438	0.445	0.443

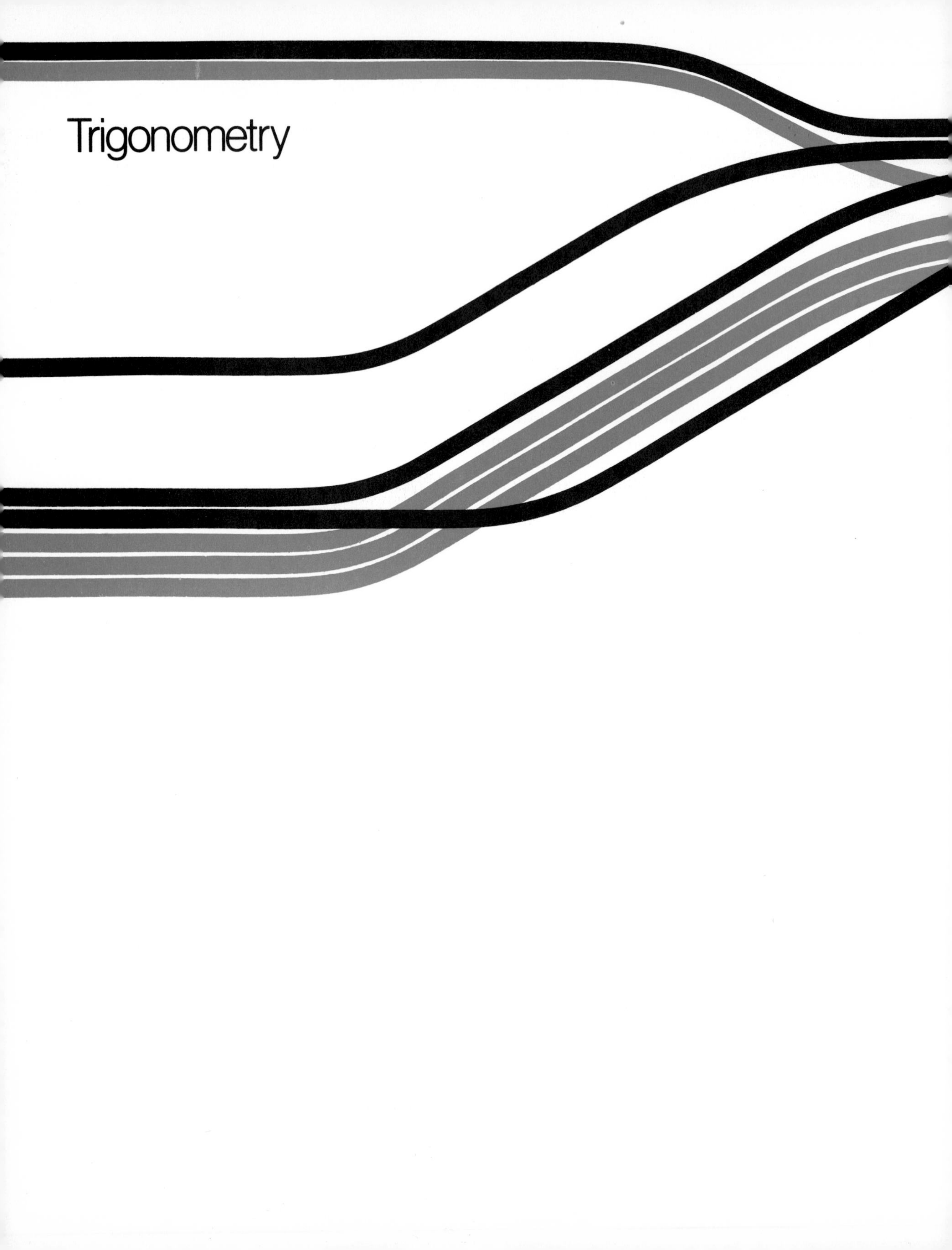
Trigonometry

APPENDIX B

B.1 Introduction

This material is intended to be a brief review of concepts from plane trigonometry that are commonly used in engineering calculations. The section deals only with plane trigonometry and furnishes no information about spherical trigonometry. The reader is referred to standard texts in trigonometry for more detailed coverage and analysis.

B.2 Trigonometric function definitions

The trigonometric functions are defined for an angle contained within a right triangle, as shown in Fig. B.1.

$$\text{sine}\,\theta = \sin\theta = \frac{\text{opposite side}}{\text{hypotenuse}} = \frac{y}{r}$$

$$\text{cosine}\,\theta = \cos\theta = \frac{\text{adjacent side}}{\text{hypotenuse}} = \frac{x}{r}$$

$$\text{tangent}\,\theta = \tan\theta = \frac{\text{opposite side}}{\text{adjacent side}} = \frac{y}{x}$$

$$\text{cotangent}\,\theta = \cot\theta = \frac{\text{adjacent side}}{\text{opposite side}} = \frac{x}{y} = \frac{1}{\tan\theta}$$

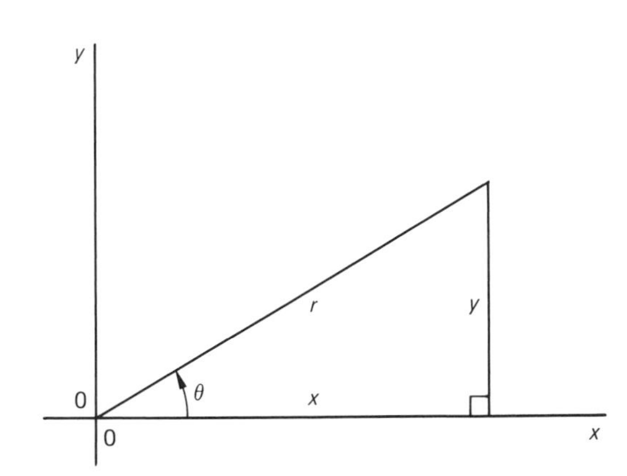

Fig. B.1 Coordinate definition.

$$\text{secant } \theta = \sec\theta = \frac{\text{hypotenuse}}{\text{adjacent side}} = \frac{r}{x} = \frac{1}{\cos\theta}$$

$$\text{cosecant } \theta = \csc\theta = \frac{\text{hypotenuse}}{\text{opposite side}} = \frac{r}{y} = \frac{1}{\sin\theta}$$

The angle θ is by convention measured positive in the counter-clockwise direction from the positive x axis.

B.3 Signs of trigonometric functions by quadrant

Table B.1

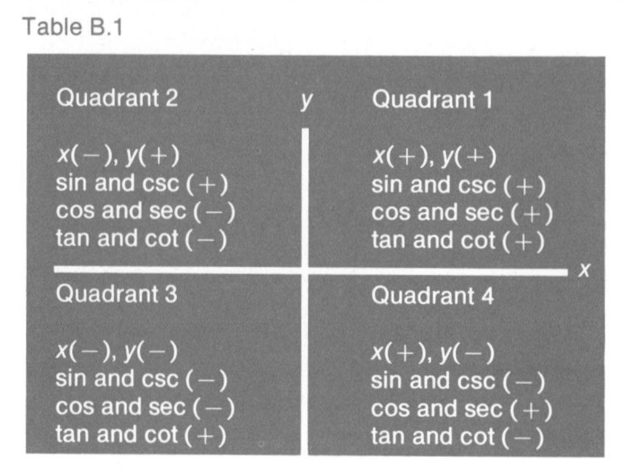

Quadrant 2	Quadrant 1
x(−), y(+)	x(+), y(+)
sin and csc (+)	sin and csc (+)
cos and sec (−)	cos and sec (+)
tan and cot (−)	tan and cot (+)
Quadrant 3	**Quadrant 4**
x(−), y(−)	x(+), y(−)
sin and csc (−)	sin and csc (−)
cos and sec (−)	cos and sec (+)
tan and cot (+)	tan and cot (−)

B.4 Radians and degrees

Angles may be measured in either degrees or radians (see Fig. B.2). By definition,

$$1 \text{ degree } (°) = \frac{1}{360} \text{ of the central angle of a circle}$$

1 radian (rad) = angle subtended at center 0 of a circle by an arc equal to the radius

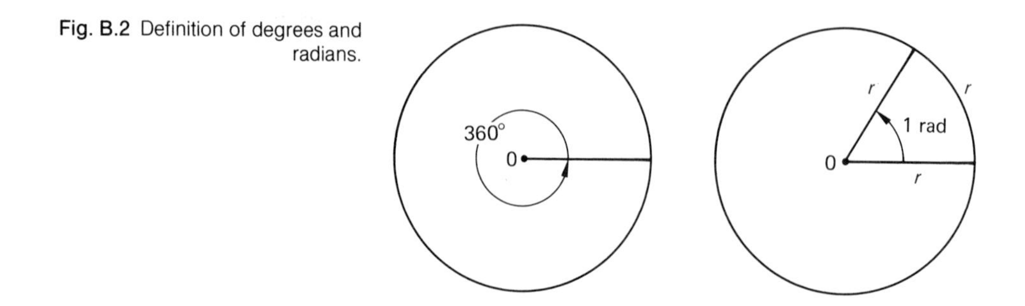

Fig. B.2 Definition of degrees and radians.

The central angle of a circle is 2π rad or 360°. Therefore,

$$1° = \frac{2\pi}{360°} = \frac{\pi}{180°} = 0.01745329 \ldots \text{ rad}$$

and

$$1 \text{ rad} = \frac{360°}{2\pi} = \frac{180°}{\pi} = 57.29578 \ldots °$$

It follows that the conversion of θ in degrees to θ in radians is given by

$$\theta\,(\text{rad}) = \theta\,(°)\frac{\pi}{180°}$$

and in like manner,

$$\theta\,(°) = \theta\,(\text{rad})\frac{180°}{\pi}$$

B.5 Plots of trigonometric functions

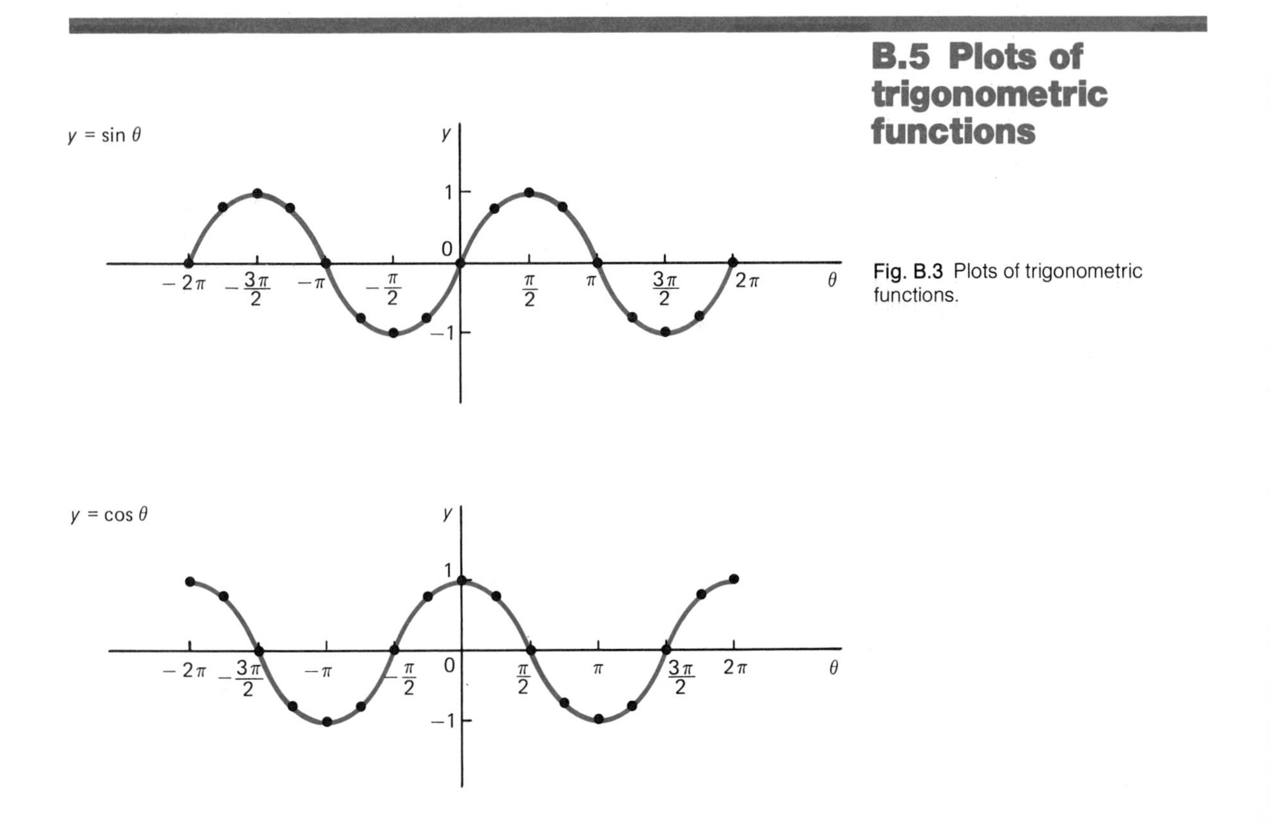

Fig. B.3 Plots of trigonometric functions.

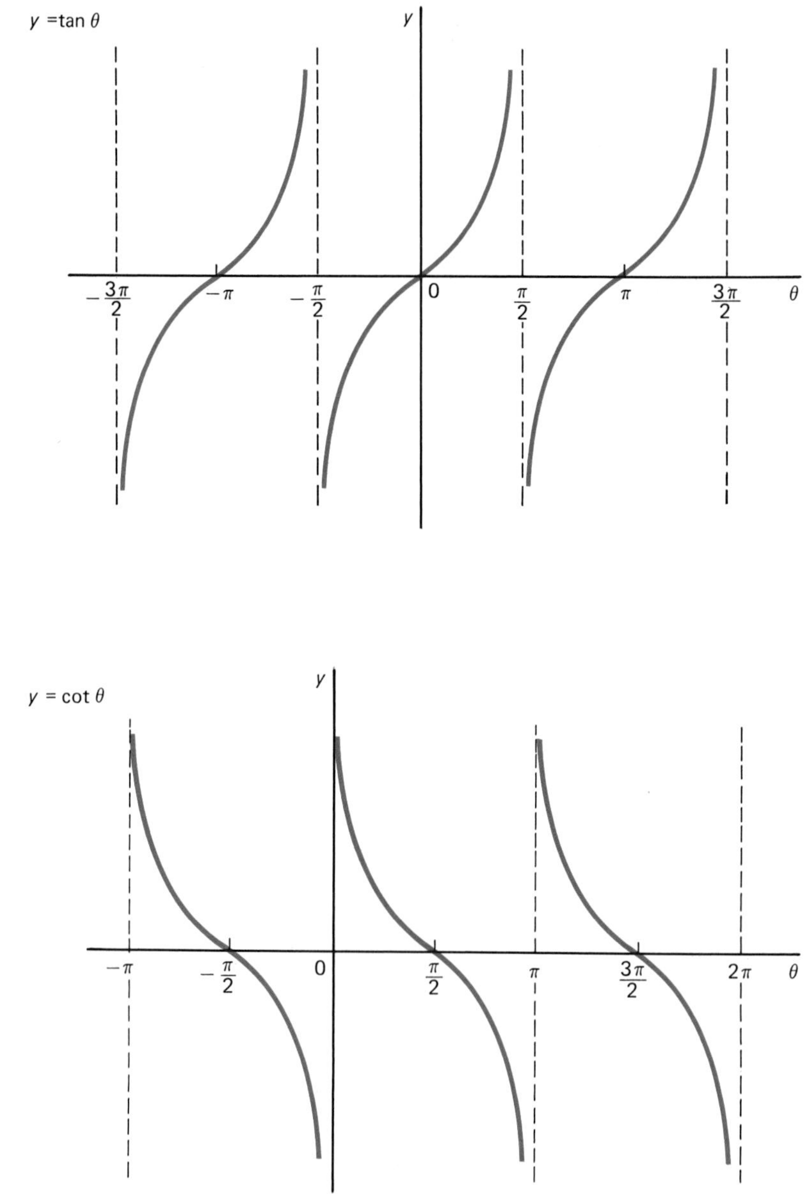

Fig. B.3 *(continued)*

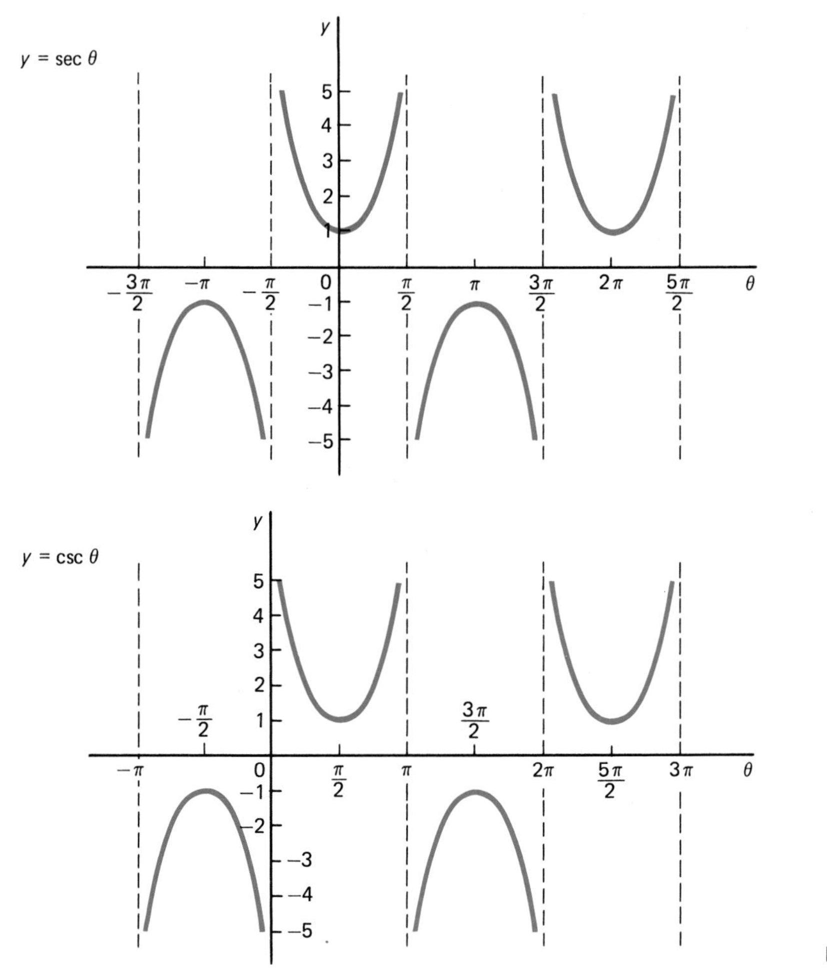

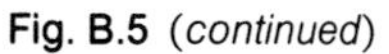
Fig. B.5 (*continued*)

B.6 Standard values of often-used angles

From three basic triangles, it is possible to compute the values of the trigonometric functions for many standard angles such as 30°, 45°, 60°, 120°, 135°, etc. It is only necessary for us to recall that

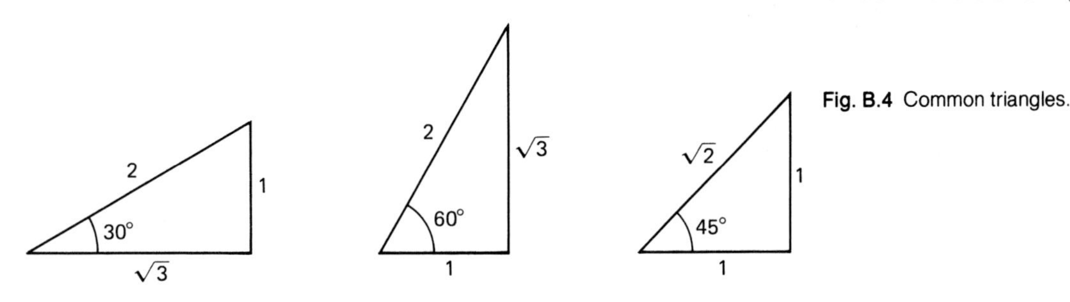

Fig. B.4 Common triangles.

Table B.2 Functions of Common Angles

Function \ Angle	0°	30°	45°	60°	90°
sin	0	1/2	$\sqrt{2}/2$	$\sqrt{3}/2$	1
cos	1	$\sqrt{3}/2$	$\sqrt{2}/2$	1/2	0
tan	0	$\sqrt{3}/3$	1	$\sqrt{3}$	∞
cot	∞	$\sqrt{3}$	1	$\sqrt{3}/3$	0
sec	1	$2\sqrt{3}/3$	$\sqrt{2}$	2	∞
csc	∞	2	$\sqrt{2}$	$2\sqrt{3}/3$	1

$\sin 30° = \cos 60° = \frac{1}{2}$, and $\tan 45° = 1$ to construct the necessary triangles from which values can be taken to obtain the other functions.

The functions for 0°, 90°, 180°, etc., can be found directly from the function definitions and a simple line sketch. See Table B.2.

B.7 Inverse trigonometric functions

Definition

If $y = \sin \theta$, then θ is an angle whose sine is y. The symbols ordinarily used to denote an inverse function are

$$\theta = \arcsin y$$

or

$$\theta = \sin^{-1} y$$

Note:

$$\sin^{-1} y \neq \frac{1}{\sin y}$$

This is an exception to the conventional use of exponents.

Inverse functions $\cos^{-1} y$, $\tan^{-1} y$, $\cot^{-1} y$, $\sec^{-1} y$, and $\csc^{-1} y$ are similarly defined. Each of these is a many-valued function of y. The values are grouped into collections called branches. One of these branches is defined to be the principal branch, and the values found there are the principal values.

The principal values are as follows:

$$-\frac{\pi}{2} \leqslant \sin^{-1} y \leqslant \frac{\pi}{2}$$

$$0 \leqslant \cos^{-1} y \leqslant \pi$$

$$-\frac{\pi}{2} < \tan^{-1} y < \frac{\pi}{2}$$

$$0 < \cot^{-1} y < \pi$$

$$0 \leqslant \sec^{-1} y \leqslant \pi \qquad \left(\sec^{-1} y \neq \frac{\pi}{2}\right)$$

$$-\frac{\pi}{2} \leqslant \csc^{-1} y \leqslant \frac{\pi}{2} \qquad (\csc^{-1} y \neq 0)$$

B.8 Plots of inverse trigonometric functions

All angles are given in radians. Principal branches are shown as solid lines. See Fig. B.5.

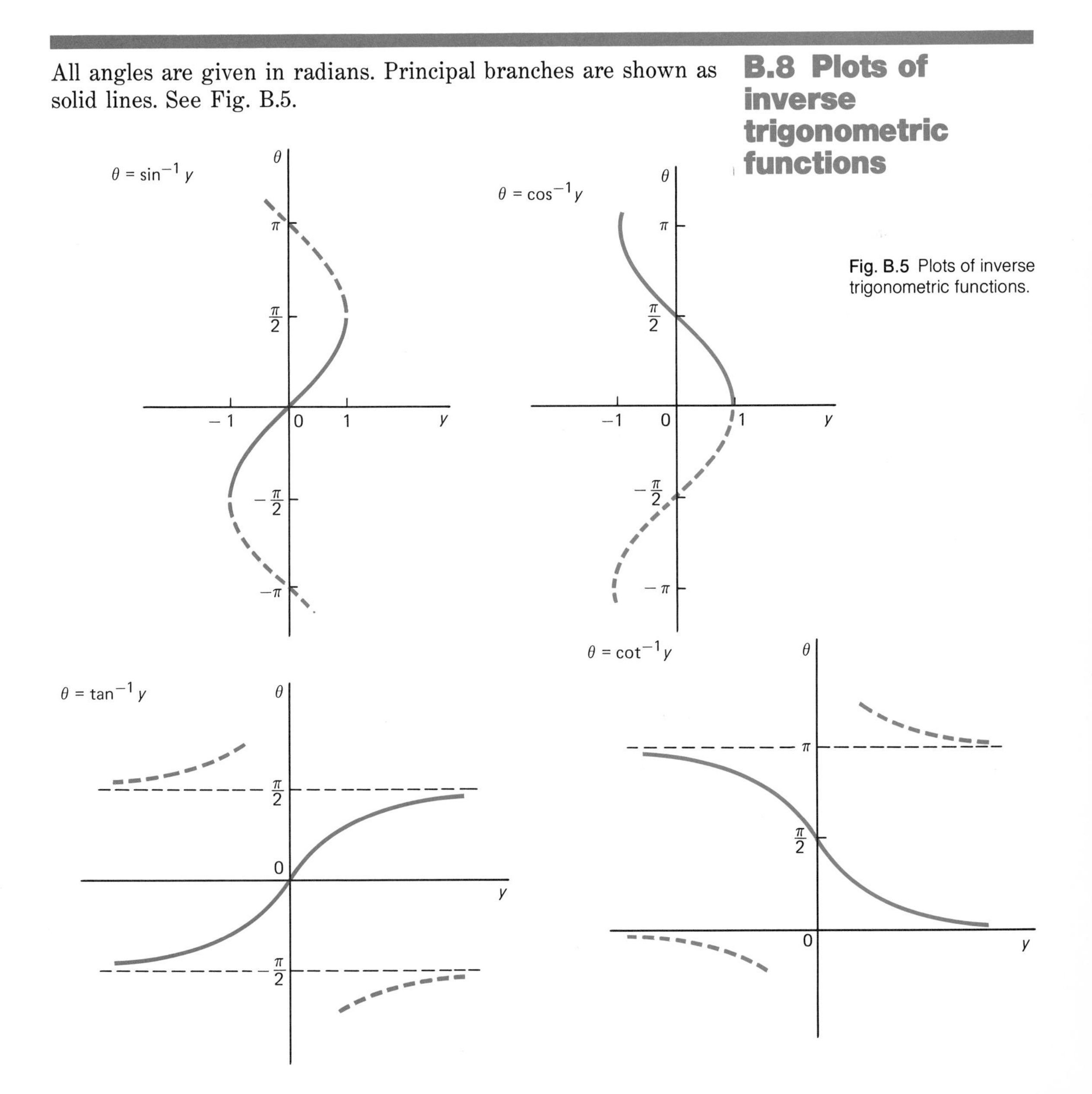

Fig. B.5 Plots of inverse trigonometric functions.

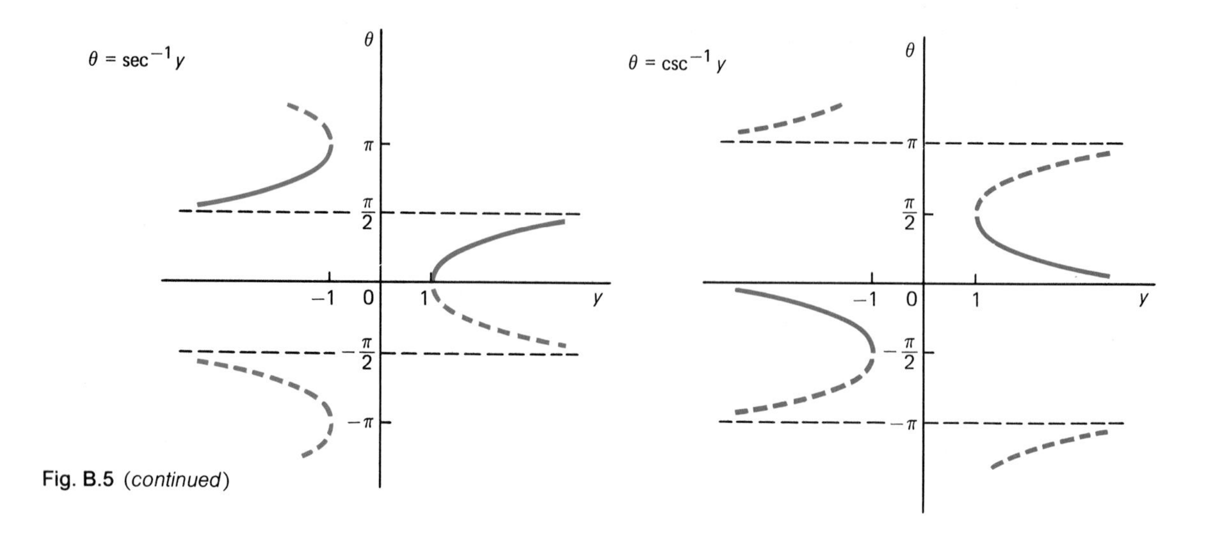

Fig. B.5 (*continued*)

B.9 Polar-rectangular coordinate conversion

See Fig. B.6.

Conversion from polar to rectangular coordinates $(r,\theta) \rightarrow (x,y)$ is given by the following equations:

$$x = r \cos \theta$$

$$y = r \sin \theta$$

Conversion from rectangular to polar coordinates $(x,y) \rightarrow (r,\theta)$ requires the following equations:

$$r = [x^2 + y^2]^{1/2}$$

$$\theta = \tan^{-1}\left(\frac{y}{x}\right)$$

The conversion from polar to rectangular coordinates can also be thought of as the determination of the x and y components of a vector (r,θ). Likewise, conversion from rectangular to polar coor-

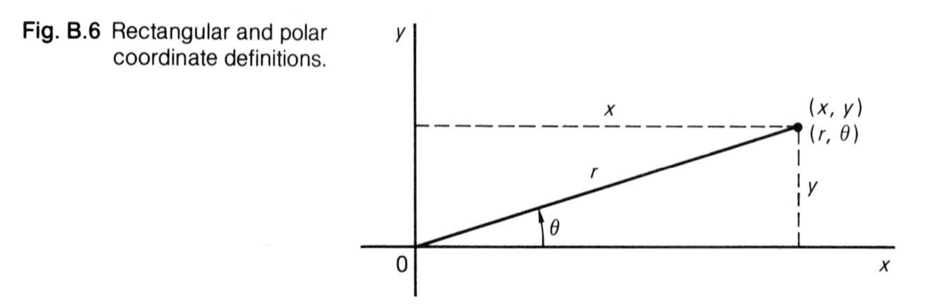

Fig. B.6 Rectangular and polar coordinate definitions.

dinates is the same as finding the resultant vector (r,θ) from its x and y components.

B.10 Laws of sines and cosines

The fundamental definitions of sine, cosine, etc., apply strictly to right triangles. Solutions needed for oblique triangles must then be accomplished by appropriate constructions that reduce the problem to a series of solutions to right triangles.

Two formulas have been derived for oblique triangles that are much more convenient to use than the construction technique. They are the law of sines and the law of cosines, which apply to any plane triangle. See Fig. B.7.

Fig. B.7 Angle and side designations.

The *law of sines* states

$$\frac{\sin A}{a} = \frac{\sin B}{b} = \frac{\sin C}{c}$$

The *law of cosines* is

$$a^2 = b^2 + c^2 - 2bc \cos A$$

or

$$b^2 = a^2 + c^2 - 2ac \cos B$$

or

$$c^2 = a^2 + b^2 - 2ab \cos C$$

Application of the law of sines is most convenient in the case where two angles and one side are known and a second side is to be found.

Example problem B.1 Determine the length of side x for the triangle with base 12.0 m as shown in Fig. B.8.

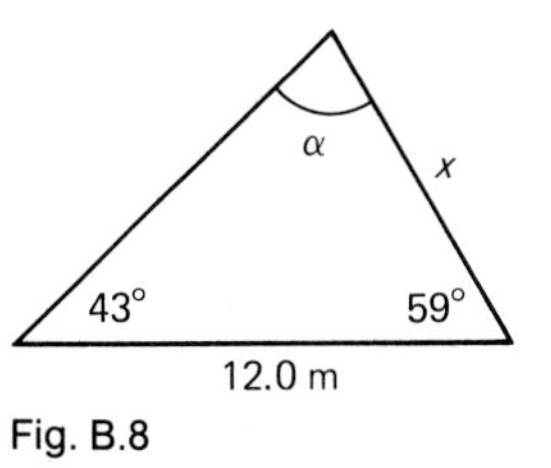

Fig. B.8

Solution The sum of the interior angles must be 180°; therefore,

$\alpha = 180° - 43° - 59° = 78°$.

Applying the law of sines,

$$\frac{\sin 78°}{12.0 \text{ m}} = \frac{\sin 43°}{x}$$

$$x = \frac{\sin 43°}{\sin 78°} 12.0 \text{ m}$$

$$= 8.37 \text{ m}$$

The law of cosines is most convenient to use when two sides and the included angle are known for a triangle.

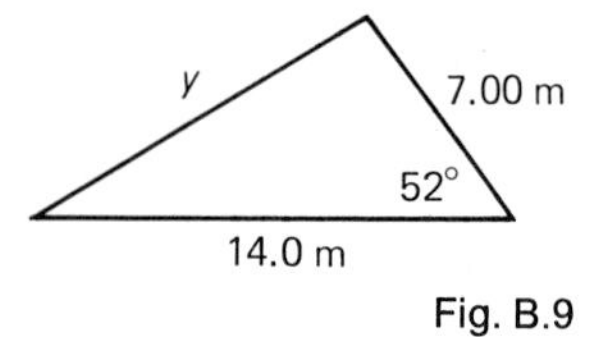

Fig. B.9

Example problem B.2 Calculate the length of side y of the triangle shown in Fig. B.9. Its base is 14.0 m.

Solution Substitute into the law of cosines.

$$y^2 = (14.0 \text{ m})^2 + (7.00 \text{ m})^2 - 2(14.0 \text{ m})(7.00 \text{ m})(\cos 52°)$$

$$= 124.33 \text{ m}^2$$

$$y = 11.2 \text{ m}$$

B.11 Area of a triangle

Formulas for the area of a triangle (see Fig. B.7) in terms of two sides and their included angle are

$$\text{Area} = \frac{1}{2}ab \sin C$$

$$\text{Area} = \frac{1}{2}ac \sin B$$

$$\text{Area} = \frac{1}{2}bc \sin A$$

Formulas written in terms of one side and three angles are

$$\text{Area} = \frac{1}{2}a^2 \frac{\sin B \sin C}{\sin A}$$

$$\text{Area} = \frac{1}{2}b^2 \frac{\sin A \sin C}{\sin B}$$

$$\text{Area} = \frac{1}{2}c^2 \frac{\sin A \sin B}{\sin C}$$

Formula for the area in terms of the sides is

$$\text{Area} = [s(s - a)(s - b)(s - c)]^{1/2}$$

$$\text{where } s = \frac{1}{2}(a + b + c)$$

Example problem B.3 Determine the areas of the two triangles defined in Sec. B.10.

Solution For the 12.0 m base triangle,

$$\text{Area} = \frac{1}{2}(12.0\ \text{m})^2 \frac{\sin 59^\circ \sin 43^\circ}{\sin 78^\circ}$$

$$= 43.0\ \text{m}^2$$

For the 14.0 m base triangle,

$$\text{Area} = \frac{1}{2}(7.00\ \text{m})(14.0\ \text{m})\sin 52^\circ$$

$$= 38.6\ \text{m}^2$$

B.12 Series representation of trigonometric functions

Infinite-series representations exist for each of the trigonometric functions. Those for sine, cosine, and tangent are

$$\sin\theta = \theta - \frac{\theta^3}{3!} + \frac{\theta^5}{5!} + \cdots + (-1)^{n-1}\frac{\theta^{2n-1}}{(2n-1)!} + \cdots$$

$$\theta \text{ in radians} \qquad (-\infty < \theta < +\infty)$$

$$\cos\theta = 1 - \frac{\theta^2}{2!} + \frac{\theta^4}{4!} + \cdots + (-1)^{n-1}\frac{\theta^{2n-2}}{(2n-2)!} + \cdots$$

$$\theta \text{ in radians} \qquad (-\infty < \theta < +\infty)$$

$$\tan\theta = \theta + \frac{\theta^3}{3} + \frac{2\theta^5}{15} + \cdots + \frac{2^{2n}(2^{2n}-1)\,B_n\,\theta^{2n-1}}{(2n)!} + \cdots$$

$$\theta \text{ in radians} \qquad \left(-\frac{\pi}{2} < \theta < \frac{\pi}{2}\right)$$

where B_n are the Bernoulli numbers

$$B_1 = \frac{1}{6}$$

$$B_2 = \frac{1}{30}$$

$$B_3 = \frac{1}{42}$$

$$B_4 = \frac{1}{30}$$

$$B_5 = \frac{5}{66}$$

$$\cdots$$

$$B_n = \frac{(2n)!}{2^{2n-1}(\pi)^{2n}}\left(1 + \frac{1}{2^{2n}} + \frac{1}{3^{2n}} + \cdots\right)$$

and where the factorial symbol (!) is defined as

$$n! = n(n-1)(n-2)\cdots(3)(2)(1)$$

Trigonometric functions to any accuracy can be calculated from the series if enough terms are used.

For small angles, on the order of five degrees or less, it may be sufficient to use only the first term of each series.

$$\sin\theta \cong \theta$$

$$\cos\theta \cong 1$$

$$\tan\theta \cong \theta$$

B.13 Trigonometric relationships

Functional relationships

$$\tan\theta = \frac{\sin\theta}{\cos\theta}$$

$$\sin^2\theta + \cos^2\theta = 1$$

$$\sec^2\theta - \tan^2\theta = 1$$

$$\csc^2\theta - \cot^2\theta = 1$$

$$\sin\theta = \cos(90^\circ - \theta) = \sin(180^\circ - \theta)$$

$$\cos\theta = \sin(90^\circ - \theta) = -\cos(180^\circ - \theta)$$

$$\tan\theta = \cot(90^\circ - \theta) = -\tan(180^\circ - \theta)$$

$$\sin(-\theta) = -\sin\theta$$

$$\cos(-\theta) = \cos\theta$$

$$\tan(-\theta) = -\tan\theta$$

Sum of angles formulas

$$\sin(\theta \pm \alpha) = \sin\theta\cos\alpha \pm \cos\theta\sin\alpha$$

$$\cos(\theta \pm \alpha) = \cos\theta\cos\alpha \mp \sin\theta\sin\alpha$$

$$\tan(\theta \pm \alpha) = \frac{\tan\theta \pm \tan\alpha}{1 \mp \tan\theta\tan\alpha}$$

Multiple-angle formulas

$$\sin 2\theta = 2\sin\theta\cos\theta$$

$$\cos 2\theta = \cos^2\theta - \sin^2\theta = 2\cos^2\theta - 1 = 1 - 2\sin^2\theta$$

$$\tan 2\theta = \frac{2\tan\theta}{1 - \tan^2\theta}$$

$$\sin 3\theta = 3\sin\theta - 4\sin^3\theta$$

$$\cos 3\theta = 4\cos^3\theta - 3\cos\theta$$

$$\tan 3\theta = \frac{3\tan\theta - \tan^3\theta}{1 - 3\tan^2\theta}$$

$$\sin\frac{\theta}{2} = \pm\sqrt{\frac{1-\cos\theta}{2}} \qquad \left(\text{sign depends on quadrant of } \frac{\theta}{2}\right)$$

$$\cos\frac{\theta}{2} = \pm\sqrt{\frac{1+\cos\theta}{2}} \qquad \left(\text{sign depends on quadrant of } \frac{\theta}{2}\right)$$

$$\tan\frac{\theta}{2} = \pm\sqrt{\frac{1-\cos\theta}{1+\cos\theta}} = \frac{1-\cos\theta}{\sin\theta} = \frac{\sin\theta}{1+\cos\theta} = \csc\theta - \cot\theta$$

$$\left(\text{sign depends on quadrant of } \frac{\theta}{2}\right)$$

Sum, difference, and product formulas

$$\sin\theta + \sin\alpha = 2\sin\left(\frac{\theta+\alpha}{2}\right)\cos\left(\frac{\theta-\alpha}{2}\right)$$

$$\sin\theta - \sin\alpha = 2\cos\left(\frac{\theta+\alpha}{2}\right)\sin\left(\frac{\theta-\alpha}{2}\right)$$

$$\cos\theta + \cos\alpha = 2\cos\left(\frac{\theta+\alpha}{2}\right)\cos\left(\frac{\theta-\alpha}{2}\right)$$

$$\cos\theta - \cos\alpha = 2\sin\left(\frac{\theta+\alpha}{2}\right)\sin\left(\frac{\alpha-\theta}{2}\right)$$

$$\sin\theta\sin\alpha = \frac{1}{2}\left[\cos(\theta-\alpha) - \cos(\theta+\alpha)\right]$$

$$\cos\theta\cos\alpha = \frac{1}{2}\left[\cos(\theta-\alpha) + \cos(\theta+\alpha)\right]$$

$$\sin\theta\cos\alpha = \frac{1}{2}\left[\sin(\theta-\alpha) + \sin(\theta+\alpha)\right]$$

Power formulas

$$\sin^2\theta = \frac{1}{2} - \frac{1}{2}\cos 2\theta$$

$$\cos^2\theta = \frac{1}{2} + \frac{1}{2}\cos 2\theta$$

$$\sin^3\theta = \frac{3}{4}\sin\theta - \frac{1}{4}\sin 3\theta$$

$$\cos^3\theta = \frac{3}{4}\cos\theta + \frac{1}{4}\cos 3\theta$$

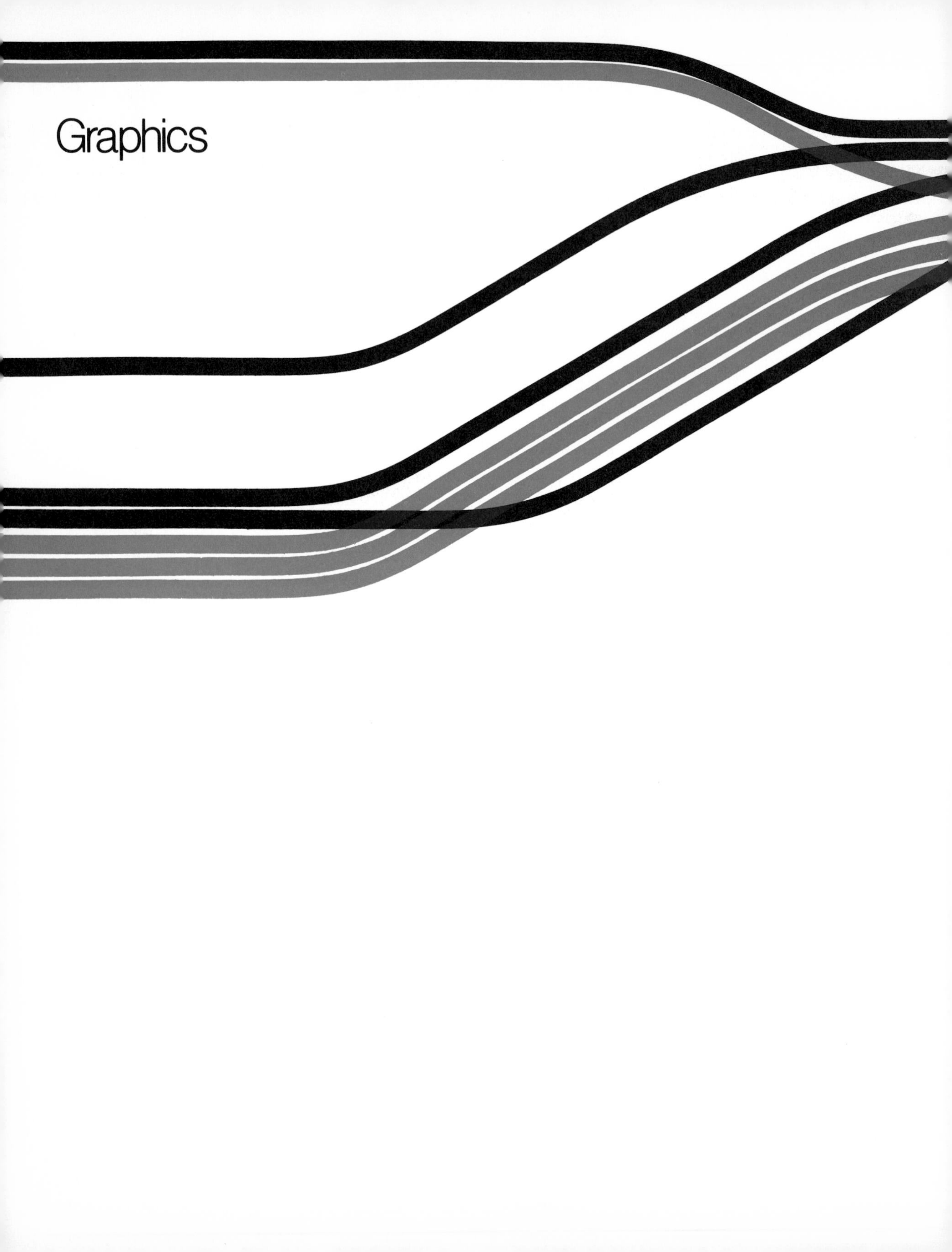

Graphics

APPENDIX C

C.1 Introduction

As an engineer you will find that the ability to communicate graphically is an essential part of your professional activity. The old proverb "A picture is worth a thousand words" demonstrates graphics to be a valuable resource in the solution of engineering problems.

Graphics is the foundation of design and as such it is an essential method of communication from conceptualization to manufacture. The topics presented in Appendix C are intended to supplement Chap. 15. Each subject area is briefly and concisely outlined with illustrations to demonstrate fundamental concepts.

C.2 Engineering lettering

Engineers in their daily professional activities find it necessary to maintain legible permanent records. Much of the work that they do must be communicated to others. This includes not only drawings

Fig. C.1 Gothic letters.

but also notes, calculations, etc. It is important that engineers develop a lettering style and an ability to render that is simple, fast, and functional.

There are many types of letters, but perhaps the most universally accepted is the gothic type illustrated in Fig. C.1. Either uppercase or lowercase, vertical or slant letters are satisfactory. Perhaps it is wise at the start of an engineering education to select a lettering style and use it consistently, since practice develops the skill.

Guidelines are very helpful when lettering, since they ensure uniform height of both letters and numbers. Guidelines can be constructed freehand for many types of work.

C.3 Freehand drawing

Freehand drawing, often referred to as sketching, differs from instrument work primarily in the amount of time and accuracy required. The end result, however, should be a clear, concise illustration suitable for the intended purpose. The ultimate use of the sketch will dictate the construction time and degree of accuracy needed.

The equipment necessary to do freehand work is quite simple: a pencil, soft eraser, and paper. For presentations or other specific communication tasks, for example, transparencies, flip charts, etc., the equipment needed may vary, but it is usually not very complex.

C.3.1 Key elements in sketching

Independent of the type of sketch or its eventual purpose there are a number of steps that must be completed. Each step may be emphasized differently depending on the purpose of the drawing, but all steps are to be considered any time a sketch is constructed.

Plan
: Conceive the sketch with respect to size, orientation, location on the paper, and degree of accuracy.

Skeleton
: Begin by making a light, overall skeleton box construction of the object.

Proportions
: Carefully check all geometric features as they are lightly constructed. Necessary changes should be made at each stage of the construction process.

Details
: Add important details that are needed to identify and distinguish the object.

Darken
: Carefully darken the object's outlines to accentuate and clarify important features and characteristics.

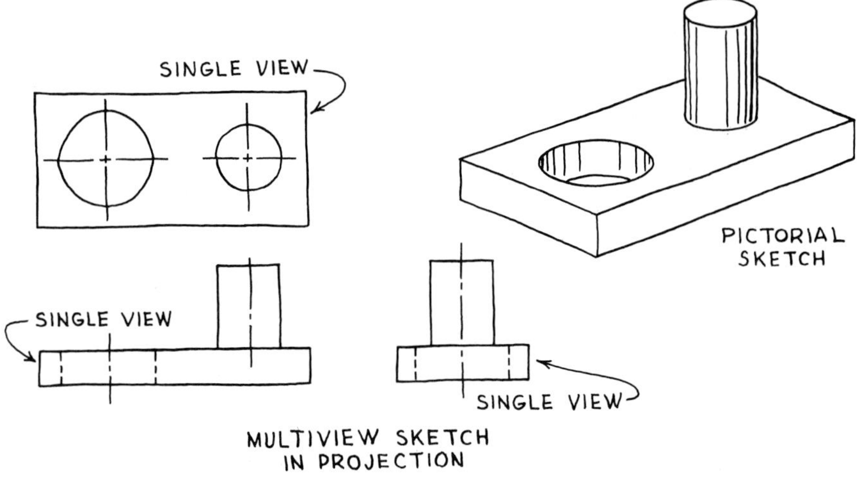

Fig. C.2 Freehand drawings.

Label
Letter titles and notes in a neat, legible fashion with standard engineering numbers and letters.

These steps are general and apply to all types of freehand construction.

C.3.2. Types of freehand drawing

Three types of freehand drawings will be considered, as illustrated in Fig. C.2.

1. Single view: Objects with primarily two dimensions: that is, maps, charts, diagrams, graphs, etc. Included in this definition are single orthographic views or one view of multiview drawings.
2. Pictorial: Objects with three dimensions illustrated: that is, length, width, and height. It is an attempt to show in a single drawing what the eye would see.
3. Multiview: Separate, single orthographic views oriented in adjacent related positions to describe an object completely. Multiview drawings are discussed in Sec. C.4.

C.3.3. Construction of single views

The construction of single views primarily requires the ability to draw parallel lines and circles.

Circles should be sketched without construction lines only if the sketch is very rough or if the circles are small. See Fig. C.3.

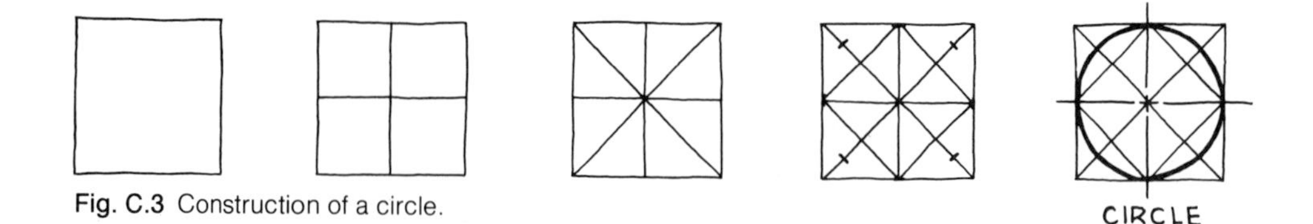

Fig. C.3 Construction of a circle.

A helpful construction technique is illustrated in Fig. C.4. The center of the circle is at point 0, so an arc can be drawn from A through E to B as follows. Construct a diagonal from A to B and then divide the line DC again. This will locate point E, which is very close to the exact location for the circular arc.

C.3.4 Construction of pictorials

The correct selection of the axes is critical when constructing a pictorial. If all faces require equal emphasis, an isometric selection would be appropriate, as seen in Fig. C.5*a*, whereas oblique would better illustrate major features in one or parallel planes (Fig. C.5*b*).

A key element in the construction of pictorial sketches is the ellipse. Circles in the major plane of an oblique drawing are circles, but in the receding planes of most pictorials they are elliptical. The construction of an ellipse is illustrated in Fig. C.6.

Note the construction of step 4 in Fig. C.6. This procedure is identical to the construction technique developed in Fig. C.4. Although it is not precisely correct mathematically, it is helpful when sketching an ellipse.

The construction of many pictorials are simple adaptations of the four basic shapes illustrated in Fig. C.7.

Practical application of the foregoing construction principles and six steps of sketching are demonstrated in the examples of Fig. C.8.

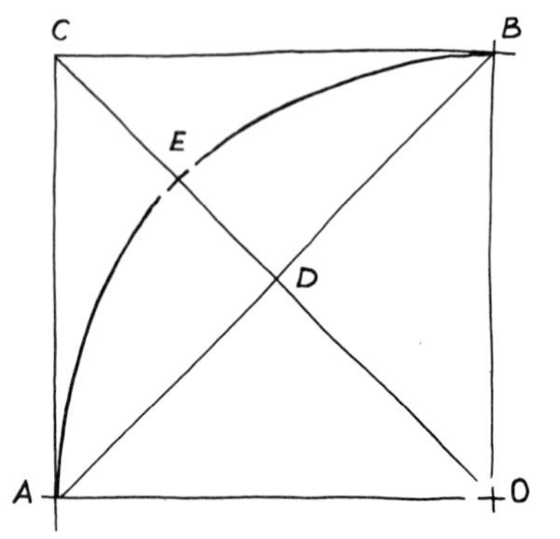

Fig. C.4 Circle construction technique.

Fig. C.5 Axes selections.

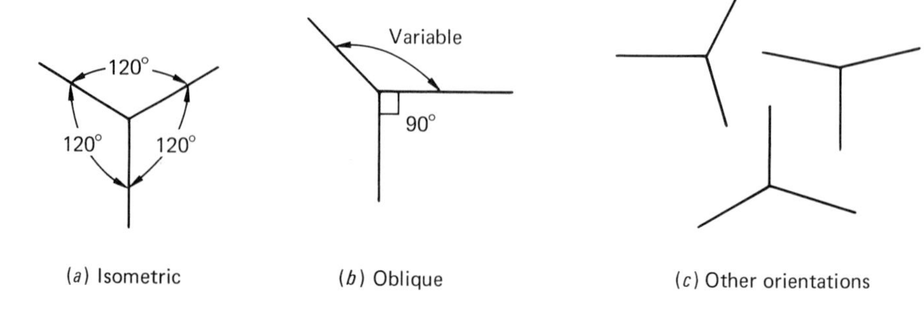

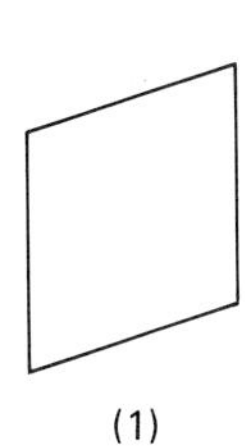

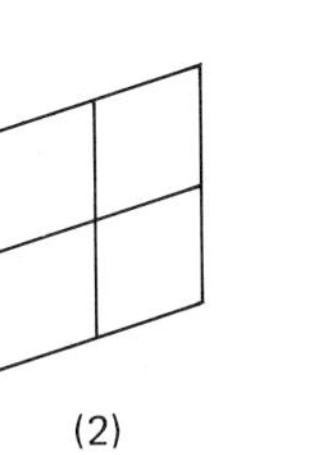

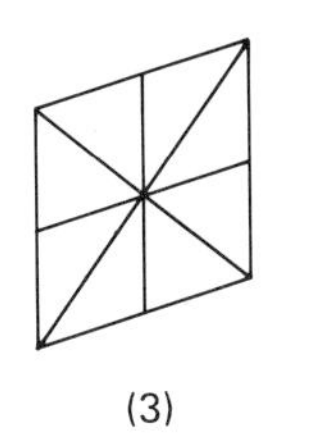

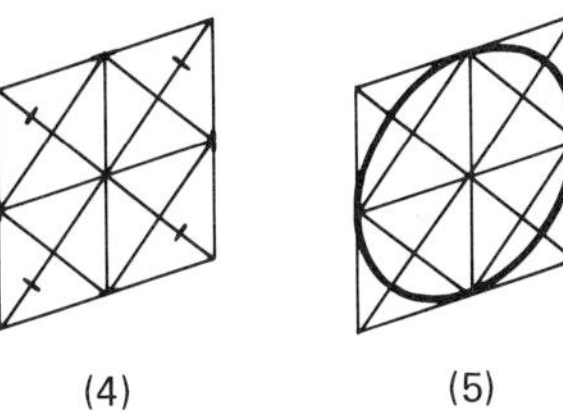

Fig. C.6 Ellipse construction.

Fig. C.7 Basic shapes.

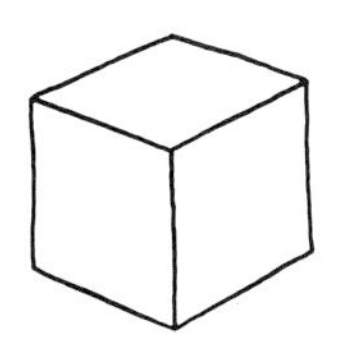

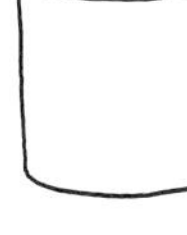

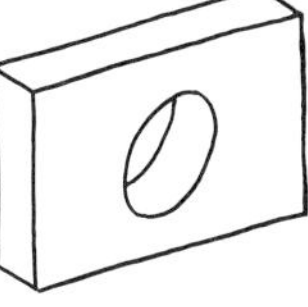
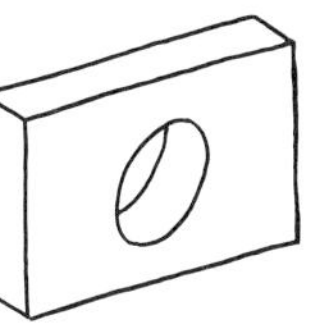
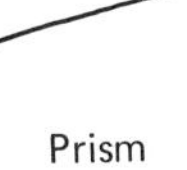

Fig. C.8 Pictorials.

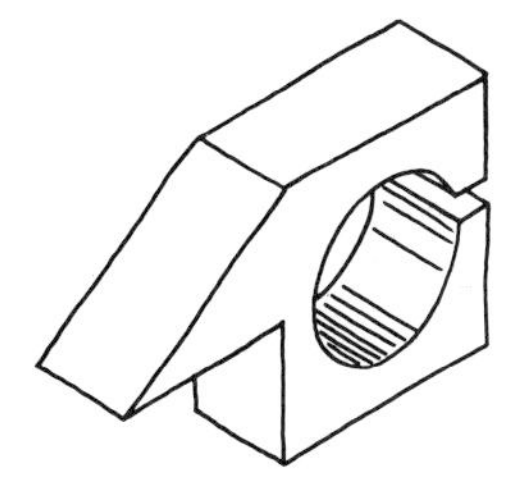

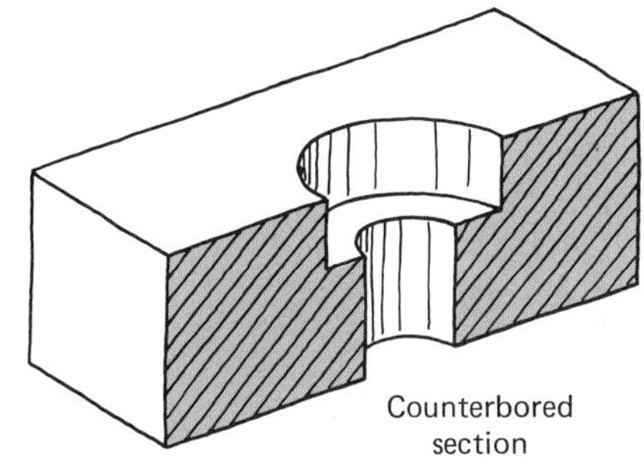

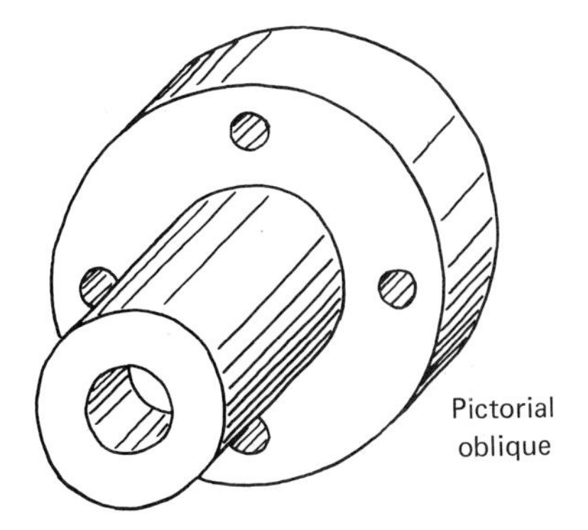

C.4 Multiview drawings

The precise definition and delineation of objects can best be represented by a series of carefully selected single views. Although pictorials are an excellent method of conveying a visual image of the object, they do not provide the detail needed for manufacturing.

Multiview drawings define an object by placing the correct number of properly constructed single views in correct orthographic alignment. Figure C.9 illustrates a simple object showing four of its six principle views, two of which are not essential for complete description of the object but are shown only to clarify various correct orthographic positions.

The number of views necessary to describe any object obviously varies with the complexity of the object, but constructing views that are not needed is a waste of time and money.

Objects with more detail may require more careful study because all contours must be represented in the correct construction of multiviews. Figure C.10*a* illustrates the use of centerlines; and Fig. C.10*b* shows proper use of both hidden lines and centerlines. Figure C.11 illustrates object lines, hidden lines, and centerlines on a single object.

The correct or necessary number of views that adequately describe an object is illustrated in Fig. C.12. The top and front views, in this case, do not completely describe a single object, but the five different right profile views, each taken independently with the given front view, should adequately describe five different objects.

An optimum number of properly constructed views is the way to communicate objects graphically without confusion and misunderstanding.

Fig. C.9 Multiview drawings.

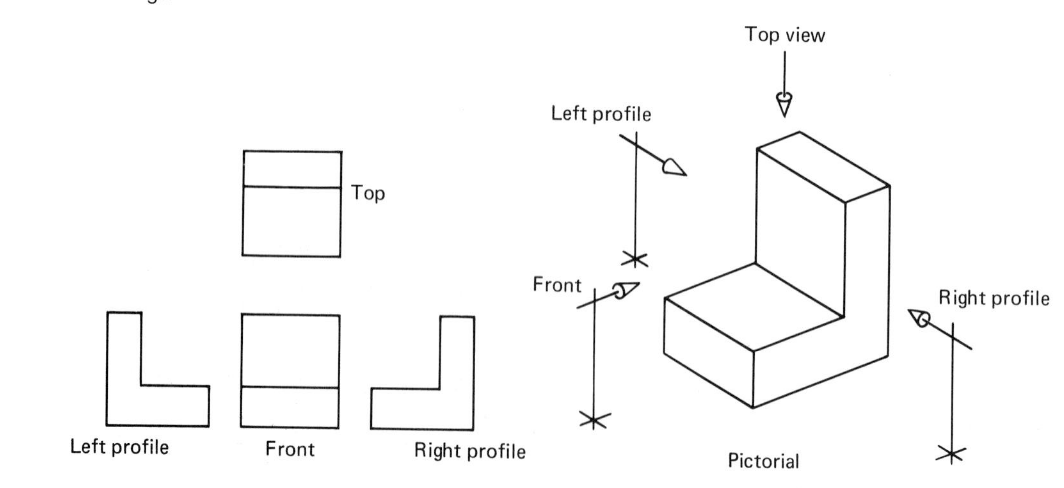

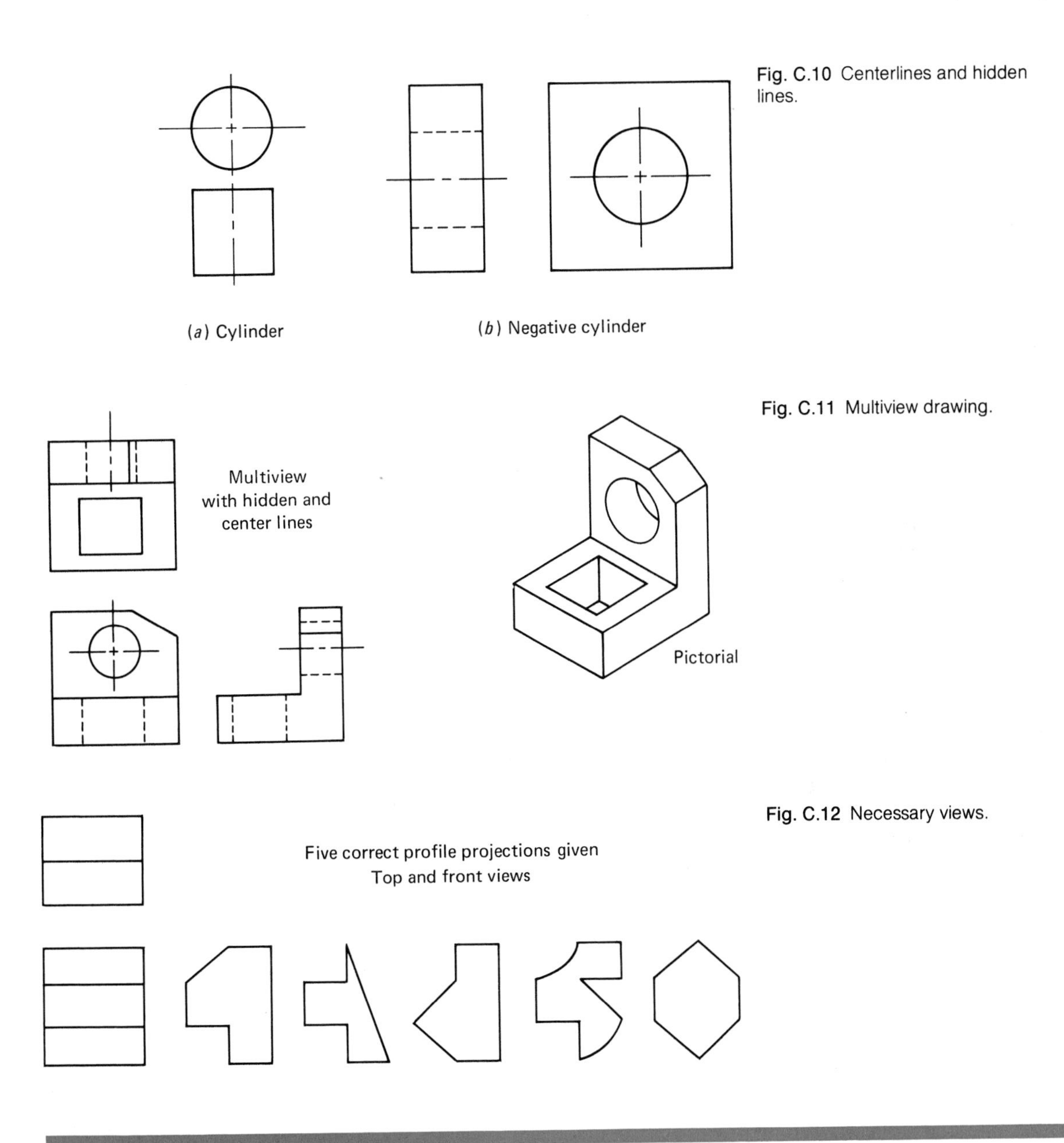

Fig. C.10 Centerlines and hidden lines.

Fig. C.11 Multiview drawing.

Fig. C.12 Necessary views.

C.5 Scales

Graduations on metric scales are identified as a ratio: 1:100, 1:20, etc. This ratio signifies the drawing reduction from actual size; for example, 1:100 indicates that 1 unit on the drawing represents 100 units on the object.

A metric scale labeled with a ratio of 1:100 signifies that the distance from 0 to the number 1 is to represent a meter, as illustrated in Fig. C.13.

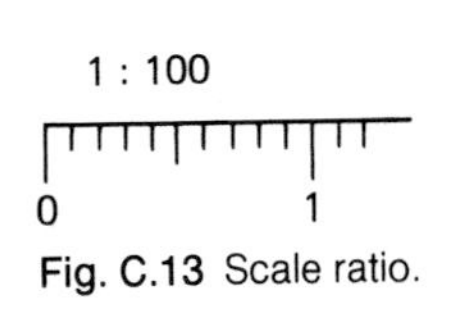

Fig. C.13 Scale ratio.

Table C.1

Ratio	Distance from 0 to 1.0 is equal to			
1:100	1 m =	10 dm =	100 cm =	1 000 mm
1:10	0.1 m =	1 dm =	10 cm =	100 mm
1:1	0.01 m =	0.1 dm =	1 cm =	10 mm
1:0.1	0.001 m =	0.01 dm =	0.1 cm =	1 mm

If you change the ratio from 1:100 to 1:10 but use the same scale, the distance from the 0 to the 1 now represents 0.1 m. Table C.1 demonstrates this functional characteristic of metric scales.

Table C.1 can be constructed for any metric scale or ratio. Figure C.14 illustrates how the scale varies with different ratios.

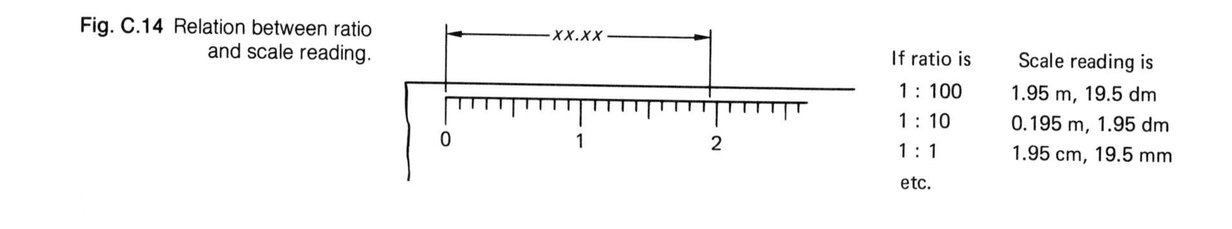

Fig. C.14 Relation between ratio and scale reading.

C.6 Dimensions

In order that objects might be perceived in terms of precise physical measurement, dimensions must be correctly indicated. The size of each geometric shape together with necessary location dimensions has to be clearly understood.

Many standards that provide considerable uniformity to dimensioning practice have been recognized and adopted. One of the most widely used is that issued by the American National Standards Institute.

C.6.1 Definition of terms

Dimensions are the numbers expressed in consistent units and used to indicate the physical lengths.

Dimension lines and arrowheads indicate the extent of the measurement. Arrowheads should be closed and consistently uniform in size and shape. Dimension lines are placed about 10 mm from the object and 6 mm from each other, as illustrated in Fig. C.15.

Extension lines are used to locate the extension of the surface, and leaders are used to dimension circles or to direct notes to a specific place.

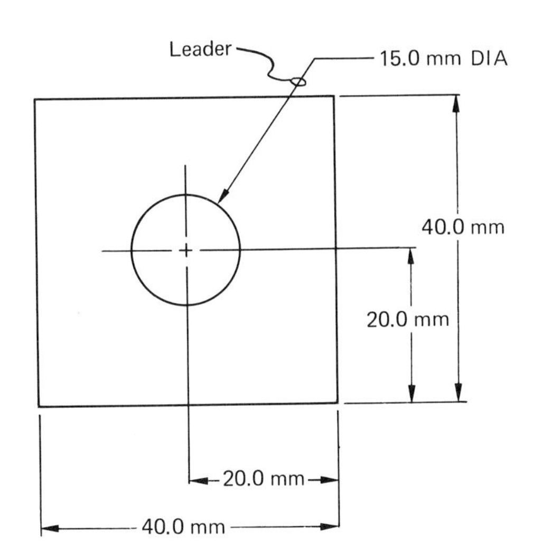

Fig. C.15 Proper spacing.

Geometric shapes

A prism, cylinder, cone, and right pyramid are correctly dimensioned in Fig. C.16.

Holes, radii, and arcs

The illustrations in Fig. C.17 demonstrate correct dimensioning procedures for *through* holes and *blind* holes.

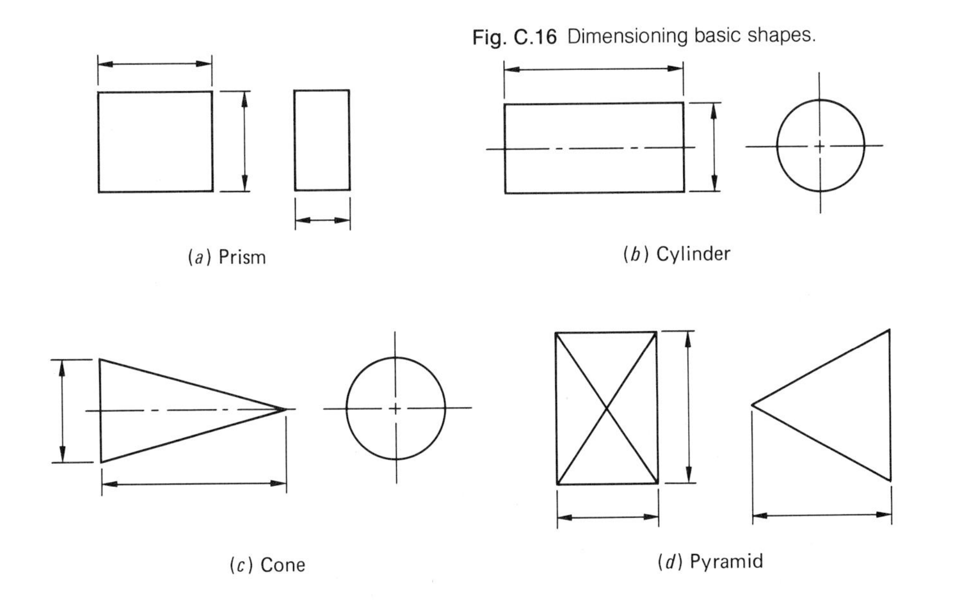

Fig. C.16 Dimensioning basic shapes.

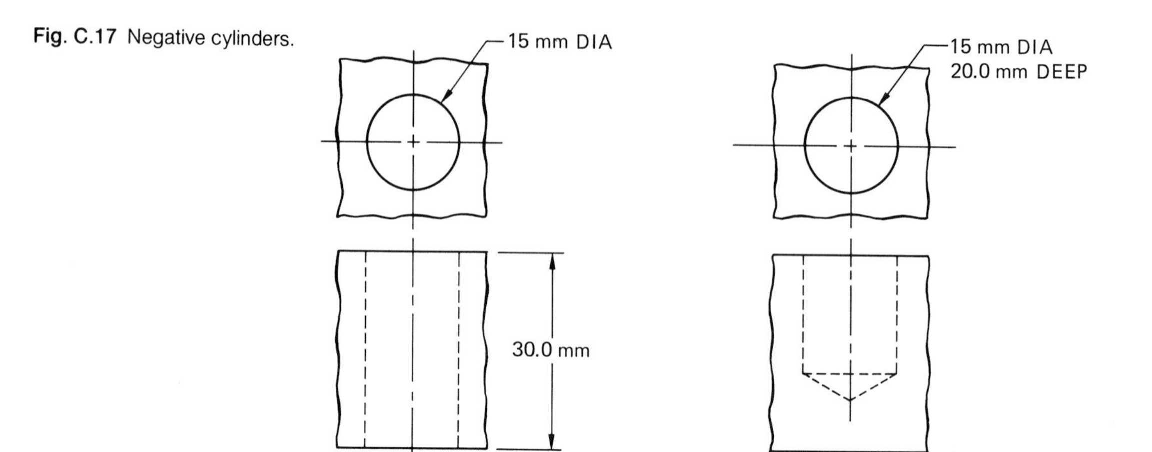

Fig. C.17 Negative cylinders.

Circular arcs are dimensioned as shown in Fig. C.18. For castings with a large number of fillets and rounds, it is customary to indicate all radii by use of a general note: for example, "all radii 8 mm unless otherwise specified."

Counterbored and countersunk holes (see Fig. C.19)

One of the most important guides to follow when dimensioning is *always to place the dimension in the view that shows the most characteristic feature.* Select a scale for the drawing, and use a minimum but sufficient number of views to completely describe and accurately dimension the object. Generally, three overall dimensions are included on all parts.

Figure C.20 illustrates three correctly dimensioned views of an object.

C.6.2 Recommended dimensioning practices

1. Never duplicate dimensions.
2. Don't crowd dimensions.
3. Place dimensions *between* views, not *on* views.
4. Avoid crossing dimension lines.

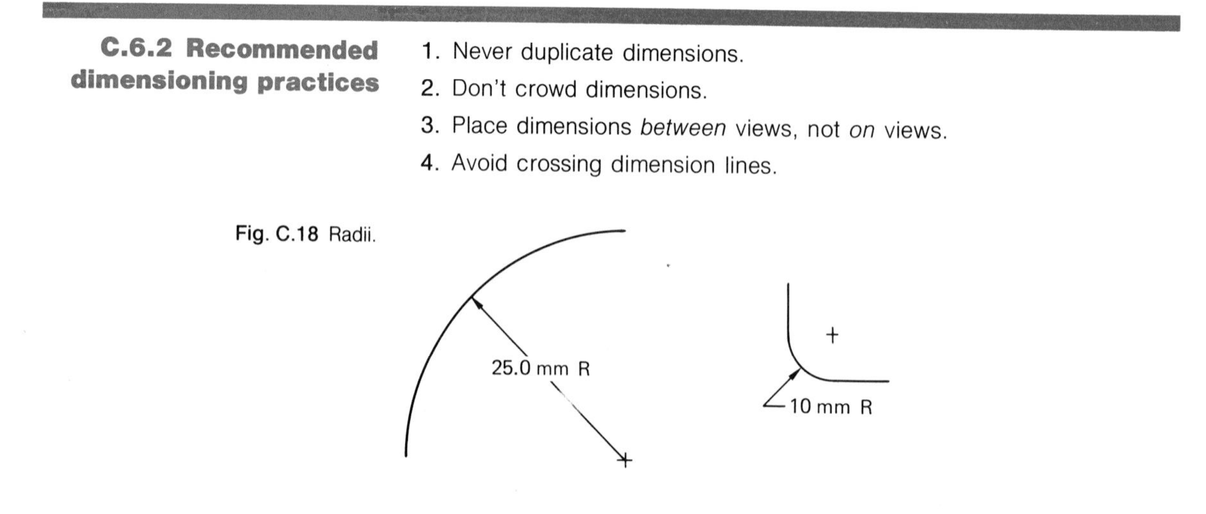

Fig. C.18 Radii.

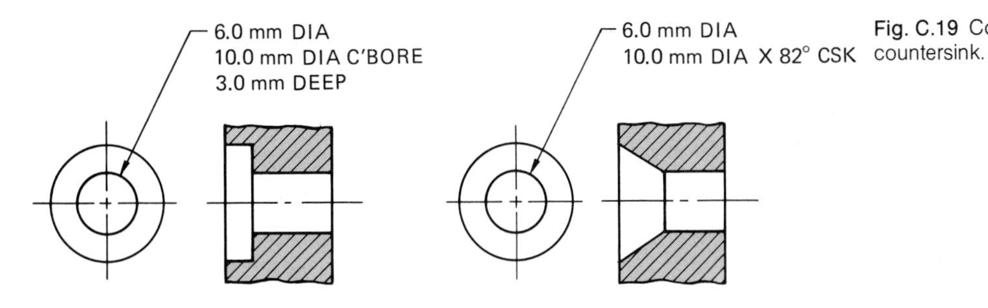

Fig. C.19 Counterbore and countersink.

5. Break dimension lines at numerals.
6. Don't dimension to hidden lines.
7. Arrange numerals and notes to be read from bottom of sheet.
8. Use standard height for letters and numerals.
9. Provide dimensions so that calculations are not necessary.
10. Measure lengths in metric units.
11. Don't allow extension lines to touch object.
12. Use centerlines as extension lines when appropriate.

C.7 Sections

At times, visualization of an object that has considerable interior complexity can be enhanced by taking an appropriate section view. An imaginary plane, as illustrated in Fig. C.21, is cut through the

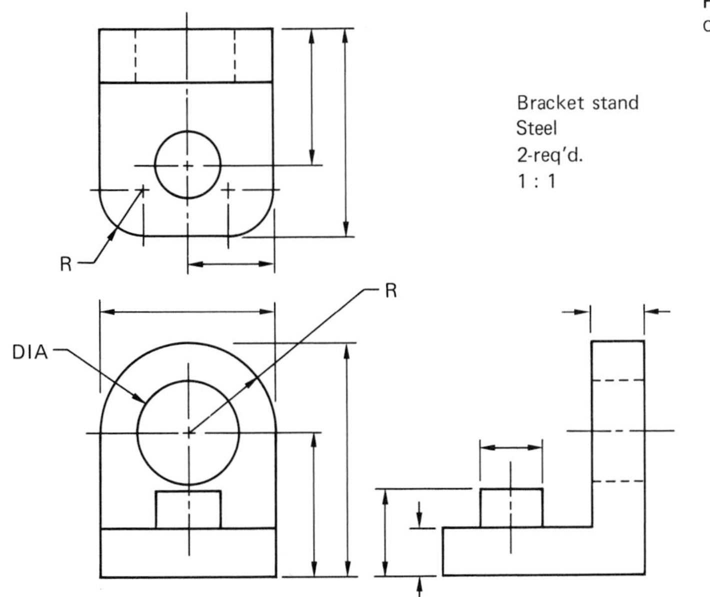

Fig. C.20 Detail drawing with complete title.

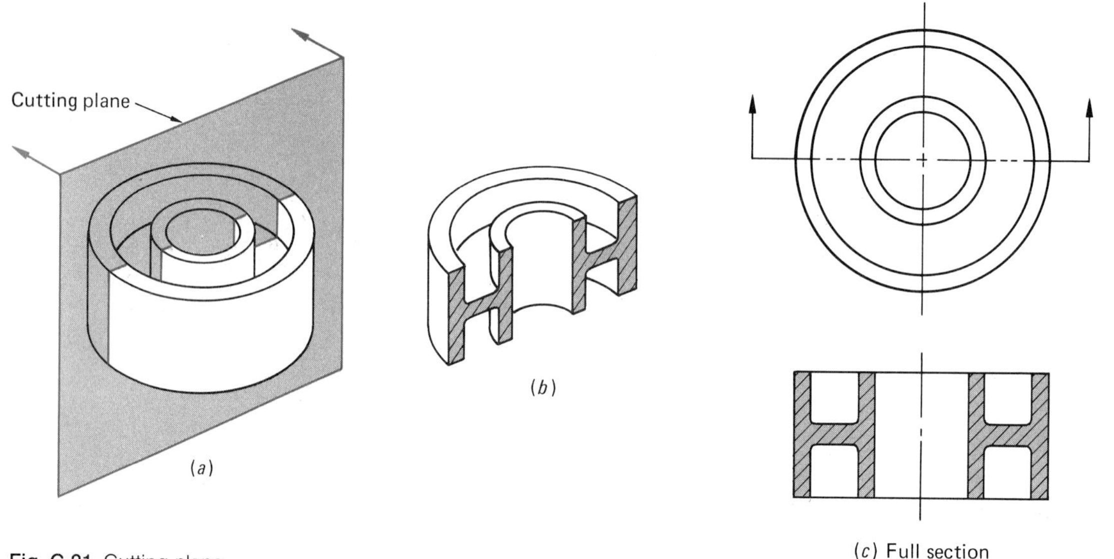

Fig. C.21 Cutting plane.

Fig. C.22 Section lining.

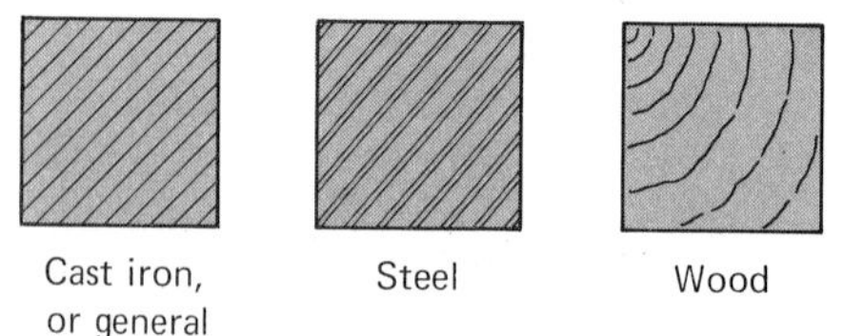

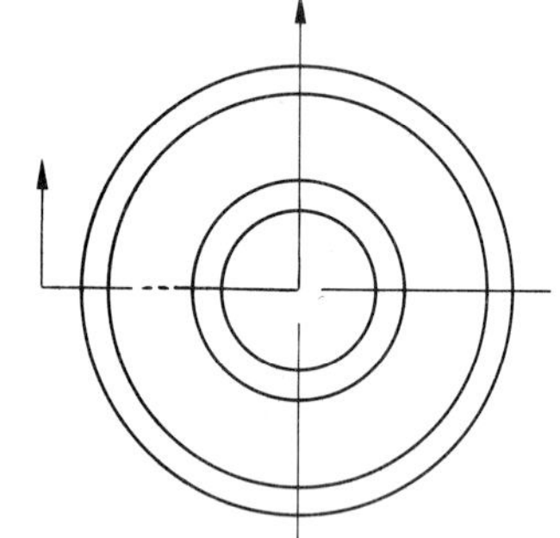

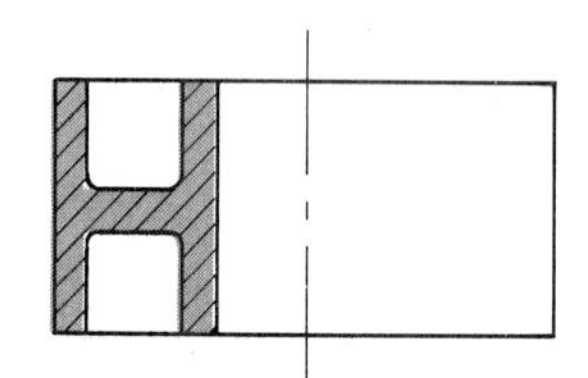

Fig. C.23 Half sections.

object and the near portion is taken away, thereby exposing the interior (Fig. C.21*b*).

Solids that are cut by the imaginary plane are cross-hatched according to preestablished standards. Three of these are illustrated in Fig. C.22.

Six different types of section views are defined and illustrated here.

Full section

A full section results when the cutting plane passes completely through the object, as illustrated in Fig. C.21*c*. Hidden lines behind the cutting plane are normally omitted unless essential.

Half sections

Half sections can be used on symmetrical objects when it is desirable to show an internal detail as well as a view of the exterior. Note that separation of the inside and outside is by a centerline, as seen in Fig. C.23.

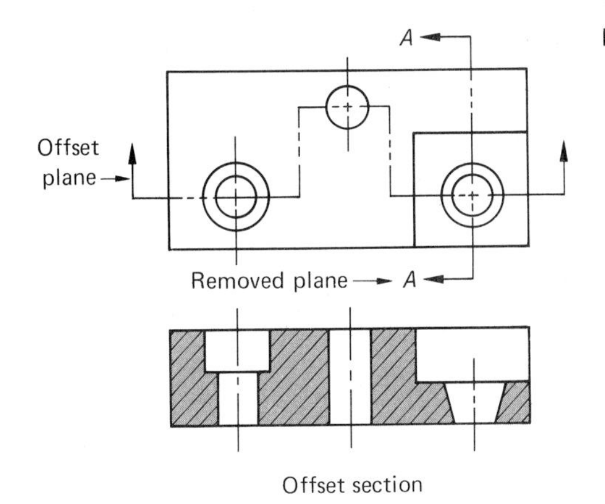

Fig. C.24 Offset section.

Offset section

An offset plane goes completely through the object but is staggered rather than straight. A convention normally practiced allows the lines of demarcation (change in direction of cutting plane) to be omitted in the sectioned view. The offset section is illustrated in Fig. C.24.

Revolved section

Revolved sections are used to show the shape and contour of ribs, spokes, etc. The plane is passed through the object and the cut area is revolved 90° and cross-hatched. Examples of this section can be seen in Fig. C.25*a*, *b*, and *c*.

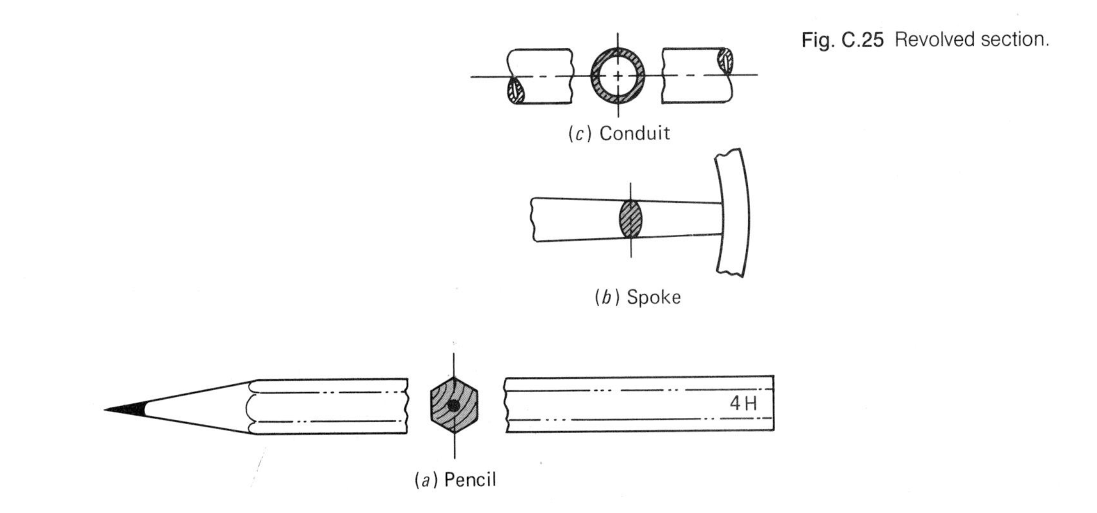

Fig. C.25 Revolved section.

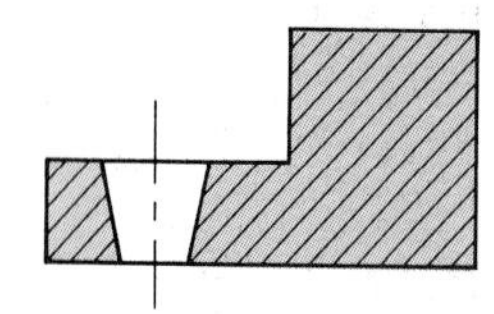

Fig. C.26 Removed section.

Removed section

There are two significant advantages to a removed section. First, it can be moved to a separate location. A second advantage is that the scale can be changed, e.g., drawn at a larger scale. The cutting plane for the removed section illustrated in Fig. C.26 is found in Fig. C.24.

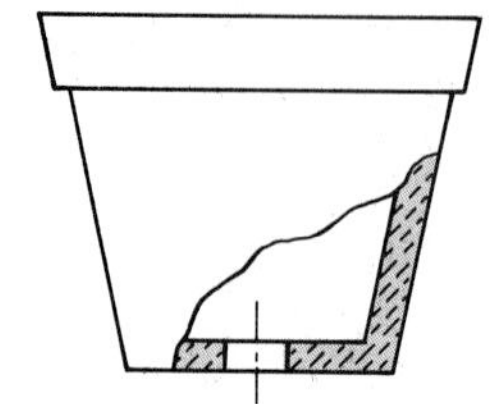
Fig. C.27 Partial section.

Partial section

A partial section is used when a portion of an orthographic view or pictorial is broken away to expose the interior (see Fig. C.27). The cross-hatching indicates the type of material from which the object is constructed.

C.8 Design drawings

Although each of the previous sections can be used at different times in the design process, they must be tied together to describe a total system more completely. It is the purpose of this section to outline some of the graphics necessary for a design to be delineated.

C.8.1. Layout drawings

A layout is a very accurate, scaled instrument drawing used to determine operational characteristics, such as clearances, and the relation of one part to another.

Fig. C.28 Layout drawing.

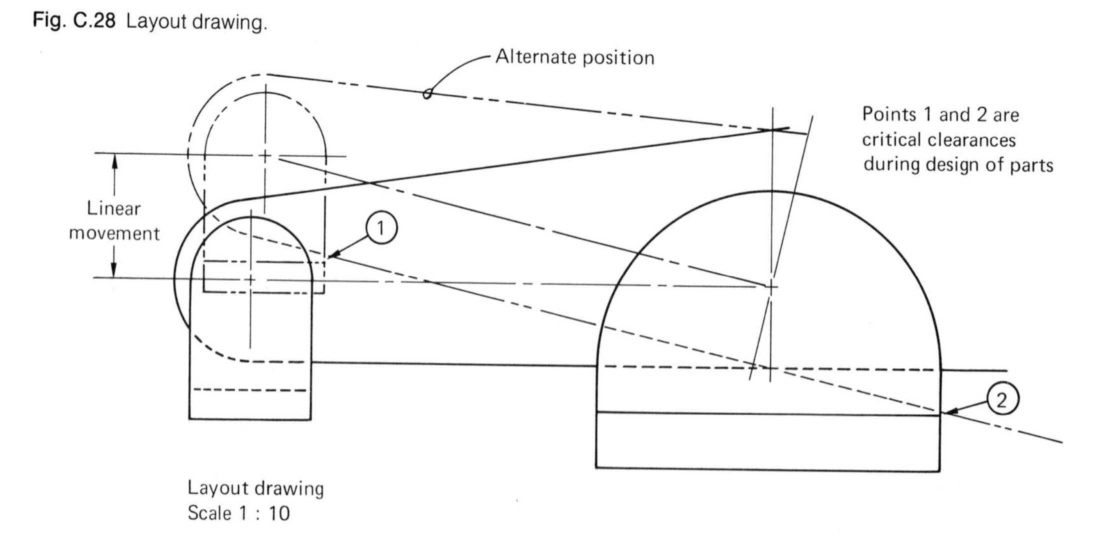

Figure C.29 is a pictorial of an assembly, whereas Fig. C.28 is a layout to determine critical clearances at points 1 and 2. Because of the drawing time involved, the layout drawing shows a minimum of information. Centerlines and key features are used to verify operation.

C.8.2 Assembly drawings

Assembly drawings illustrate, either in pictorial or orthographic form, the individual parts assembled. Figures C.29 and C.30 are both examples of assembly drawings. This type of drawing has several important functions. It demonstrates how the entire collection of individual items fit together. Critical dimensions, centerline dimensions, and overall dimensions are normally included to specify clearances, etc. The drawing provides an opportunity to investigate the relationship of individual parts as they move or rotate to alternate positions.

Ballons (numbers within circles) and leaders pointing to the individual parts establish an identification system for the assembly of parts. Detail drawings (Sec. C.8.3) are keyed to the assembly by use of these identification numbers. The bill of material (Sec. C.8.4) also uses the same identification system.

Fig. C.29 Pictorial assembly.

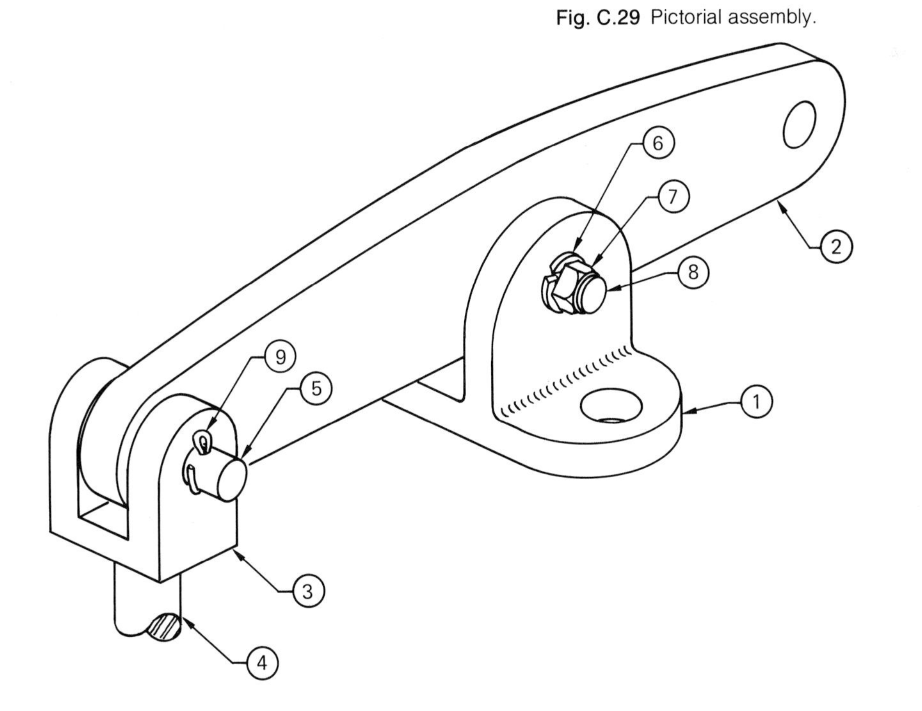

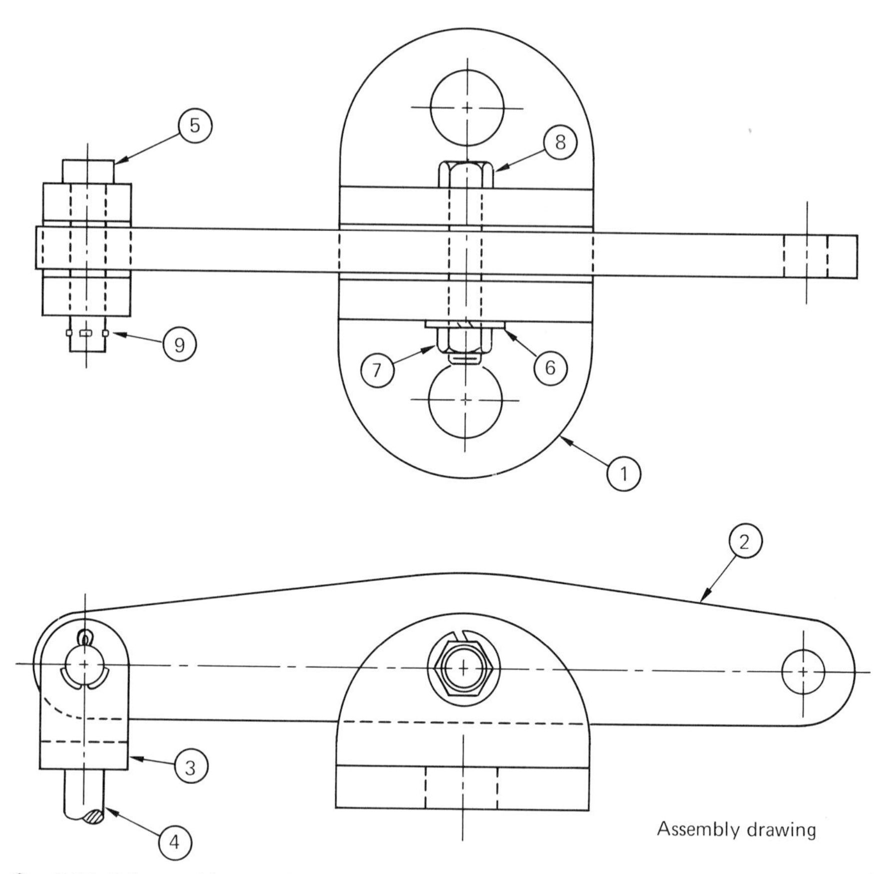

Fig. C.30 Orthographic assembly.

C.8.3 Detail drawings A detail drawing, as illustrated in Fig. C.31, is a drawing of a *single* part completely specified. By working with the assembly and layout drawings the detail can be completed so that the individual part can be made. Standard parts, i.e., nuts, bolts, etc., are not customarily detailed.

A detail drawing will consist of sufficient views, completely dimensioned, with appropriate section views, and a title that includes the identification number, name, number required in assembly, material, scale, and necessary notes.

C.8.4 Bill of material Every item in the assembly should appear in the bill of materials (see Table C.2). Materials, sizes, notes, etc., are added as the assembly is formalized.

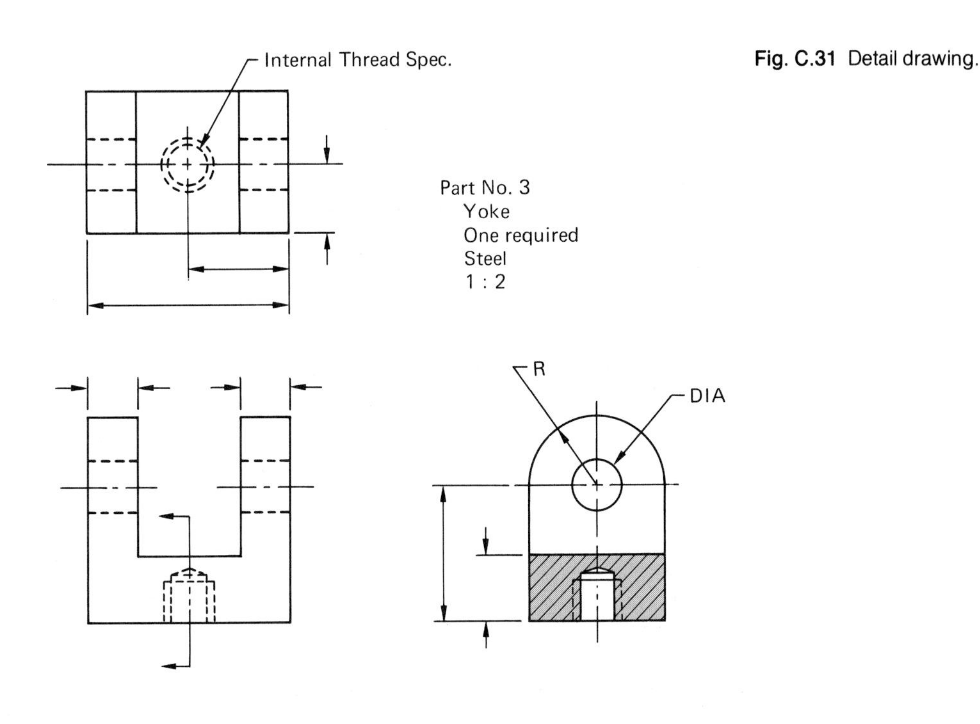

Fig. C.31 Detail drawing.

C.9 Presentation drawings

It is often necessary for the engineer to present data to people who are not familiar with technical graphs. A few of the many methods available to do this are included below.

Pie diagrams

Pie diagrams are most popular when representing items that total 100 percent. All lettering and percentages should be placed on the sector or immediately adjacent. The circle should not contain more than five or six categories and each should be cross-hatched or

Table C.2

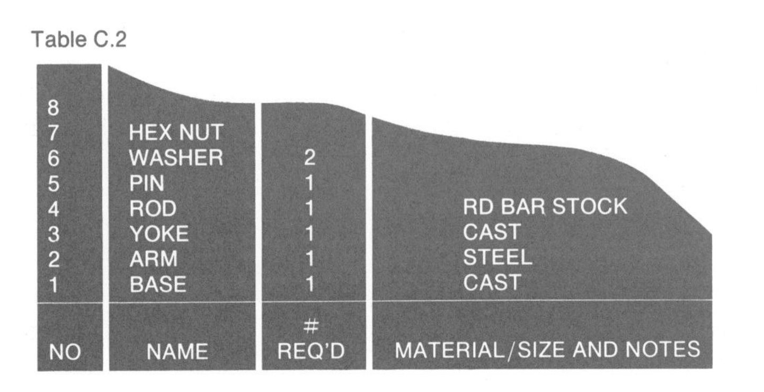

NO	NAME	# REQ'D	MATERIAL/SIZE AND NOTES
8			
7	HEX NUT		
6	WASHER	2	
5	PIN	1	
4	ROD	1	RD BAR STOCK
3	YOKE	1	CAST
2	ARM	1	STEEL
1	BASE	1	CAST

Fig. C.32 Pie diagram. (Data collected at Iowa State University (student opinion).

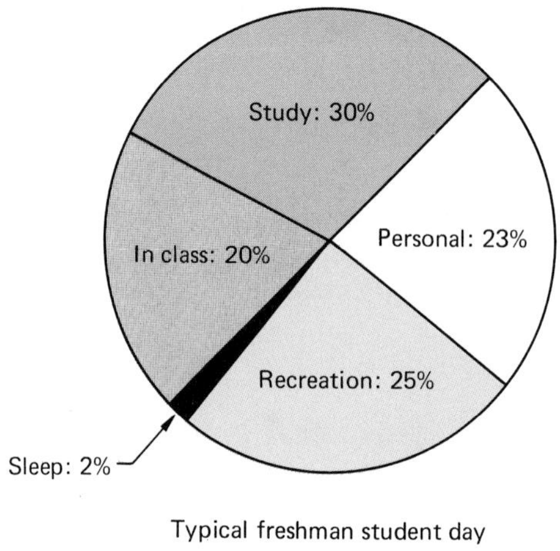

Study: 30%
Personal: 23%
In class: 20%
Recreation: 25%
Sleep: 2%

Typical freshman student day

marked differently. A title with pertinent information concerning source, etc., should always be included. See Figure C.32.

Column charts

One of the most common nontechnical methods of representing information is the column chart, illustrated in Fig. C.33.

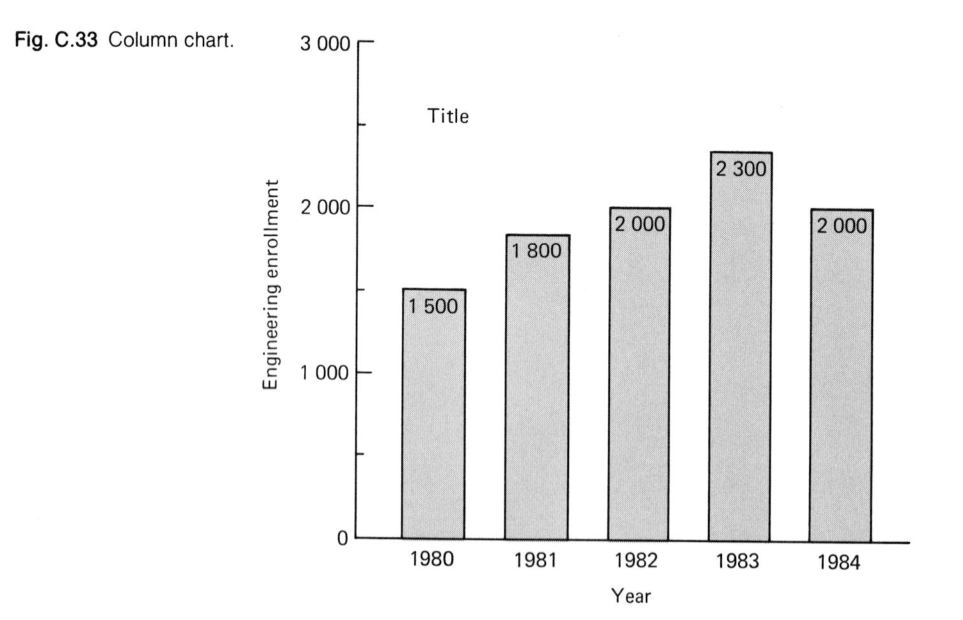

Fig. C.33 Column chart.

The quantity to be graphed is illustrated by bars, each of whose length is proportional to the value represented.

Block diagrams

A block diagram, as represented in Fig. C.34, is an excellent technique to show in a simple fashion the overall process. Different symbols denote certain processes with size and balance considerations important for ease of understanding.

Schematic diagrams

Combination block and schematic diagrams, as illustrated in Figure C.35, and the schematic diagram in Fig. C.36 convey the operational characteristics of electrical as well as other complex systems. Symbols, layout, and labels provide a graphical expression of the functional relationship or interconnection of component parts.

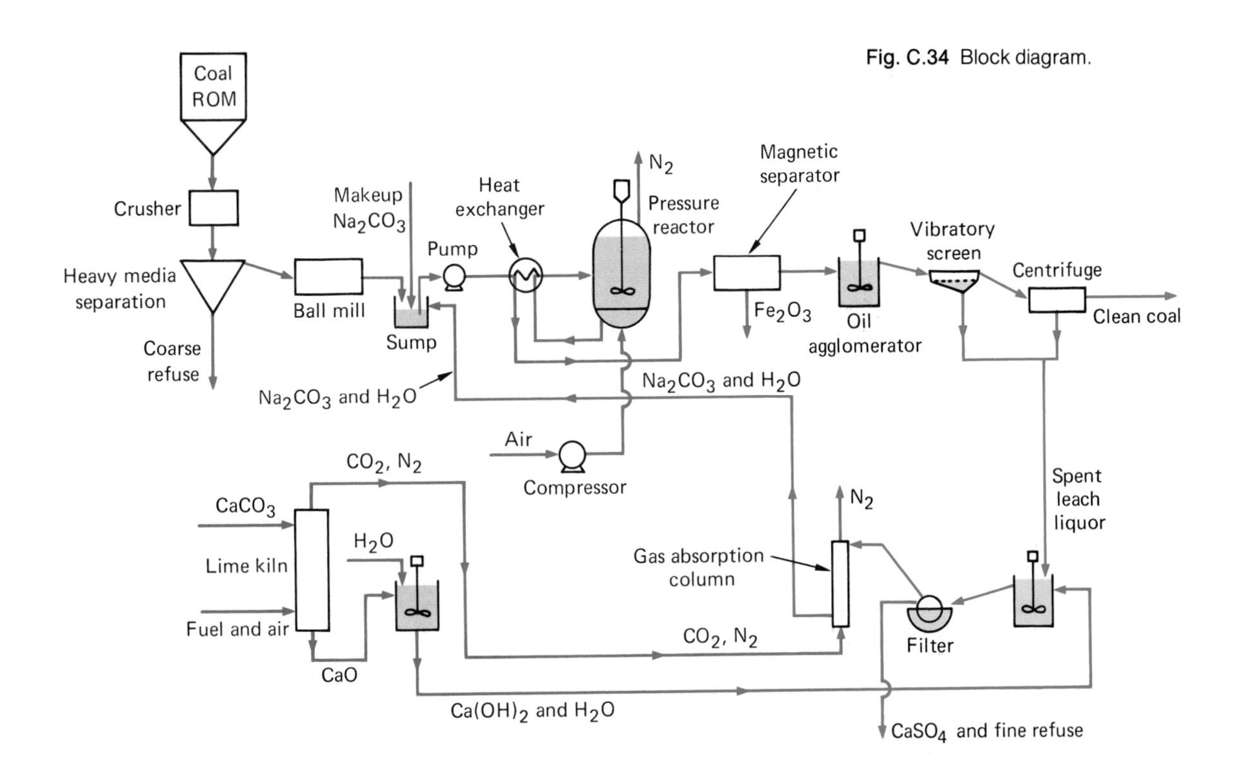

Fig. C.34 Block diagram.

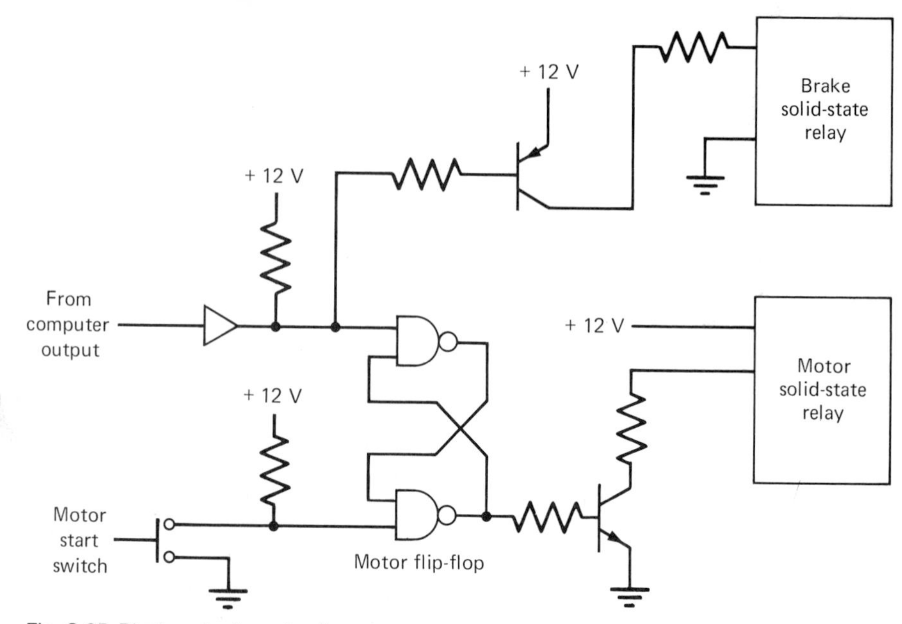

Fig. C.35 Block and schematic diagram.

Fig. C.36 Schematic diagram.

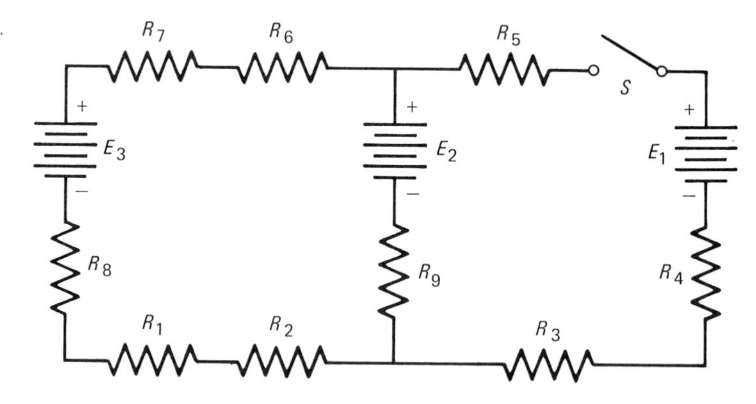

Schematic diagram

General

APPENDIX D

Approximate specific gravities and densities

Material	Specific gravity	Average density lb/ft^3	Average density kg/m^3
Gases (0°C and 1 atm)			
Air		0.080 18	1.284
Ammonia		0.048 13	0.771 0
Carbon dioxide		0.123 4	1.977
Carbon monoxide		0.078 06	1.251
Ethane		0.084 69	1.357
Helium		0.011 14	0.178 4
Hydrogen		0.005 611	0.089 88
Methane		0.044 80	0.717 6
Nitrogen		0.078 07	1.251
Oxygen		0.089 21	1.429
Sulfur dioxide		0.182 7	2.927
Liquids (20°C)			
Alcohol, ethyl	0.79	49	790
Alcohol, methyl	0.80	50	800
Benzene	0.88	55	880
Gasoline	0.67	42	670
Heptane	0.68	42	680
Hexane	0.66	41	660
Octane	0.71	44	710
Oil	0.88	55	880
Toluene	0.87	54	870
Water	1.00	62.4	1000
Metals (20°C)			
Aluminum	2.55– 2.80	165	2640
Brass, cast	8.4 – 8.7	535	8570
Bronze	7.4 – 8.7	510	8170
Copper, cast	8.9	555	8900
Gold, cast	19.3	1210	19 300
Iron, cast	7.04– 7.12	440	7050
Iron, wrought	7.6 – 7.9	485	7770
Iron ore	5.2	325	5210
Lead	11.3	705	11 300
Manganese	7.4	462	7400
Mercury	13.6	849	13 600
Nickel	8.9	556	8900
Silver	10.4 –10.6	655	10 500
Steel, cold drawn	7.83	489	7830

Material	Specific gravity	Average density lb/ft³	kg/m³
Steel, machine	7.80	487	7800
Steel, tool	7.70	481	7700
Tin, cast	7.30	456	7300
Titanium	4.5	281	4500
Uranium	18.7	1170	18 700
Zinc, cast	6.9 – 7.2	440	7050

Nonmetallic solids (20°C)

Material	Specific gravity	lb/ft³	kg/m³
Brick, common	1.80	112	1800
Cedar	0.35	22	350
Clay, damp	1.8 – 2.6	137	2200
Coal, bituminous	1.2 – 1.5	84	1350
Concrete	2.30	144	2300
Douglas fir	0.50	31	500
Earth, loose	1.2	75	1200
Glass, common	2.5 – 2.8	165	2650
Gravel, loose	1.4 – 1.7	97	1550
Gypsum	2.31	144	2310
Limestone	2.0 – 2.9	153	2450
Mahogany	0.54	34	540
Marble	2.6 – 2.9	172	2750
Oak	0.64– 0.87	47	750
Paper	0.7 – 1.2	58	925
Rubber	0.92– 0.96	59	940
Salt	0.8 – 1.2	62	1000
Sand, loose	1.4 – 1.7	97	1550
Sugar	1.61	101	1610
Sulfur	2.1	131	2100

Unit prefixes

Multiple and submultiple	Prefix	Symbol
1,000,000,000,000 = 10^{12}	tera	T
1,000,000,000 = 10^{9}	giga	G
1,000,000 = 10^{6}	mega	M
1,000 = 10^{3}	kilo	k
100 = 10^{2}	hecto	h
10 = 10	deka	da
0.1 = 10^{-1}	deci	d
0.01 = 10^{-2}	centi	c
0.001 = 10^{-3}	milli	m
0.000 001 = 10^{-6}	micro	μ
0.000 000 001 = 10^{-9}	nano	n
0.000 000 000 001 = 10^{-12}	pico	p
0.000 000 000 000 001 = 10^{-15}	femto	f
0.000 000 000 000 000 001 = 10^{-18}	atto	a

Chemical elements

Element	Symbol	Atomic No.	Atomic Weight
Actinium	Ac	89	
Aluminum	Al	13	26.9815
Americium	Am	95	
Antimony	Sb	51	121.750
Argon	Ar	18	39.948
Arsenic	As	33	74.9216
Astatine	At	85	
Barium	Ba	56	137.34
Berkelium	Bk	97	
Beryllium	Be	4	9.0122
Bismuth	Bi	83	208.980
Boron	B	5	10.811
Bromine	Br	35	79.904
Cadmium	Cd	48	112.40
Calcium	Ca	20	40.08
Californium	Cf	98	
Carbon	C	6	12.01115
Cerium	Ce	58	140.12
Cesium	Cs	55	132.905
Chlorine	Cl	17	35.453
Chromium	Cr	24	51.996
Cobalt	Co	27	58.9332
Columbium (see Niobium)			
Copper	Cu	29	63.546
Curium	Cm	96	
Dysprosium	Dy	66	162.50
Einsteinium	Es	99	
Erbium	Er	68	167.26
Europium	Eu	63	151.96
Fermium	Fm	100	
Fluorine	F	9	18.9984
Francium	Fr	87	
Gadolinium	Gd	64	157.25
Gallium	Ga	31	69.72
Germanium	Ge	32	72.59
Gold	Au	79	196.967
Hafnium	Hf	72	178.49
Helium	He	2	4.0026
Holmium	Ho	67	164.930
Hydrogen	H	1	1.00797
Indium	In	49	114.82
Iodine	I	53	126.9044
Iridium	Ir	77	192.2
Iron	Fe	26	55.847
Krypton	Kr	36	83.80
Lanthanum	La	57	138.91
Lead	Pb	82	207.19
Lithium	Li	3	6.939
Lutetium	Lu	71	174.97
Magnesium	Mg	12	24.312
Manganese	Mn	25	54.9380
Mendelevium	Md	101	
Mercury	Hg	80	200.59
Molybdenum	Mo	42	95.94
Neodymium	Nd	60	144.24
Neon	Ne	10	20.183
Neptunium	Np	93	
Nickel	Ni	28	58.71
Niobium	Nb	41	92.906
Nitrogen	N	7	14.0067
Nobelium	No	102	
Osmium	Os	76	109.2
Oxygen	O	8	15.9994
Palladium	Pd	46	106.4
Phosphorus	P	15	30.9738
Platinum	Pt	78	195.09
Plutonium	Pu	94	
Polonium	Po	84	
Potassium	K	19	39.102
Praseodymium	Pr	59	140.907
Promethium	Pm	61	
Protactinium	Pa	91	
Radium	Ra	88	
Radon	Rn	86	
Rhenium	Re	75	186.2
Rhodium	Rh	45	102.905
Rubidium	Rb	37	85.47
Ruthenium	Ru	44	101.07
Samarium	Sm	62	150.35
Scandium	Sc	21	44.956
Selenium	Se	34	78.96
Silicon	Si	14	28.086
Silver	Ag	47	107.868
Sodium	Na	11	22.9898
Strontium	Sr	38	87.62
Sulphur	S	16	32.064
Tantalum	Ta	73	180.948
Technetium	Tc	43	
Tellurium	Te	52	127.60
Terbium	Tb	65	158.924
Thallium	Tl	81	204.37
Thorium	Th	90	232.038
Thulium	Tm	69	168.934
Tin	Sn	50	118.69
Titanium	Ti	22	47.90
Tungsten	W	74	183.85
Uranium	U	92	238.03
Vanadium	V	23	50.942
Xenon	Xe	54	131.30
Ytterbium	Yb	70	173.04
Yttrium	Y	39	88.905
Zinc	Zn	30	65.37
Zirconium	Zr	40	91.22

Greek alphabet

Alpha	A	α
Beta	B	β
Gamma	Γ	γ
Delta	Δ	δ
Epsilon	E	ϵ
Zeta	Z	ζ
Eta	H	η
Theta	Θ	θ
Iota	I	ι
Kappa	K	κ
Lambda	Λ	λ
Mu	M	μ
Nu	N	ν
Xi	Ξ	ξ
Omicron	O	o
Pi	Π	π
Rho	P	ρ
Sigma	Σ	σ
Tau	T	τ
Upsilon	Υ	υ
Phi	Φ	ϕ
Chi	X	χ
Psi	Ψ	ψ
Omega	Ω	ω

Physical constants

Avogadro's number = $6.022\ 57 \times 10^{23}$/mol

Density of dry air at 0°C, 1 atm = 1.293 kg/m^3

Density of water at 3.98°C = $9.999\ 973 \times 10^2$ kg/m^3

Equatorial radius of the earth = 6378.39 km = 3963.34 mi

Gravitational acceleration (standard) at sea level = 9.806 65 m/s^2 = 32.174 ft/s^2

Gravitational constant = 6.672×10^{-11} N · m^2/kg^2

Heat of fusion of water, 0°C = 3.3375×10^5 J/kg = 143.48 Btu/lb

Heat of vaporization of water, 100°C = 2.2591×10^6 J/kg = 971.19 Btu/lb

Mass of hydrogen atom = $1.673\ 39 \times 10^{-27}$ kg

Mean density of the earth = 5.522×10^3 kg/m^3 = 344.7 lb/ft^3

Molar gas constant = 8.3144 J/(mol · K)

Planck's constant = $6.625\ 54 \times 10^{-34}$ J/Hz

Polar radius of the earth = 6356.91 km = 3949.99 mi

Velocity of light in a vacuum = 2.9979×10^8 m/s

Velocity of sound in dry air at 0°C = 331.36 m/s = 1087.1 ft/s

Code of ethics for engineers

Preamble

The Engineer, to uphold and advance the honor and dignity of the engineering profession and in keeping with high standards of ethical conduct:

Will be honest and impartial, and will serve with devotion his employer, his clients, and the public;

Will strive to increase the competence and prestige of the engineering profession;

Will use his knowledge and skill for the advancement of human welfare.

Section 1

The Engineer will be guided in all his professional relations by the highest standards of integrity, and will act in professional matters for each client or employer as a faithful agent or trustee.

a. He will be realistic and honest in all estimates, reports, statements, and testimony.

b. He will admit and accept his own errors when proven wrong and refrain from distorting or altering the facts in an attempt to justify his decision.

c. He will advise his client or employer when he believes a project will not be successful.

d. He will not accept outside employment to the detriment of his regular work or interest, or without the consent of his employer.

e. He will not attempt to attract an engineer from another employer by false or misleading pretenses.

f. He will not actively participate in strikes, picket lines, or other collective coercive actions.

g. He will avoid any act tending to promote his own interest at the expense of the dignity and integrity of the profession.

Section 2

The Engineer will have proper regard for the safety, health, and welfare of the public in the performance of his professional duties. If his engineering judgment is overruled by nontechnical authority, he will clearly point out the consequences. He will notify the proper authority of any observed conditions which endanger public safety and health.

a. He will regard his duty to the public welfare as paramount.

b. He shall seek opportunities to be of constructive service in civic affairs and work for the advancement of the safety, health and well-being of his community.

c. He will not complete, sign, or seal plans and/or specifications that are not of a design safe to the public health and welfare and in conformity with accepted engineering standards. If the client or employer insists on such unprofessional conduct, he shall notify the proper authorities and withdraw from further service on the project.

Section 3 **The Engineer will avoid all conduct or practice likely to discredit or unfavorably reflect upon the dignity or honor of the profession.**

a. The Engineer shall not advertise his professional services but may utilize the following means of identification:

(1) Professional cards and listings in recognized and dignified publications, provided they are consistent in size and are in a section of the publication regularly devoted to such professional cards and listings. The information displayed must be restricted to firm name, address, telephone number, appropriate symbol, name of principal participants and the fields of practice in which the firm is qualified.

(2) Signs on equipment, offices and at the site of projects for which he renders services, limited to firm name, address, telephone number and type of services, as appropriate.

(3) Brochures, business cards, letterheads and other factual representations of experience, facilities, personnel and capacity to render service, providing the same are not misleading relative to the extent of participation in the projects cited, and provided the same are not indiscriminately distributed.

(4) Listings in the classified section of telephone directories, limited to name, address, telephone number and specialties in which the firm is qualified.

b. The Engineer may advertise for recruitment of personnel in appropriate publications or by special distribution. The information presented must be displayed in a dignified manner, restricted to firm name, address, telephone number, appropriate symbol, name of principal participants, the fields of practice in which the firm is qualified and factual descriptions of positions available, qualifications required and benefits available.

c. The Engineer may prepare articles for the lay or technical press which are factual, dignified and free from ostentations or laudatory implications. Such articles shall not imply other than his direct participation in the work described unless credit is given to others for their share of the work.

d. The Engineer may extend permission for his name to be used in commercial advertisements, such as may be published by manufacturers, contractors, material suppliers, etc., only by means of a modest dignified notation acknowledging his participation and the scope thereof in the project or product described. Such permission shall not include public endorsement of proprietary products.

e. The Engineer will not allow himself to be listed for employment using exaggerated statements of his qualifications.

Section 4

The Engineer will endeavor to extend public knowledge and appreciation of engineering and its achievements and to protect the engineering profession from misrepresentation and misunderstanding.

a. He shall not issue statements, criticisms, or arguments on matters connected with public policy which are inspired or paid for by private interests, unless he indicates on whose behalf he is making the statement.

Section 5

The Engineer will express an opinion of an engineering subject only when founded on adequate knowledge and honest conviction.

a. The Engineer will insist on the use of facts in reference to an engineering project in a group discussion, public forum or publication of articles.

Section 6

The Engineer will undertake engineering assignments for which he will be responsible only when qualified by training or experience; and he will engage, or advise engaging, experts and specialists whenever the client's or employer's interests are best served by such service.

Section 7

The Engineer will not disclose confidential information concerning the business affairs or technical processes of any present or former client or employer without his consent.

a. While in the employ of others, he will not enter promotional efforts or negotiations for work or make arrangements for other employment as a principal or to practice in connection with a specific project for which he has gained particular and specialized knowledge without the consent of all interested parties.

Section 8

The Engineer will endeavor to avoid a conflict of interest with his employer or client, but when unavoidable, the Engineer shall fully disclose the circumstances to his employer or client.

a. The Engineer will inform his client or employer of any business connections, interests, or circumstances which may be deemed as influencing his judgment or the quality of his services to his client or employer.

b. When in public service as a member, advisor, or employee of a governmental body or department, an Engineer shall not participate in considerations or actions with respect to services provided by him or his organization in private engineering practice.

c. An Engineer shall not solicit or accept an engineering contract from a governmental body on which a principal or officer of his organization serves a member.

Section 9 The Engineer will uphold the principle of appropriate and adequate compensation for those engaged in engineering work.

a. He will not undertake or agree to perform any engineering service on a free basis, except for civic, charitable, religious, or eleemosynary nonprofit organizations when the professional services are advisory in nature.

b. He will not undertake work at a fee or salary below the accepted standards of the profession in the area.

c. He will not accept remuneration from either an employee or employment agency for giving employment.

d. When hiring other engineers, he shall offer a salary according to the engineer's qualifications and the recognized standards in the particular geographical area.

e. If, in sales employ, he will not offer, or give engineering consultation, or designs, or advice other than specifically applying to the equipment being sold.

Section 10 The Engineer will not accept compensation, financial or otherwise, from more than one interested party for the same service, or for services pertaining to the same work, unless there is full disclosure to and consent of all interested parties.

a. He will not accept financial or other considerations, including free engineering designs, from material or equipment suppliers for specifying their product.

b. He will not accept commissions or allowances, directly or indirectly, from contractors or other parties dealing with his clients or employer in connection with work for which he is responsible.

Section 11 The Engineer will not compete unfairly with another engineer by attempting to obtain employment or advancement or professional engagements by competitive bidding, by taking advantage of a salaried position, by criticizing other engineers, or by other improper or questionable methods.

a. The Engineer will not attempt to supplant another engineer in a particular employment after becoming aware that definite steps have been taken toward the other's employment.

b. He will not pay, or offer to pay, either directly or indirectly, any commission, political contribution, or a gift, or other consideration in order to secure work, exclusive of securing salaried positions through employment agencies.

c. He shall not solicit or submit engineering proposals on the basis of competitive bidding. Competitive bidding for professional engineering services is defined as the formal or informal submission, or receipt, of verbal or written estimates of cost or proposals in terms of dollars, man days of work required, percentage of construction cost, or any other measure of compensation whereby the prospective client may compare engineering services on a price basis prior to the time that one engineer, or one engineering organization, has been selected for negotiations. The disclosure of recommended fee schedules prepared by various engineering societies is not considered to constitute competitive bidding. An Engineer requested to submit a fee proposal or bid prior to the selection of an engineer or firm subject to the negotiation of a satisfactory contract, shall attempt to have the procedure changed to conform to ethical practices, but if not successful he shall withdraw from consideration for the proposed work. These principles shall be applied by the Engineer in obtaining the services of other professionals.

d. An engineer shall not request, propose, or accept a professional commission on a contingent basis under circumstances in which his professional judgment may be compromised, or when a contingency provision is used as a device for promoting or securing a professional commission.

e. While in a salaried position, he will accept part-time engineering work only at a salary or fee not less than that recognized as standard in the area.

f. An Engineer will not use equipment, supplies, laboratory, or office facilities of his employer to carry on outside private practice without consent.

Section 12

The Engineer will not attempt to injure, maliciously or falsely, directly or indirectly, the professional reputation, prospects, practice or employment of another engineer, nor will he indiscriminately criticize another engineer's work. If he believes that another engineer is guilty of unethical or illegal practice, he shall present such information to the proper authority for action.

a. An Engineer in private practice will not review the work of another engineer for the same client, except with the knowledge of such engineer, or unless the connection of such engineer with the work has been terminated.

b. An Engineer in governmental, industrial or educational employ is entitled to review and evaluate the work of other engineers when so required by his employment duties.

c. An Engineer in sales or industrial employ is entitled to make engineering comparisons of his products with products by other suppliers.

Section 13 The Engineer will not associate with or allow the use of his name by an enterprise of questionable character, nor will he become professionally associated with engineers who do not conform to ethical practices, or with persons not legally qualified to render the professional services for which the association is intended.

a. He will conform with registration laws in his practice of engineering.

b. He will not use association with a nonengineer, a corporation, or partnership, as a "cloak" for unethical acts, but must accept personal responsibility for his professional acts.

Section 14 The Engineer will give credit for engineering work to those to whom credit is due, and will recognize the proprietary interests of others.

a. Whenever possible, he will name the person or persons who may be individually responsible for designs, inventions, writings, or other accomplishments.

b. When an Engineer uses designs supplied to him by a client, the designs remain the property of the client and should not be duplicated by the Engineer for others without express permission.

c. Before undertaking work for others in connection with which he may make improvements, plans, designs, inventions, or other records which may justify copyrights or patents, the Engineer should enter into a positive agreement regarding the ownership.

d. Designs, data, records, and notes made by an engineer and referring exclusively to his employer's work are his employer's property.

Section 15 The Engineer will cooperate in extending the effectiveness of the profession by interchanging information and experience with other engineers and students, and will endeavor to provide opportunity for the professional development and advancement of engineers under his supervision.

a. He will encourage his engineering employees' efforts to improve their education.

b. He will encourage engineering employees to attend and present papers at professional and technical society meetings.

c. He will urge his engineering employees to become registered at the earliest possible date.

d. He will assign a professional engineer duties of a nature to utilize his full training and experience, insofar as possible, and delegate lesser functions to subprofessionals or to technicians.

e. He will provide a prospective engineering employee with complete information on working conditions and his proposed status of employment, and after employment will keep him informed of any changes in them.

Note

In regard to the question of application of the Code to corporations vis-a-vis real persons, business form or type should not negate nor influence conformance of individuals to the Code. The Code deals with professional services, which services must be performed by real persons. Real persons in turn establish and implement policies within business structures. The Code is clearly written to apply to the Engineer and it is incumbent on a member of NSPE to endeavor to live up to its provisions. This applies to all pertinent sections of the Code.

NSPE Publication 1102, as revised January 1974. Reprinted by permission.

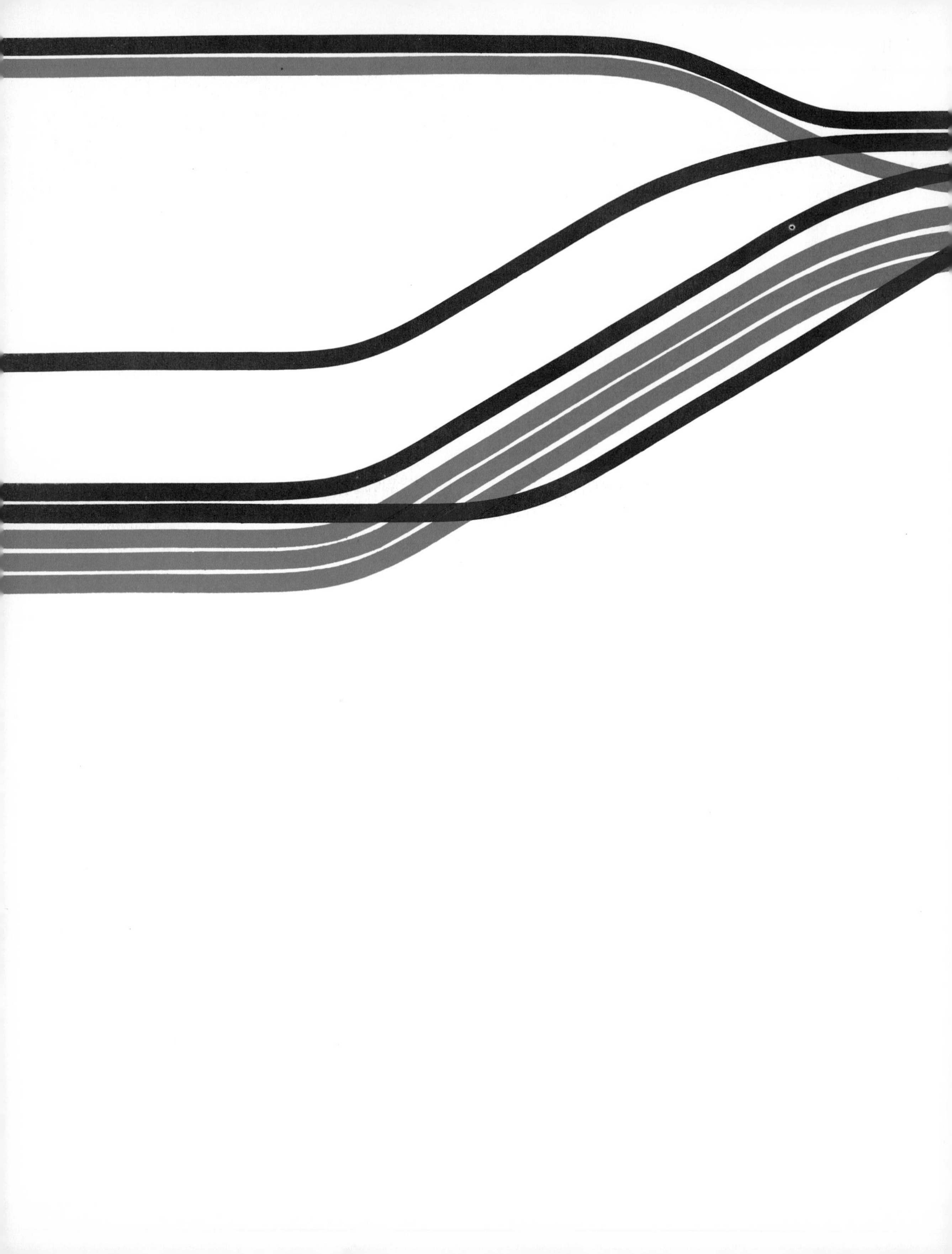

UNIT CONVERSIONS

To convert from	To	Multiply by
acres	ft^2	4.356×10^4
acres	ha	4.0469×10^{-1}
acres	m^2	4.0469×10^3
amperes	C/s	1
ampere hours	C	3.6×10^3
angstroms	cm	1×10^{-8}
angstroms	in	3.9370×10^{-9}
atmospheres	bars	1.0133
atmospheres	in of Hg	2.9921×10^1
atmospheres	kg/cm^2	1.0332
atmospheres	lb/in^2	1.4696×10^1
atmospheres	mm of Hg	7.6×10^2
atmospheres	Pa	1.0133×10^5
barrels (petroleum, US)	gal (US liquid)	4.2×10^1
bars	atm	9.8692×10^{-1}
bars	in of Hg	2.9530×10^1
bars	kg/cm^2	1.0197
bars	lb/in^2	1.4504×10^1
bars	Pa	1×10^5
Btu	ft · lb	7.7765×10^2
Btu	hp · h	3.9275×10^{-4}
Btu	J	1.0551×10^3
Btu	kg · m	1.0751×10^2
Btu	kWh	2.9288×10^{-4}
Btu per hour	ft · lb/s	2.1601×10^{-1}
Btu per hour	hp	3.9275×10^{-4}
Btu per hour	W	2.9288×10^{-1}
Btu per minute	ft · lb/min	7.7765×10^2
Btu per minute	hp	2.3565×10^{-2}
Btu per minute	kg · m/min	1.0751×10^2
Btu per minute	kW	1.7573×10^{-2}
bushels (US)	ft^3	1.2445
bushels (US)	L	3.5239×10^1
bushels (US)	m^3	3.5239×10^{-2}
candelas	lm/sr	1
candelas per square foot	lamberts	3.3816×10^{-3}
centimeters	Å	1×10^8
centimeters	ft	3.2808×10^{-2}
centimeters	in	3.9370×10^{-1}
centipoises	g/(cm · s)	1×10^{-2}
circular mils	cm^2	5.0671×10^{-6}
circular mils	in^2	7.8540×10^{-7}

To convert from	To	Multiply by
coulombs	A·s	1
cubic centimeters	in^3	6.1024×10^{-2}
cubic centimeters	ft^3	3.5315×10^{-5}
cubic centimeters	gal (US liquid)	2.6417×10^{-4}
cubic centimeters	L	1×10^{-3}
cubic centimeters	oz (US fluid)	3.3814×10^{-2}
cubic centimeters per gram	ft^3/lb	1.6018×10^{-2}
cubic centimeters per second	ft^3/min	2.1189×10^{-3}
cubic centimeters per second	gal (US liquid)/min	1.5850×10^{-2}
cubic feet	acre·ft	2.2957×10^{-5}
cubic feet	bushels (US)	8.0356×10^{-1}
cubic feet	gal (US liquid)	7.4805
cubic feet	in^3	1.728×10^{3}
cubic feet	L	2.8317×10^{1}
cubic feet	m^3	2.8317×10^{-2}
cubic feet per minute	gal (US liquid)/min	7.4805
cubic feet per minute	L/s	4.7195×10^{-1}
cubic feet per pound	cm^3/g	6.2428×10^{1}
cubic feet per second	gal (US liquid)/min	4.4883×10^{2}
cubic feet per second	L/s	2.8317×10^{1}
cubic inches	bushels (US)	4.6503×10^{-4}
cubic inches	cm^3	1.6387×10^{1}
cubic inches	gal (US liquid)	4.3290×10^{-3}
cubic inches	L	1.6387×10^{-2}
cubic inches	m^3	1.6387×10^{-5}
cubic inches	oz (US fluid)	5.5411×10^{-1}
cubic meters	acre·ft	8.1071×10^{-4}
cubic meters	bushels (US)	2.8378×10^{1}
cubic meters	ft^3	3.5315×10^{1}
cubic meters	gal (US liquid)	2.6417×10^{2}
cubic meters	L	1×10^{3}
cubic yards	bushels (US)	2.1696×10^{1}
cubic yards	gal (US liquid)	2.0197×10^{2}
cubic yards	L	7.6455×10^{2}
cubic yards	m^3	7.6455×10^{-1}
dynes	g	1.0197×10^{-3}
dynes	N	1×10^{-5}
dynes per square centimeter	atm	9.8692×10^{-7}
dynes per square centimeter	bars	1×10^{-6}
dynes per square centimeter	kg/m^2	1.0197×10^{-2}
dynes per square centimeter	lb/in^2	1.4504×10^{-5}
dyne centimeters	ft·lb	7.3756×10^{-8}
dyne centimeters	kg·m	1.0197×10^{-8}
dyne centimeters	N·m	1×10^{-7}
ergs	dyne·cm	1
fathoms	ft	6
feet	cm	3.048×10^{1}
feet	in	1.2×10^{1}
feet	km	3.048×10^{-4}
feet	m	3.048×10^{-1}
feet	mi	1.8939×10^{-4}
feet	rods	6.0606×10^{-2}
feet per second	km/h	1.0973
feet per second	m/min	1.8288×10^{1}
feet per second	mi/h	6.8182×10^{-1}
feet per second squared	m/s^2	3.048×10^{-1}
foot-candles	lm/ft^2	1
foot-candles	lux	1.0764×10^{1}

To convert from	To	Multiply by
foot pounds	Btu	1.2859×10^{-3}
foot pounds	dyne · cm	1.3558×10^{7}
foot pounds	hp · h	5.0505×10^{-7}
foot pounds	J	1.3558
foot pounds	kg · m	1.3825×10^{-1}
foot pounds	kWh	3.7662×10^{-7}
foot pounds	N · m	1.3558
foot pounds per hour	Btu/min	2.1432×10^{-5}
foot pounds per hour	ergs/min	2.2597×10^{5}
foot pounds per hour	hp	5.0505×10^{-7}
foot pounds per hour	kW	3.7662×10^{-7}
furlongs	ft	6.6×10^{2}
furlongs	m	2.0117×10^{2}
gallons (US liquid)	ft^3	1.3368×10^{-1}
gallons (US liquid)	in^3	2.31×10^{2}
gallons (US liquid)	L	3.7854
gallons (US liquid)	m^3	3.7854×10^{-3}
gallons (US liquid)	oz (US fluid)	1.28×10^{2}
gallons (US liquid)	pt (US liquid)	8
gallons (US liquid)	qt (US liquid)	4
grams	lb	2.2046×10^{-3}
grams per centimeter second	poises	1
grams per cubic centimeter	lb/ft^3	6.2428×10^{1}
gram centimeters	Btu	9.3011×10^{-8}
gram centimeters	dyne · cm	9.8067×10^{2}
gram centimeters	ft · lb	7.2330×10^{-5}
gram centimeters	hp · h	3.6530×10^{-11}
gram centimeters	J	9.8067×10^{-5}
gram centimeters	kWh	2.7241×10^{-11}
hectares	acres	2.4711
hectares	ares	1×10^{2}
hectares	ft^2	1.0764×10^{5}
hectares	m^2	1×10^{4}
horsepower	Btu/h	2.5461×10^{3}
horsepower	ft · lb/s	5.5×10^{2}
horsepower	kW	7.4570×10^{-1}
horsepower	W	7.4570×10^{2}
horsepower hours	Btu	2.5461×10^{3}
horsepower hours	ft · lb	1.98×10^{6}
horsepower hours	J	2.6845×10^{6}
horsepower hours	kg · m	2.7375×10^{5}
horsepower hours	kWh	7.4570×10^{-1}
hours	min	6×10^{1}
hours	s	3.6×10^{3}
inches	Å	2.54×10^{8}
inches	cm	2.54
inches	ft	8.3333×10^{-2}
inches	mils	1×10^{3}
inches	yd	2.7778×10^{-2}
joules	Btu	9.4845×10^{-4}
joules	ft · lb	7.3756×10^{-1}
joules	hp · h	3.7251×10^{-7}
joules	kg · m	1.0197×10^{-1}
joules	kWh	2.7778×10^{-7}
joules	W · s	1
joules per second	Btu/min	5.6907×10^{-2}
joules per second	ergs/s	1×10^{7}
joules per second	ft · lb/s	7.3756×10^{-1}

To convert from	To	Multiply by
joules per second	g · cm/s	1.0197×10^{4}
joules per second	hp	1.3410×10^{-3}
joules per second	W	1
kilograms	lb	2.2046
kilograms	N	9.8067
kilograms	slugs	6.8522×10^{-2}
kilograms	t	1×10^{-3}
kilogram meters	Btu	9.3011×10^{-3}
kilogram meters	ft · lb	7.2330
kilogram meters	J	9.8067
kilogram meters	kWh	2.7241×10^{-6}
kilogram meters	N · m	9.8067
kilometers	ft	3.2808×10^{3}
kilometers	mi	6.2137×10^{-1}
kilometers	nmi (nautical mile)	5.3996×10^{-1}
kilometers per hour	ft/min	5.4681×10^{1}
kilometers per hour	ft/s	9.1134×10^{-1}
kilometers per hour	knots	5.3996×10^{-1}
kilometers per hour	m/s	2.7778×10^{-1}
kilometers per hour	mi/h	6.2137×10^{-1}
kilowatts	Btu/h	3.4144×10^{3}
kilowatts	ergs/s	1×10^{10}
kilowatts	ft · lb/s	7.3756×10^{2}
kilowatts	hp	1.3410
kilowatts	J/s	1×10^{3}
kilowatt hours	Btu	3.4144×10^{3}
kilowatt hours	ft · lb	2.6552×10^{6}
kilowatt hours	hp · h	1.3410
kilowatt hours	J	3.6×10^{6}
knots	ft/s	1.6878
knots	mi/h	1.1508
liters	bushels (US)	2.8378×10^{-2}
liters	ft³	3.5315×10^{-2}
liters	gal (US liquid)	2.6417×10^{-1}
liters	in³	6.1024×10^{1}
liters per second	ft³/min	2.1189
liters per second	gal (US liquid)/min	1.5850×10^{1}
lumens	candle power	7.9577×10^{-2}
lumens per square foot	foot-candles	1
lumens per square meter	foot-candles	9.2903×10^{-2}
lux	lm/m²	1
meters	Å	1×10^{10}
meters	ft	3.2808
meters	in	3.9370×10^{1}
meters	mi	6.2137×10^{-4}
meters per minute	cm/s	1.6667
meters per minute	ft/s	5.4681×10^{-2}
meters per minute	km/h	6×10^{-2}
meters per minute	knots	3.2397×10^{-2}
meters per minute	mi/h	3.7282×10^{-2}
microns	Å	1×10^{4}
microns	ft	3.2808×10^{-6}
microns	m	1×10^{-6}
miles	ft	5.28×10^{3}
miles	furlongs	8
miles	km	1.6093
miles	nmi (nautical mile)	8.6898×10^{-1}
miles per hour	cm/s	4.4704×10^{1}

To convert from	To	Multiply by
miles per hour	ft/min	8.8×10^{1}
miles per hour	ft/s	1.4667
miles per hour	km/h	1.6093
miles per hour	knots	8.6898×10^{-1}
miles per hour	m/min	2.6822×10^{1}
nautical miles	mi	1.1508
newtons	dynes	1×10^{5}
newtons	kg	1.0197×10^{-1}
newtons	lb	2.2481×10^{-1}
newton meters	dyne · cm	1×10^{7}
newton meters	ft · lb	7.3756×10^{-1}
newton meters	kg · m	1.0197×10^{-1}
ounces	g	2.8350×10^{1}
ounces	lb	6.25×10^{-2}
ounces	oz (troy)	9.1146×10^{-1}
ounces (US fluid)	cm^3	2.9574×10^{1}
ounces (US fluid)	gal (US liquid)	7.8125×10^{-3}
ounces (US fluid)	in^3	1.8047
ounces (US fluid)	L	2.9574×10^{-2}
pascals	atm	9.8692×10^{-6}
pascals	lb/ft^2	2.0885×10^{-2}
pascals	lb/in^2	1.4504×10^{-4}
poises	g/(cm · s)	1
pounds	g	4.5359×10^{2}
pounds	kg	4.5359×10^{-1}
pounds	lb (troy)	1.2153
pounds	N	4.4482
pounds	oz	1.6×10^{1}
pounds	slugs	3.1081×10^{-2}
pounds	t	4.5359×10^{-4}
pounds	tons (short)	5×10^{-4}
pounds per cubic foot	g/cm^3	1.6018×10^{-2}
pounds per cubic foot	kg/m^3	1.6018×10^{1}
pounds per square foot	atm	4.7254×10^{-3}
pounds per square foot	Pa	4.7880×10^{1}
pounds per square inch	atm	6.8046×10^{-2}
pounds per square inch	bars	6.8948×10^{-2}
pounds per square inch	in of Hg	2.0360
pounds per square inch	kg/cm^2	7.0307×10^{-2}
pounds per square inch	mm of Hg	5.1715×10^{1}
pounds per square inch	Pa	6.8948×10^{3}
radians	°	5.7296×10^{1}
radians	r (revolutions)	1.5915×10^{-1}
radians per second	r/min	9.5493
slugs	kg	1.4594×10^{1}
slugs	lb	3.2174×10^{1}
square centimeters	ft^2	1.0764×10^{-3}
square centimeters	in^2	1.5500×10^{-1}
square feet	acre	2.2957×10^{-5}
square feet	cm^2	9.2903×10^{2}
square feet	gal (US liquid)	7.4805
square feet	ha	9.2903×10^{-6}
square feet	m^2	9.2903×10^{-2}
square meters	ft^2	1.0764×10^{1}
square meters	in^2	1.5500×10^{3}
square miles	acres	6.4×10^{2}
square miles	ft^2	2.7878×10^{7}
square miles	ha	2.5900×10^{2}

To convert from	To	Multiply by
square miles	km^2	2.5900
square millimeters	ft^2	1.0764×10^{-5}
square millimeters	in^2	1.5500×10^{-3}
stokes	cm^2/s	1
stokes	in^2/s	1.5500×10^{-1}
tons (long)	lb	2.24×10^3
tons (long)	t	1.0160
tons (long)	tons (short)	1.12
watts	Btu/h	3.4144
watts	ergs/s	1×10^7
watts	ft · lb/min	4.4254×10^1
watts	hp	1.3410×10^{-3}
watts	J/s	1
watt hours	Btu	3.4144
watt hours	ft · lb	2.6552×10^3
watt hours	hp · h	1.3410×10^{-3}

ANSWERS TO SELECTED PROBLEMS

Chapter 2

2.1 $\alpha = 55°$ $B_y = 5.0$ m $B = 8.8$ m

2.5 $\overline{A} = 40$ m ∠53° $\overline{C} = 62$ m ∠31°

2.7 6.8×10^2 m

2.11 $AB = 156.8$ m $\angle ABC = 130.0°$

2.15 1.8×10^2 cm

2.18 (*a*) 320 km/h N 35°E
(*b*) N 24°E

Chapter 3

3.7 (*b*) $V = 2.1t + 1.7$
(*c*) Acceleration

3.10 (*b*) $y = 65\, e^{-0.41x}$

3.13 (*b*) $I = 2\, e^{-1.04t}$

3.16 (*b*) $Q = 1.5\, h^{2.47}$

3.17 (*b*) $V = 4.5\, h^{0.5}$

Chapter 4

4.2 5.3×10^{11} m³/year

4.8 170 kL (Assume 45 in class, quarter system)

4.10 3×10^{10} sticks/year

4.11 143×10^6 tires/year

Chapter 5

5.7 (*a*) 9.90×10^2 N
(*b*) 1.7×10^2 N

5.11 Yes, 3.27 m/s

5.14 710 m³, 710 Mg

5.17 (*a*) 103 cm²
(*b*) 0.460 kg
(*c*) Will not float

Chapter 6

6.6 (*a*) $x = 1.497$
(*b*) $x = 2.506$

Chapter 8

8.2 Tension in short length = 404.7 N
Tension in long length = 173.4 N

8.6 10.9 N 76.1°

8.8 68.0 N 54.3°

8.13 757 N · m

8.20 721 N · m

8.25 $\bar{D} = 49.6$ N ↑ $\bar{E} = 121$ N ↑

8.29 $\bar{A} = 2.9$ N ↑ $\bar{C} = 17$ N ↑

8.34 $\bar{D} = 1.30 \times 10^2$ N ↑ $\bar{E}_x = 4.7$ N ← $\bar{E}_y = 27$ N ↓

Chapter 9

9.1 22°C 295 K

9.8 (*a*) 18.015 3
(*b*) 78.114 72
(*c*) 138.213
(*d*) 171.35

9.13 2.70 g/cm^3

9.16 2.5

9.21 87 kPa

9.25 −12.3°C

9.31 (*a*) $2\ C_8H_{18} + 25\ O_2 \rightarrow 16\ CO_2 + 18\ H_2O$
(*b*) 5.7 kg

9.37 (*a*) $CaH_2 + 2\ H_2O \rightarrow Ca(OH)_2 + 2\ H_2$
(*b*) 9.58 g

9.41 38.82 percent

9.45 $B_4Na_2O_7$

Chapter 10

10.2 13.7 g

10.6 29 percent

10.8 Flow rate = 1350 kg/h
Processed livers: 52% inert, 46% ether, 2.0% oil

10.12 Leftover amount = 260 kg
Concentrated H_2SO_4 = 1200 kg
Concentrated HNO_3 = 590 kg

10.14 62 percent

10.20 Coal feed = $8.3t/h$
Water requirement = $3.46t/h$

Chapter 11

11.1 20 A

11.3 $0.90

11.5 1.0×10^2 A

11.8 (*b*) 52 Ω
(*c*) 1.8 A
(*d*) $E_1 = 22$ V, $E_2 = 27$ V, $E_3 = 46$ V

11.11 (*a*) 2.0×10^1 V
(*b*) 880 W

11.12 (*a*) 110 V
(*b*) 6.0×10^3 W
(*c*) 6.0×10^2 W

11.14 Charging current = 9 A $V_m = 1.2 \times 10^2$ V

Chapter 12

12.1 0.41 MJ

12.2 4.1 (10^3) N

12.3 281 m

12.7 (*a*) 177 TJ
(*b*) 22 m^3/s

12.9 (*a*) 7.14 kJ
(*b*) 0.798 hp

12.14 27 kJ

12.17 (*a*) 78 kJ/min
(*b*) 33 percent

12.21 316°C

12.23 0.36 kW

12.25 (*a*) 430 W
(*b*) 5.2 (10^4) L

12.29 $7568.64

12.31 25 percent

Chapter 13

13.1 (*a*) $783.53
(*b*) $497.18
(*c*) $327.68

13.5 $7996.80

13.9 $4212.36

13.12 20.15 years

13.17 $74 514.74

13.21 $514.60

13.26 (*a*) 5.5 percent per year
(*b*) 1.51 percent per quarter

Chapter 14

14.2 (*c*) Median = 67.5 mode = 68.2 mean = 66.39

14.5 (*c*) Median = 72 mode = 70 mean = 73

14.7 (*a*) 13.46
(*d*) One deviation (59.18 – 86.10)
Two deviations (45.72 – 99.56)

14.8 (*a*) $y = 0.7955x + 14.71$
(*d*) 0.863

Selected Bibliography

Allen, Myron S.: *Morphological Creativity,* Prentice-Hall, Englewood Cliffs, N.J., 1962.

Assaf, Karen, and Said A. Assaf: *Handbook of Mathematical Calculations,* Iowa State University Press, Ames, Iowa, 1974.

Bassin, Milton G., Stanley M. Brodsky, and Harold Woloff: *Statics and Strengths of Materials,* McGraw-Hill, New York, 1969.

Beakley, George C., and H. W. Leach: *Engineering: An Introduction of a Creative Profession,* 3d ed., Macmillan, New York, 1977.

Bullinger, Clarence E.: *Engineering Economy,* McGraw-Hill, New York, 1958.

Drago, Russell S.: *Principles of Chemistry with Practical Perspective,* Allyn and Bacon, Boston, 1974.

Erickson, William H., and Nelson H. Bryant: *Electrical Engineering, Theory and Practice,* 2d ed., Wiley, New York, 1959.

Fletcher, Leroy S., and Terry E. Shoup: *Introduction to Engineering Including Fortran Programming,* Prentice-Hall, Englewood Cliffs, N.J., 1978.

Freund, John: *Modern Elementary Statistics,* 3d ed., Prentice-Hall, Englewood Cliffs, N.J., 1967.

Gajda, Walter J., Jr., and William E. Biles: *Engineering: Modeling and Computation,* Houghton Mifflin, Boston, 1978.

Gibson, John E.: *Introduction to Engineering Design,* Holt, New York, 1968.

Giesicke, Frederick E., Alva Mitchell, Henry Cecil Spencer, Ivan Leroy Hill, and Robert Olin Loving: *Engineering Graphics,* 2d ed., Macmillan, New York, 1975.

Glorioso, Robert M., and Francis S. Hill, Jr. (eds.): *Introduction to Engineering,* Prentice-Hall, Englewood Cliffs, N.J., 1975.

Gordon, William J. J.: *Synectics,* Harper, New York, 1961.

Grant, Eugene L., W. Grant Ireson, and Richard S. Leavenworth: *Principles of Engineering Economy,* 6th ed., Ronald, New York, 1976.

Hill, Percy H.: *The Science of Engineering,* Holt, New York, 1970.

Jensen, C. H.: *Engineering Drawing and Design,* McGraw-Hill, New York, 1968.

Johnston, David O., John T. Netterville, James L. Wood, and Mark M. Jones: *Chemistry and the Environment,* Saunders, Philadelphia, 1973.

Katz, Donald L., Robert O. Goetz, Edward R. Lady, and Dale C. Ray: *Engineering Concepts and Perspectives,* Wiley, New York, 1968.

Krick, Edward V.: *An Introduction to Engineering: Methods, Concepts, and Issues,* Wiley, New York, 1976.

Langford, Cooper H., and Ralph A. Beebe: *The Development of Chemical Principles,* Addison-Wesley, Reading, Mass., 1969.

Longo, Frederick R.: *General Chemistry—Interaction of Matter, Energy, and Mass,* McGraw-Hill, New York, 1974.

Luzadder, Warren J.: *Fundamentals of Engineering Drawing for Design, Product Development, and Numerical Control,* 7th ed., Prentice-Hall, Englewood Cliffs, N.J., 1977.

Mahan, Bruce H.: *University Chemistry,* Addison-Wesley, Reading, Mass., 1965.

"Metric Editorial Guide," 3d ed., American National Metric Council, Washington, D.C., 1978.

"Metric Guide for Educational Materials," American National Metric Council, Washington, D.C., 1977.

Osborn, Alex F.: *Applied Imagination,* Scribner, New York, 1963.

Peterson, Ottis: *Your Future in Engineering Careers,* R. Rosen, New York, 1975.

"Reference Manual for SI (Metric)," Inland Steel Company, Steel Division, Chicago, 1976.

Rising, James S., Maurice W. Almfeldt, and Paul S. DeJong: *Engineering Graphics,* 5th ed., Kendall/Hunt, Dubuque, Iowa, 1977.

Seeley, Fred B., and Newton E. Ensign: *Analytical Mechanics for Engineers,* Wiley, New York, 1957.

Smith, Gerald W.: *Engineering Economy: Analysis of Capital Expenditures,* Iowa State University Press, Ames, Iowa, 1973.

Smith, Ralph J.: *Engineering as a Career,* 2d ed., McGraw-Hill, New York, 1962.

"Standard for Metric Practice," American National Standards Institute E 388-76 268-1976.

Thuesen, H. G., W. J. Fabrycky, and G. L. Thuesen: *Engineering Economy,* Prentice-Hall, Englewood Cliffs, N.J., 1977.

Wasserman, Leonard S.: *Chemistry: Basic Concepts and Contemporary Applications,* Wadsworth, Belmont, Calif., 1974.

Woodson, Thomas T.: *Introduction to Engineering Design,* McGraw-Hill, New York, 1966.

Index

Abacus, 126
Abscissa, 56
Absolute temperature scales,
    195–196
Absolute zero, 195
Accidental errors, 80–81
Accuracy, 84
Adiabatic process, 264
Algorithm, 124–125, 151–152
Alternative solutions, 325, 327–328,
    341–345
Analog computers, 127, 137–139
Analysis, 325, 327–328, 346–352
Angles, standard trig values, 393–394
Annuity, 287–302
Approximation, 85–94
Area of basic shapes *(see front endpapers)*
Arithmetic mean, 309–310
Arrowheads, 410–411
Assembly drawing, 417–418
Atomic number, 197
  table of, 427
Atomic weight, 198
  table of, 427
Atoms, 197, 233–234
Avogadro's number, 198
Axes breaks, 57
Axis identification, 63

Babbage, Charles, 127
Balloons, 417–418
Base units, definition, 103–104
Batch process, 218–221
Battery, 238
Bill of material, 418–419
Block diagram, 421
Boyle's law, 202
Brainstorming, 343–345
Briggs, Henry, 127
Buffer, 131

Calibrations, 58–62
Carnot efficiency, 265
Celsius scale, 194
Celsius unit, 194–196
Center lines, 408–409
Central processing unit (CPU), 129
Charles' law, 202
Checkoff lists, 343
Chemical elements:
  atomic number of, 427
  atomic weights of, 427
  symbols for, table, 427
Chemical energy, 257
Chemical equations, balancing of,
    203–205
Circle construction, 405–406
Closed system, 261–262
Code of ethics, 29
Coefficient of correlation, 318–320
Colinear force, 167
Column chart, 420
Communications, 325, 327–328,
    361–370
Compound interest, 283–302
Computer language, 155
Computers:
  analog, 127, 137–139
  definition, 124
  digital, 126–137
  hybrid, 139–140
  micro-, 140–141
Conceptualization, 353–357
Conclusions and recommendations,
    366
Concurrent force, 167
Conditional test, 151–159
Conductance, 238
Conservation laws:
  energy, 261
  mass, 216–218
Constraints, 325, 327–328, 333,
    337–341
Conversion of units, 114–116
  table, 437–441
Coordinate axes, 57
Coplanar force, 167
Correlation, coefficient of, 318–320
Coulomb, 234–235
Counterbore, 412–413
Countersink, 412–413
Craftsman, 10
Cramer's rule, 383–384
Criteria, 325, 327–328, 337–341, 349,
    351
Current, 235–237
Curve fitting, 66–67
Cutting plane, 414

Data sheets, 52–53
Data symbols, 64
Decimal multiples, 107
Decision, 325, 327–328, 330, 350,
    352–360
Deductive reasoning, 36–37
Definition of problem, 325, 327–328,
    332–333
Degree (angle):
  conversion, 391
  definition, 390
Density:
  definition, 200
  table of, 425–426
Depreciation, 291–299
  Matheson formula, 294–297
  sinking-fund, 293–297
  straight-line, 292–297
  sum of years' digits, 293–297
Derived dimensions, 101–102
Derived units, 103, 105
Descriptive statistics, 306–307
Design, 7, 325–375
  drawing, 416
  process, 325–326, 346

Design:
  project, 325
  team, 326, 329, 344, 357
Detail drawing, 418–419
Digital computers, 126–137
Dimension lines, 410–411
Dimension system, absolute, 101–102
Dimensioning, 410–413
  practices, 412–413
Dimensions, 101–102
Dynamics, 164

e, 382
Economic selection, 296–298
Edison, Thomas Alva, 4
Electric current, 235
Electric energy, 231, 244, 258
Electric power, 244
Electromotive force (emf), 235, 246
Electron drift, 237
Electrostatics, 234–235
Elements:
  atomic number of, 427
  atomic weights of, 427
  symbols for, 427
Empirical data, plotting of, 63
Empirical equations, 67
Empirical formula, 198–199
  determination of, 207–209
Energy:
  chemical, 257
  electric, 231, 244, 258
  internal, 257
  kinetic, 256–257
  nuclear, 257–258
  potential, 256
  radiant, 258
  stored, 256–258
Energy efficiency ratio, 268
Engineer:
  definition, 6
  description of, 6–9
Engineering:
  definition, 6
  economy, 281–302
  education, 22–26
  functions, 12–22
  lettering, 403–404
  method, 37–39, 124
Engineering functions, 12–22
  construction, 17–18
  consulting, 20–21
  design, 14–16
  development, 12–14
Engineering functions:
  management, 20
  operations, 18–19
  production, 16–17
  research, 12
  sales, 19
  teaching, 21–22
  testing, 16–17
Engineers' Creed, 30
Equilibrium, 171–178
Errors, 80–81
  accidental, 80–81
  systematic, 80
Estimation, 85–94
Exponential functions, 380
Extension lines, 410–411

Fahrenheit scale, 194
Fahrenheit unit, 194–196
Fillets and rounds, 412
First law of thermodynamics, 261
Flowchart, 150–159
Force, 165–178
Free-body diagram, 171–177
Freehand drawing, 404–407
Frequency distribution, 307–308
Full section, 414
Functional scale, 58–60
Fundamental dimensions, 101–102

Gay-Lussac's law, 202
Gothic letters, 403–404
Graduations, 58–62
Graph paper, 55–56
Graphic display, 347–368
Graphical analysis, 51–52
Graphical presentation of data, 51–52
Graphing procedure, steps in, 54, 65
Gravitational acceleration, 112–113
Gravitational dimension system, 102
Greek alphabet, 428
Gunter, Edmund, 127

Half section, 414
Heart pacemaker, 13–14
Heat, 258
Hidden lines, 408–409
Histogram, 308–309
Hollerith, Herman, 132
Hybrid computers, 139–140

Inductive reasoning, 36–37
Input, 151
Installment loan, 289–291
Interest, 282–302
Internal energy, 257
Invention, 342
Inverse trigonometric functions:
  plots, 395–396
  principal values, 394–395
  symbols, 394
Isothermal process, 264
Isotope, definition, 197

Kelvin scale, 195–196
Kinetic energy, 256–257
Kirchhoff's laws, 248–251

Law of cosines, 397–398
Law of sines, 397
Laws of exponents, 379–380
Layout drawings, 416–417
Leibniz, Baron von, 127
Letter of transmittal, 366
Line printers, 132
Linear regression, 314–318
Log-log graph paper, 55–56
Logarithms, 380–383
Looping, 153–159

Magnetic core, 137
Magnetic tape, 132–133
Mass, 112–113
Mass number, 197
Matheson formula, 294–297
Mean, 309–310
Measurements, 79–80
Mechanics, 164–189
  of materials, 164
Median, 310
Mensuration formulas table *(see front and back endpapers)*
Method:
  of least squares, 67, 316–318
  of selected points, 67
Microcomputer, 140–141
Mode, 310
Models, 340, 347–348
Modulus of elasticity, 181–183
Mole, 198
Molecular formula, 199
Molecules, 233–234
Moments, 170–178
Morphological listing, 343
Multiview drawings, 405, 408–409

Napier, John, 127
Need, 325, 327–328, 333
Neutrons, 233–234
Nonuniform scale, 58–60
Normal distribution, 313–314
Notebook of original entry, 52
Nuclear energy, 257–258

Observed data plot, 63
Offset section, 415
Ohm, 237
One, two, five rule, 60
Optical character readers, 133
Optimization, 357–358
Ordinate, 56
Overall efficiency, 266

Parallel circuit, 241–243
Parallelogram, 169
Partial section, 416
Pascal, Blaise, 126
Perfect gas law, 202–203
Physical constants, table of, 428
Pictorials, 405–407
  construction of, 406
Pie diagram, 419–420
Polar-rectangular conversion, 396–397
Population, 309
Potential energy, 256
Power, 268
Power functions, 380
Precision, 84
Present worth, 285–302
Presentation of problems:
  guidelines, 40–41
  steps in, 41–46
Pressure, definition, 201
Problem layout:
  guidelines, 40–41
  steps in, 41–46
Problem solution, engineering method of, 37–39
Problem solving, 149–150
Process:
  batch, 218–221
  rate, 218–221
Professional ethics, 28
Professional registration, 28
Professionalism, 27–28
Program, computer, 149–151
Programmable calculators, 134–135
Programming, 150
Protons, 233–234
Punched cards, 132

Radian:
  conversion, 391
  definition, 390
Radiant energy, 258
Radicals, 379
Rankine scale, 195–196
Rate of return, 282
Rate process, 218–221
Rectangular-polar conversion, 396–397
Rectilinear paper, 55–56
Refrigeration efficiency, 268
Removed section, 415–416
Reports, 365–370
Resistance, 237–238
Resistor, 238
Restrictions (constraints), 333
Retirement plan, 291
Revolved section, 415

Sample, 309
Scalar quantity, 165
Scale, graphical, 58, 60–62
Scale multipler, 58–60
Scales, 409–410
Schematic diagram, 421–422
Scientist, 5–6
Search (research), 325, 327–328, 333–337
Second law of thermodynamics, 265
Section lining, 414
Sections, 413–416
Semilog graph paper, 55–56
Series circuit, 239–240, 243
Shear, 179
SI rules, 109–112
Significant figures, 81–83
Single views, 405
Sinking fund, 287–302
Sketches, examples of, 336, 338, 346, 354–357
Sketching, 404–407
Slide rule, 127
Specific gravity, 201
  table of, 425–426
Specification, 325, 327–328, 360–363
Standard deviation, 310–312
Statics, 164–178
Statistical inference, 306–307
Stored energy, 256–258
Strain, 180–181
Stress, 178–180
Subroutine, 150–151
Sum of years' digits, 293–297
Supplementary units, 103–104
Switch, 238
System, 216, 261–262
Systematic error, 80
Système International d'Unites (SI), 99

Technician, 9–10
Technologist, 9–10
Temperature:
  conversion formulas, 195–196
  definition, 194
  scales, 194–196
  units, 194–196
Tensile strength, 181–183
Tension, 172, 175–176
Testing, 340, 351
Theoretical data plot, 63
Time sharing, 130–131
Tradeoff, 352–353
Trigonometric series, 399–400
Trigonometry, 389–401
  function definitions, 389–390
  plots of functions, 391–393
  relationships, 400–401

Unified atomic mass unit, 197
Uniform scale, 58
Units, 102–116
  conversion table, 437–441
  prefixes, 426

Variance, 311
Vector, 164
Velocity of sound, 116
Visual aid, 368–369
Volume of basic solids *(see back endpapers)*

Watt, 245–246
Weight, 112–113
Work, 258

Yield strength, 181–183

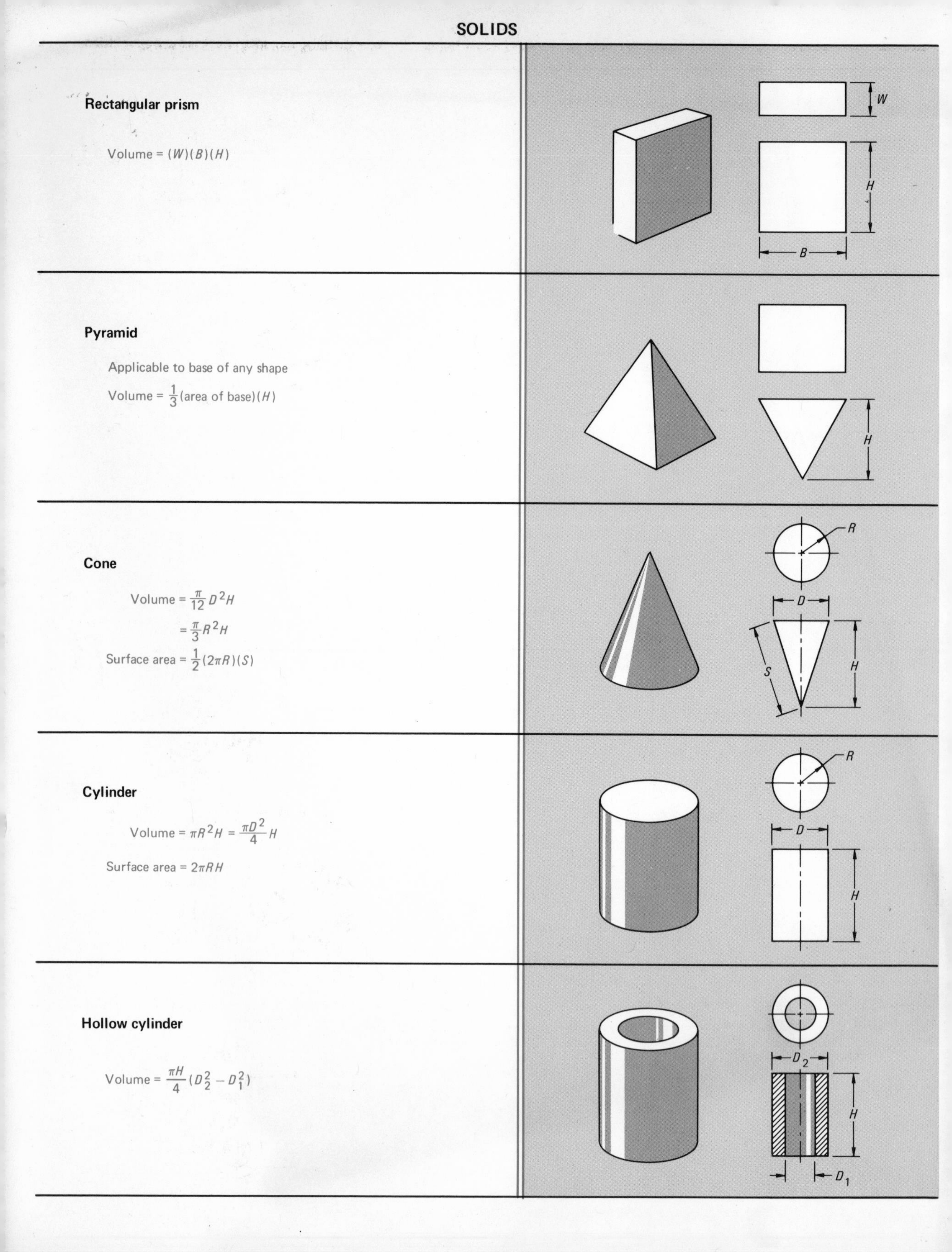

SOLIDS

Rectangular prism

Volume = $(W)(B)(H)$

Pyramid

Applicable to base of any shape

Volume = $\frac{1}{3}$(area of base)(H)

Cone

Volume = $\frac{\pi}{12} D^2 H$

$= \frac{\pi}{3} R^2 H$

Surface area = $\frac{1}{2}(2\pi R)(S)$

Cylinder

Volume = $\pi R^2 H = \frac{\pi D^2}{4} H$

Surface area = $2\pi R H$

Hollow cylinder

Volume = $\frac{\pi H}{4}(D_2^2 - D_1^2)$